Lecture Notes in Physics

The Lecture Notes in Physics

The series Lecture Notes in Physics (LNP), founded in 1969, reports new developments in physics research and teaching – quickly and informally, but with a high quality and the explicit aim to summarize and communicate current knowledge in an accessible way. Books published in this series are conceived as bridging material between advanced graduate textbooks and the forefront of research and to serve three purposes:

- to be a compact and modern up-to-date source of reference on a well-defined topic

- to serve as an accessible introduction to the field to postgraduate students and nonspecialist researchers from related areas

- to be a source of advanced teaching material for specialized seminars, courses and schools

Both monographs and multi-author volumes will be considered for publication. Edited volumes should, however, consist of a very limited number of contributions only. Proceedings will not be considered for LNP.

Volumes published in LNP are disseminated both in print and in electronic formats, the electronic archive being available at springerlink.com. The series content is indexed, abstracted and referenced by many abstracting and information services, bibliographic networks, subscription agencies, library networks, and consortia.

Proposals should be sent to a member of the Editorial Board, or directly to the managing editor at Springer:

Christian Caron
Springer Heidelberg
Physics Editorial Department I
Tiergartenstrasse 17
69121 Heidelberg / Germany
christian.caron@springer.com

J.-P. Rozelot
C. Neiner (Eds.)

The Rotation of Sun and Stars

 Springer

Jean-Pierre Rozelot
Université de Nice Sophia-Antipolis
CNRS-OCA
Dept. Fizeau
av. Copernic
06130 Grasse
France
rozelot@obs-azur.fr

Coralie Neiner
Observatoire de Paris
Section de Meudon
GEPI
5 place Jules Janssen
92195 Meudon CX
France
coralie.neiner@obspm.fr

Rozelot, J.-P., Neiner, C. (Eds.), *The Rotation of Sun and Stars*, Lect. Notes Phys. 765 (Springer, Berlin Heidelberg 2009), DOI 10.1007/ 978-3-540-87831-5

ISBN: 978-3-540-87830-8   e-ISBN: 978-3-540-87831-5

DOI 10.1007/978-3-540-87831-5

Lecture Notes in Physics ISSN: 0075-8450   e-ISSN: 1616-6361

Library of Congress Control Number: 2008936495

Cover design: Integra Software Services Pvt Ltd.

Printed on acid-free paper

9 8 7 6 5 4 3 2 1

springer.com

Preface

The Sun and stars rotate in different ways and at different velocity rates. The knowledge of how they rotate is important in understanding the formation and evolution of stars and their structure. The closest star to our Earth, the Sun, is a good laboratory to study in detail the rotation of a G star and allows to test new ideas and develop new techniques to study stellar rotation. More or less massive, more or less evolved objects, however, can have a very different rotation rate, structure and history.

In recent years our understanding of the rotation of the Sun has greatly improved. The Sun has a well-known large-scale rotation, which can be measured thanks to visible features across the solar disk, such as sunspots, or via spectroscopy. In addition, several studies cast light on differential rotation in the convective zone and on meridional circulation in the radiative zone of the Sun. Even the rotation of the core of the Sun can now be studied thanks to various methods, such as dynamics of the gravitational moments and of course, helioseismology, through g-modes analysis.

Moreover, the magnetic field is strongly linked to the matter motions in the solar plasma. The solar magnetic field can be measured only at the surface or in the upper layers. It is the product of the internal dynamo or of the local dynamos if they exist – in any case magnetic field and rotation cannot thus be separated.

The wide variety of stars, however, offers an equally wide variety of rotation rates and rotational evolution. From the slowly rotating stars to stars rotating close to their breakup velocity (such as Be stars), different techniques and models have to be developed to study rotation and its effects on physical aspects of stars.

This book, while not attempting to answer all questions about rotation – given that many issues still have to be further investigated, focuses on the basic and some particular aspects and aims to show why it is important, from a physical point of view, to study stellar rotation. Specifically

- The first chapter (J.P. Zahn) compares the Sun to other slowly rotating stars, investigates the angular momentum history of the Sun and reviews the physical processes responsible for its internal rotation profile.

- The second chapter (J.P. Rozelot) develops the current issues raised from observation of the shape of the Sun and shows the interest of the sub-surface layers.
- The third chapter (M.J. Goupil) explains effects of rotation on p-modes of pulsations.
- The fourth chapter (M. Rieutord) develops the basic knowledge needed to understand the properties of the low-frequency spectrum of rotating stars.
- The fifth chapter (S. Turck-Chièze) presents the current knowledge of the rotation of the solar core, including very recent results.
- The sixth chapter (G. Meynet) reviews the effects of axial rotation in stellar interior models and their important role at low metallicity.
- The seventh chapter (A. Domiciano de Souza) presents the advent of interferometry in the study of rotation for various types of stars.
- The eighth chapter (Ph. Stee and A. Meilland) considers Be stars and the need for critical rotation to trigger the Be phenomenon.
- The ninth chapter (F. Royer) details the effects of gravity darkening and differential rotation with particular attention to the case of A-type stars.
- Finally, the last chapter (V. Bommier) presents the next step to be made after introducing the effects of rotation in stellar models: the detailed study of magnetic fields, with the Sun as a prime example. Indeed, in the most upper layers of the solar photosphere, where the magnetic field begins to play an active role well marked by prominences (or other streams), this magnetic field shows a global structure linked to the differential rotation (within the leptocline, which is the seat of the structured magnetic field in connection with the differential rotation and the change of the radial gradient of rotation).

Based on tutorial lectures given at a graduate school on the same topic held under the auspices of the CNRS (France), we foresee that our book will be of interest and useful to a rather broad audience of scientists and students – in particular for the latter as a kind of high-level, yet accessible introduction – as we currently witness a complete renewal of astrophysical ideas about stellar rotation, mainly due to the development of new models including high-order effects of rotation and magnetism. In this context it appeared important to confront the experience of solar astronomers with that of stellar astronomers. Transposing progresses obtained in the field of solar physics (e.g. from helioseismology to asteroseismology) has always been a fruitful way to proceed.

We thus hope that our book will contribute to get many astronomers and students interested in studying the rotation of the Sun and stars and its interaction with other physical processes. At this point, we would like to thank the authors for their commitment to this endeavour.

France   *J.P. Rozelot*
July 2008   *C. Neiner*

Contents

The Sun: A Slowly Rotating Star

J.-P. Zahn

LUTH, Observatoire de Paris, CNRS, Université Paris-Diderot,
Place Jules Janssen, 92195 Meudon, France
Jean-Paul.Zahn@obspm.fr

Abstract After a brief historical sketch, we compare the slowly rotating Sun to other stars and explain how it was able to lose so much of its angular momentum. We next discuss the physical processes that may be responsible for its internal rotation profile and conclude that the most efficient of them is the transport of angular momentum by the gravito-inertial waves excited at the base of the convection zone.

1 The Sun Rotates

As many of us have experienced, sunspots can be easily detected with the naked eye, in foggy weather for instance. But it was only at the beginning of the seventeenth century, with the invention of the refracting telescope, that their position could be recorded with sufficient precision to establish that they are moving westward across the solar disk, in about 2 weeks. After some debate on whether these spots might be planets transiting in front of the disk, Galileo confirmed in 1612 Fabricius' interpretation that they were located at the surface of the rotating Sun, as revealed by the foreshortening effect when they are close to the limb. Scheiner went a step further: by measuring the rotation period with higher precision (see Fig. 1), he was able to show that the spots located farther from the equator move slower, thus discovering the Sun's differential rotation.

For two centuries thereafter, astronomers seem to have lost interest in the solar rotation. Then, in the 1850s, Carrington and Spörer undertook systematic observations of the motions of sunspots and Faye gave the result in the form

$$\Omega = 14^0 37 - 3^0 10 \sin^2 \phi \qquad \text{deg/day}, \tag{1}$$

ϕ being the latitude. Nowadays, one often prefers to fit the data by a law that includes also a term in $\sin^4 \phi$.

Somewhat later, Dunér and Halm determined the solar rotation rate from the Doppler shift of spectral lines at opposite rims of the disk, and they concluded that Faye's formula above represented adequately the differential rotation, even up to higher latitudes than those explored by the spots. Their measurements

Zahn, J.-P.: *The Sun: A Slowly Rotating Star*. Lect. Notes Phys. **765**, 1–14 (2009)
DOI 10.1007/978-3-540-87831-5_1   © Springer-Verlag Berlin Heidelberg 2009

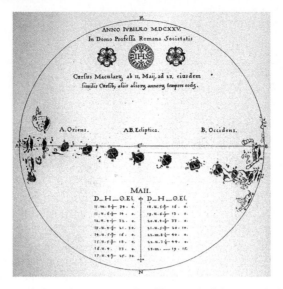

Fig. 1. Observation of the solar rotation by Christoph Scheiner: daily engravings of a sunspot group, as the Sun rotates, in May 1625. (From his *Rosa Ursini*, 1630.)

were made visually, but they were confirmed soon after by Adams' more precise spectrographic observations; he found moreover that different lines, emitted at different heights, gave somewhat different velocities.

All subsequent investigations confirmed that picture of a Sun rotating differentially both in depth and in latitude. For example, Fig. 2 clearly demonstrates that the sunspots rotate faster than the photosphere, suggesting that the rota-

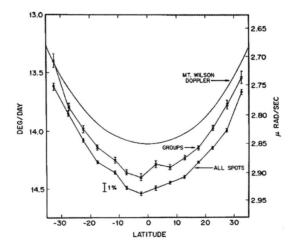

Fig. 2. Variation of the rotation in latitude of three different tracers: photospheric plasma (from Doppler shifts), sunspot groups, individual sunspots (Howard et al. [11]; courtesy ApJ)

tion rate increases with depth and that the spots are rooted below the photo-sphere [11].

2 The Sun Is a Slow Rotator

The rotation rates of other stars can be determined through methods that are similar to those used for the Sun: photometry, by detecting tracers such as bright or dark regions, or spectroscopy, by measuring the Doppler broadening of spectral lines. The first method has the advantage of yielding directly the rotation period, but it requires a spotted surface, while the second method can be applied to any star, but it delivers the projected equatorial velocity $V \sin i$; to derive the period, one needs in addition the value of the radius and that of the inclination angle.

Figure 3 summarizes what was known in 1969 at the IAU colloquium devoted to stellar rotation [20]: early-type stars on the main sequence display equatorial speeds of the order of 200 km/s, whereas the rotational velocity amounts only to a few km/s for solar-type stars. This contrasted behavior suggests that the outer convection zone, which is present in late-type stars but not in early-type stars, must play a key role in the loss (or the conservation) of angular momentum.

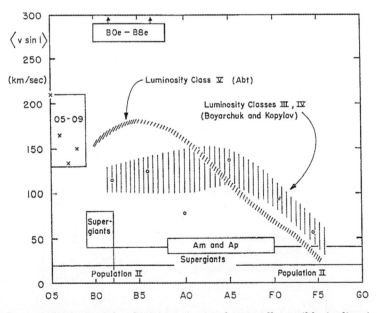

Fig. 3. Projected equatorial velocities, averaged over all possible inclinations, as a function of spectral type. On the main sequence (luminosity class V), early-type stars have rotational velocities that reach and even exceed 200 km/s; these velocities drop to a few km/s for late-type stars, such as the Sun (type G2) (Slettebak [20]; courtesy Gordon & Breach)

3 Solar Rotation in the Past

An important step forward was accomplished by Skumanich [19], when he compared the characteristic rotation period in the Pleiades and in the Hyades, whose ages were then estimated to be, respectively, 40 and 400 Myr, with that of the Sun (4.5 Gyr). He found that, over this long time span, the rotation rate declined with age t as $\Omega \propto t^{-1/2}$, a relation which is now referred to as the Skumanich law. (He went actually further, by comparing Ω with the Li abundance and the Ca^+ emission taken as an activity indicator; see Fig. 4, where age is labeled as τ.)

Such a time dependence of the rotation velocity indicates that these stars lose their angular momentum at a rate proportional to Ω^3, neglecting the variation of the moment of inertia that accompanies the main sequence evolution. Therefore

$$\frac{d\Omega}{dt} = -K\Omega^3,\qquad(2)$$

where K is a constant that can be calibrated with the observations. Integrating this equation in time, one finds that

$$\Omega^{-2} = 2Kt + \Omega_i^{-2}.\qquad(3)$$

It shows that the rotation rate loses quickly the memory of its initial value Ω_i and then follows asymptotically the Skumanich law. If one wishes to learn something about the past solar rotation, one cannot just extrapolate back the present

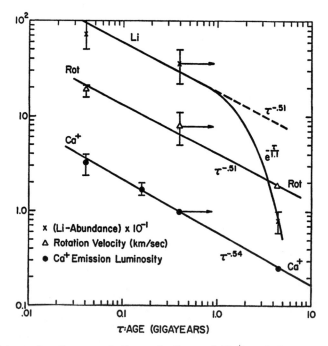

Fig. 4. Lithium abundance, rotation velocity and Ca^+ emission versus stellar age (Skumanich [19]; courtesy ApJ)

rotation rate, but one has to turn to the observation of a series of clusters of different age, as Skumanich did.

More recent investigations essentially confirm Skumanich's findings (see for example Bouvier et al. [1]). Figure 5 presents the latest compilation, which includes the results of a vast photometric monitoring survey of young open clusters, i.e., Monitor [12]. Note that the dispersion of the data exists already in the youngest clusters: presumably the stars have decoupled at different epochs from their protostellar disk. But the scatter has been much reduced at the age of the Hyades, as expected from the braking law (2). The results suggest also that this simple law holds only below some saturation velocity $\omega_{\mathrm{sat}} \approx 14\omega_\odot$. Furthermore, they are consistent with an almost uniform rotation in the bulk of the stars (in solid lines); in long-dashed lines the model makes some allowance for decoupling the rotation of the convection zone from that of the interior – clearly these curves do not fit the data.

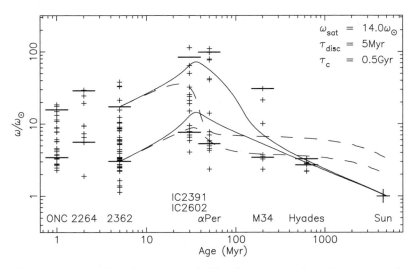

Fig. 5. Rotational angular velocity ω, scaled by the present solar value $\omega_\odot$, as a function of time, in the mass bin 0.9–1.1 $M_\odot$. *Crosses* show the data, and *short horizontal lines* the 25 and 90 percentiles of ω (Irwing et al. [12]; courtesy ApJ)

From this study we may conclude that the Sun was probably rotating between 10 and 100 times faster than at present when it arrived on the ZAMS, although one cannot exclude that it was then already a slow rotator.

4 How Did the Sun Lose Its Angular Momentum?

There is only one way for a single star to lose its angular momentum, once it is no longer locked to its protostellar disk, and that is through mass loss (wind, coronal mass ejection).[1] But as Fesenkov [6] pointed out, it is not easy to shed a large

---

[1] We ignore here the loss by gravitational waves, which is negligible in MS stars.

amount of angular momentum without spending a (relatively) large fraction of matter. To illustrate this, consider a star that loses angular momentum in the most efficient way: at the equator. The loss of mass and angular momentum are then related by

$$\frac{d}{dt}I\Omega = R^2\Omega\frac{d}{dt}M, \tag{4}$$

R being the radius and I the moment of inertia. Introducing the gyration radius kR, and replacing I by k^2MR^2, Eq. (4) is easily integrated, assuming that k remains constant. One obtains the following relation between the initial and final states:

$$\frac{(R^2\Omega)_f}{(R^2\Omega)_i} = \left[\frac{M_f}{M_i}\right]^p, \tag{5}$$

where $p = k^{-2} - 1 \approx 16$ for a solar MS model. In order to decrease its angular momentum by a factor of 100, a star would have to lose 25% of its mass on the main sequence, which is incompatible with the observed cluster isochrones.

It was Schatzman [17] who solved that puzzle: since solar-like stars possess an outer convection zone which produces a dynamo field, that magnetic field compels the matter to rotate rigidly with the star until it reaches the so-called Alfvén radius, where the wind speed overcomes the Alfvén velocity. This enhances considerably the loss of angular momentum, since Eq. (4) must be replaced by

$$\frac{d}{dt}I\Omega = D^2\Omega\frac{d}{dt}M, \tag{6}$$

where D is the Alfvén radius, which is approximately $15R_\odot$ according to Schatzman's estimate. integrating this equation as before, one obtains the same result (5), but where the exponent now is $p = (D/R)^2k^{-2} - 1 \approx 3800$. Under such conditions, the star can reduce its angular momentum by a factor of 100 while spending as little as 1.2×10^{-3} of its mass. To put this figure in perspective, let us recall that the present mass loss through the solar wind is estimated at 3×10^{-14} $M_\odot$/yr; at this rate the Sun would have lost only 1.4×10^{-4} of the initial mass, proving that its angular momentum loss has been higher in the past. But its mass decreases also through nuclear burning: the mass equivalent of the energy it has radiated since its birth is of the order of 3×10^{-4} $M_\odot$.

5 The Interior Rotation Unveiled by Helioseismology

In what precedes, we have implicitly assumed that the Sun is rotating uniformly inside, although we know already that its surface is in differential rotation. Moreover, since the angular momentum has to be carried to the surface, at which the Lorentz torque is applied, it has long been considered as plausible that the core keeps rotating faster than the surface. The advent of helioseismology, in the late 1970s, made it possible to probe the interior rotation of the Sun, since the eigenfrequencies of the solar pulsations are split in multiplets by the rotation, from which one can infer the internal rotation, in depth and in latitude (see for instance Thompson et al. [20]).

Figure 6 displays the solar rotation inferred from the data obtained by the MDI (Michelson Doppler Instrument) on board of the satellite SoHO – the ground-based networks BiSON and GONG provide similar results. The convection zone rotates in the same differential way as observed at the surface, with little dependence on radius, whereas the radiative core is in quasi-uniform rotation. The transition from one rotation regime to the other occurs in a thin layer (less than 5% in radius) that has been called the tachocline (from $\tau\alpha\chi o\sigma$: speed, and $\kappa\lambda\iota\nu\eta$: layer; see Spiegel & Zahn [21], Zahn [28]).

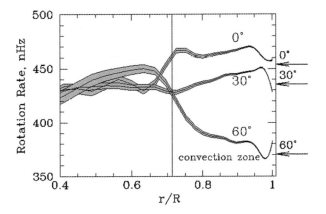

Fig. 6. Internal rotation rate of the Sun as a function of depth at three different latitudes, inferred from data obtained with the SoHO/MDI instrument. The *arrows* indicate the rotation rate measured at the surface through the Doppler shift of the spectral lines

Thus the rotation profile is rather flat overall, which implies that the angular momentum has been carried from the interior to the surface by a sufficiently efficient process. In the convection zone, that transport is mediated by the turbulent motions; however in the radiation zone, the mechanism responsible for it has been identified only recently, after a search that lasted over 40 years.

But before we address this question, let us ask why the tachocline is so thin.

6 The Thin Tachocline Puzzle

From recent 3D numerical simulations performed by Brun and Toomre [2], we learned how turbulent convection in a spherical rotating shell generates the differential rotation of the type observed in the Sun, namely with the equator spinning faster than the poles. When this differential rotation is applied at the top of the radiation zone, the stratification becomes baroclinic, i.e., the temperature varies from pole to equator on a level surface, or isobar. This horizontal temperature gradient then diffuses into the radiative interior, where it sets up a similar differential rotation. This diffusion is partly inhibited by the stable stratification, but even so, the whole Sun should be rotating differentially down to a radius of $r \approx 0.3R_\odot$ [21]. In other words, the tachocline should encompass half of the radiation zone, in contradiction with helioseismic sounding which sets an upper limit of 5% of the solar radius on the thickness of the tachocline.

What causes the thinness of that layer still remains to be elucidated. One possibility is that its penetration is prevented by shear-induced anisotropic turbulence, as was suggested initially by Spiegel and Zahn. As an alternative, Gough and McIntyre [9] suggested that a fossil magnetic field could be responsible for that; but it will be shown in Sect. 8 that this scenario is ruled out by recent numerical simulations.

More recently it has been shown by Forgács-Dajka and Petrovay [7] that the spread of the tachocline could be prevented by the alternating dynamo field generated in the convection zone, whose penetration into the radiative interior is limited by Ohmic diffusion (see Sect. 8). That poloidal dynamo field is sheared into a toroidal field, and the resulting Lorentz force acts to reduce the differential rotation. This "fast tachocline" model, as it has been named, is compatible with the observations, for reasonable values of the Ohmic turbulent diffusivity. If this mechanism operates in the Sun, we should observe a cyclic variation of the thickness of the tachocline, which could not be detected so far.

Let us come back now to the question of which physical mechanism was able to render quasi-uniform the internal rotation of the Sun.

7 Angular Momentum Transported by Internal Motions in the Radiation Zone

The first transport process that has been considered in this context is the large-scale meridional circulation generated in the radiation zone by the torque which is applied on its top (cf. Howard et al. [10]). As explained in [27], the circulation adjusts itself such as to carry precisely the amount of angular momentum that is lost by the wind; this can only be achieved when the rotation is non-uniform, and it is accompanied by a weak shear-induced turbulence. Due to the stable vertical stratification, that turbulence is probably strongly anisotropic, and it smoothes out the differential rotation in latitude; to first approximation, one may thus treat the rotation in the radiation zone as if it were only depth dependent.

Assuming that meridional circulation and turbulence are the only processes that carry the angular momentum in the radiation zone, one can solve the relevant equations (conservation of mass, momentum and energy) to predict the present rotation profile of the Sun. The result is shown in Fig. 7, where the calculations were started with a model rotating uniformly at the equatorial velocity of 100 km/s. The angular momentum loss was asserted by the Skumanich

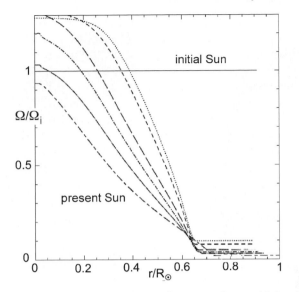

Fig. 7. Internal rotation rate of the Sun as a function of time, if the angular momentum transport is ensured only by meridional circulation and shear turbulence. Initially the model rotates uniformly with an equatorial velocity of 100 km/s. The angular momentum loss is applied at the surface according to the Skumanich law (2). The profiles are shown for the successive ages, in Myr: 0 (ZAMS), 50, 100, 500, 1000, 2000, 4550. At the age of the Sun, the core has been barely spun down [13]

law (2) calibrated such as to yield the present surface velocity of 2 km/s. One sees that the present Sun would be in very strong differential rotation, with a core still rotating approximately at its initial rate [13]. Such a steep profile is excluded by helioseismology, and therefore we have to conclude that another, more efficient process, was responsible for the extraction of angular momentum.

Two candidates have been proposed for this: an interior magnetic field or internal gravity waves; we shall now discuss them in turn.

8 Is a Magnetic Field Able to Rigidify the Solar Rotation?

At first sight the answer is yes, since even a weak poloidal field can render the rotation uniform along its field lines [5, 15]. Note that the field which is considered here cannot be the dynamo field observed at the surface: due to its cyclic behavior that field diffuses extremely little into the radiation zone, a phenomenon similar to the skin effect in electricity [8]. Thus it has to be a field residing in the radiative interior, presumably of fossil origin, which has been invoked by Gough and McIntyre [9] to prevent the penetration of the tachocline (see Sect. 6).

The problem with such a field is that it diffuses, due to Ohmic dissipation, and that it tends to connect with the convection zone, hence imprinting the

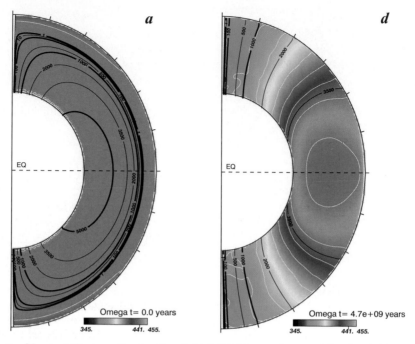

Fig. 8. Ohmic diffusion of a fossil poloidal field in a solar model. The field lines are drawn in *black*, the lines of constant rotation rate in *white*. A differential rotation is applied at the *top* of the computational domain, which extends from $r = 0.35R_\odot$ to $0.7R_\odot$, taken as base of the convection zone. Initially (*a, left*), the field is completely buried in the radiation zone, whose rotation is uniform; at the equivalent age of the Sun (*d, right*), the field has penetrated into the convection zone, and imposes the differential rotation of that zone, along the field lines, beneath on the radiative interior (Brun & Zahn [3]; courtesy A&A)

differential rotation of that region on the radiative interior – which is not observed. Gough and McIntyre were aware of this difficulty, but they suggested that the large-scale flow in the tachocline may keep the field from diffusing into the convection zone.

To check the validity of this scenario, we have performed a series of high-resolution calculations [3], of which some results are presented in Fig. 8. We used for this the 3D ASH code (anelastic spherical harmonics), which was designed to simulate a turbulent rotating convection zone, in spherical geometry, and we adapted it to the stable stratification of a radiation zone. For numerical reasons, we had to increase all diffusion coefficients, but we took care to respect their natural hierarchy (radiative diffusivity ≫ Ohmic diffusivity ≫ viscosity). We introduced an initial poloidal field, and let it evolve through Ohmic diffusion, while allowing also for the tachocline circulation, which was driven by the differential rotation applied at the top of the radiation zone.

We found that even when the calculation starts with a fossil field entirely buried in the radiation zone (left panel), this field diffuses outward and ends up connecting with the convection zone after about 1 Gyr of suitably rescaled time (with the enhanced diffusivities). The tachocline circulation is unable to prevent that inexorable diffusion, contrary to the claim made by Gough and McIntyre. Thus, either the fossil field is strong enough to enforce uniform rotation along its field lines, and then it imposes the differential rotation of the convection zone on the radiative interior, as shown in the right panel, or the field is too weak to achieve that, as it is confirmed by helioseismology, but then it cannot be invoked to rigidify the rotation in the radiation zone.

9 The Solution: Transport by Internal Gravity Waves

Since magnetic fields seem unable to enforce uniform rotation in the radiative interior of solar-type stars, as observed in the Sun, we turn to the other possible mechanism, namely the transport of angular momentum by the internal gravity waves excited by the turbulent motions at the base of the convection zone. The importance of that process had already been anticipated by Press [16], Schatzman [18] and Zahn et al. [29].

The restoring force operating on gravity waves is the buoyancy force: therefore they travel only in stably stratified regions, i.e., in radiation zones. Presumably a whole spectrum of such waves is emitted at the base of the convection zone of late-type stars; they propagate into the radiation zone, transporting angular momentum which they deposit wherever they are dissipated through radiative damping. By shaping the rotation profile, they indirectly participate in the mixing of chemicals.

Let us first examine the behavior of those waves that are of short wavelength and are therefore dissipated close to the convection zone. Prograde waves carry positive angular momentum and retrograde waves negative angular momentum. When they travel in a medium which is rotating faster than the layer where they have been emitted, their frequency is Doppler-shifted, and therefore the prograde waves are more damped than the retrograde waves. For this reason the angular velocity tends to increase where it was already high, and its slope with depth steepens until the shear becomes unstable and turbulent. That turbulent layer then merges with the convection zone, and the shear is smoothed out. But in the meanwhile the retrograde waves have deposited negative angular momentum somewhat further down, thus building there another shear layer of opposite slope, which now takes the place of the former one, and so forth. A similar phenomenon is observed in the Earth atmosphere, where it is called the quasi-biennial oscillation (cf. McIntyre [14]).

The question then arises how this very thin shear layer, located just below the convection zone, affects the waves of longer period and wavelength, which are much less damped. If there is no slope in angular velocity, the shear layer oscillation is perfectly symmetrical in time, and its effect is the same on the prograde and on the retrograde waves which cross it. But if Ω increases even slightly with depth in that layer, the prograde waves will be more dissipated, and this bias allows the waves to extract angular momentum from the deep interior.

This scenario has been tested through numerical simulations performed by Talon et al. [25], using a rather crude approach where the Coriolis force was neglected and where an arbitrary turbulent viscosity was imposed. Their calculations were one dimensional, with the flux of angular momentum averaged over level surfaces, which were assumed to rotate uniformly due to the anisotropic shear-induced turbulence discussed above. Recently the mechanism has been confirmed through more realistic calculations; besides the internal gravity waves, these include also the meridional circulation and the shear-induced turbulence [22–24].

The result fulfills all hopes: internal gravity waves succeed in enforcing nearly uniform rotation in solar-type stars at the solar age, as demonstrated in Fig. 9 by Charbonnel and Talon [4]. Furthermore their models predict the observed Li abundances, which is a crucial test to validate the mixing processes in the radiation zone. They also explain the Spite plateau for population II stars and the Li dip in galactic clusters. Clearly, the process responsible for the flat rotation profile in the Sun has now been identified.

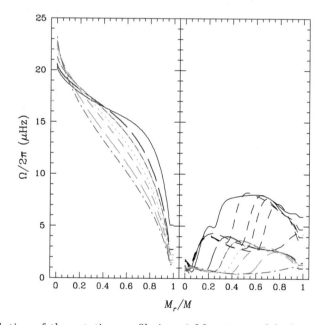

Fig. 9. Evolution of the rotation profile in a 1 $M_\odot$ star model: the angular velocity is plotted as a function of mass. When turbulence and meridional circulation are the only mechanisms which transport angular momentum (*left*), the core keeps rotating much faster than the surface, as in Fig. 7. When internal gravity waves are included (*right*), the model reaches a state of almost uniform rotation. The same braking law is applied in both cases, starting from an initial equatorial velocity of 50 km/s. The different curves correspond to ages of 0.2, 0.5, 0.7, 1, 1.5, 3.0 and 4.6 Gyr (Charbonnel & Talon [4], courtesy Science)

10 Perspectives

Many of the puzzles concerning the solar rotation have been solved, some quite recently. We now understand how the solar-type stars are spun down and how they achieve a flat internal rotation profile. But there are still some open questions.

- Presently, helioseismology is unable to decide whether the innermost core of the Sun is rotating slower or faster than the surface. The question would probably be settled by observing gravity modes, but so far there has been no firm confirmation of the candidates that have been proposed.
- We do not know the precise thickness of the solar tachocline, and its variation in time, which would help us greatly to elucidate the role of that layer in the solar dynamo and to constrain the mixing processes that occur there. It would also allow to confirm whether the dynamo field is responsible for the confinement of the tachocline.
- We are still lacking a reliable prescription of how stars lose mass and angular momentum, although much progress has been accomplished by surveys such as the Monitor project.
- We need a better description of the gravity and gravito-inertial waves in a rotating star (excitation level, spectrum, balance or eventual bias between prograde and retrograde waves).

Work is in progress on all these subjects, and we are waiting eagerly for the returns of the CoRoT mission, on the internal rotation of other stars than the Sun.

References

1. Bouvier, J., Forestini, M., Allain, S.: A&A **326**, 1023 (1997)
2. Brun, A.S., Toomre, J.: ApJ **570**, 685 (2002)
3. Brun, A.S., Zahn, J.-P.: A&A **457**, 665 (2006)
4. Charbonnel, C., Talon, S.: Science **309**, 2189 (2005)
5. Ferraro, V.C.A.: MNRAS **97**, 458 (1937)
6. Fesenkov, V.G.: AJ-USSR **26**, 62 (1949)
7. Forgács-Dajka, E., Petrovay, K.: A&A **389**, 629 (2002)
8. Garaud, P.: MNRAS **304**, 583 (1999)
9. Gough, D.O., McIntyre, M.E.: Nature **394**, 755 (1998)
10. Howard, L.N., Moore, D.W., Spiegel, E.A.: Nature **214**, 1297 (1967)
11. Howard, R., Gilman, P.I., Gilman, P.A.: ApJ **283**, 373 (1984)
12. Irwing, J., Hodgkin, S., Aigrain, S., Hebb, L., Bouvier, J., Clarke, C., Moraux, E., Bramich, D.M.: MNRAS **377**, 741 (2007)
13. Matias, J., Zahn, J.-P.: Sounding solar and stellar interiors. In: Provost, J., Schmider, F.-X. (eds.) IAU Symposium 181, poster volume, p. 103 (1997)
14. McIntyre, M.E.: NATO ASI subseries I, global environmental change, **25**, 293 (1994)
15. Mestel, L., Weiss, N.O.: MNRAS **226**, 123 (1987)

16. Press, W.H.: ApJ **245**, 286 (1981)
17. Schatzman, E.: Ann. Ap. **25**, 18 (1962)
18. Schatzman, E.: A&A **279**, 431 (1993)
19. Skumanich, A.: ApJ **171**, 565 (1972)
20. Slettebak, A.: In: Stellar Rotation (Proc. IAU Coll. 4; ed. A. Slettebak; Reidel, Dordrecht) p. 3 (1970)
21. Spiegel, E.A., Zahn, J.-P.: A&A **265**, 106 (1992)
22. Talon, S., Charbonnel, C.: A&A **405**, 1025 (2003)
23. Talon, S., Charbonnel, C.: A&A **418**, 1051(2004)
24. Talon, S., Charbonnel, C.: A&A **440**, 981 (2005)
25. Talon, S., Kumar, P., Zahn, J.-P.: ApJ **574**, L175 (2002)
26. Thompson, M.J., Tomre, J., and 24 coauthors: Science **272**, 1300 (1996)
27. Zahn, J.-P.: A&A **265**, 115 (1992)
28. Zahn, J.-P.: The Solar Tachocline. p. 89, Cambridge University Press (2007)
29. Zahn, J.-P., Talon, S., Matias, J.: A&A **322**, 320 (1997)

What Is Coming: Issues Raised from Observation of the Shape of the Sun

J.-P. Rozelot

Université de Nice-Sophia-Antipolis, CNRS –UMR 6525, Observatoire de la Côte
d'Azur (Dpt H. Fizeau), Av. Copernic, 06130 Grasse, France
Jean-Pierre.Rozelot@obs-azur.fr

Abstract Variations of the diameter, shape and irradiance are ultimately related to
solar activity, but a further investigation of how a weak magnetic field might cause
variations in the irradiance amplitude and phase, combined with a shrinking or an
expanding shape, is still needed. Indeed, accurate measurements of the solar diameter
started by Jean Picard showed that the solar diameter might be greater during the
Maunder minimum of the solar activity. After Jean Picard (and some other heirs), there
has been a lot of other measurements, ground based or from space. In this chapter we
will review the question, extending diameter variability to shape changes. We will show
how helioseismology results allow us to look at the variations below the surface, where
changes are not uniform, and putting in evidence a new shallow layer, the *leptocline*.
This layer is the seat of solar asphericities, radius variations with the 11 year cycle
and probably also the cradle of sub-layers where act complex physical processes such
as partial ionization of the light elements, opacities changes, superadiabaticity, strong
gradient of rotation and pressure. We will base our discussion on physical grounds
and show why it is important to get accurate measurements from space (SDO – Solar
Dynamics Observatory or DynaMICCS/GOLF-NG). Such measurements will provide
us a unique opportunity to study in detail changes of the global solar properties and
their relationship to changes in the Sun's interior.

1 Introduction

Since the highest Antiquity the determination of the value of the solar diameter
has been a subject widely debated. A number of historical books already treated
this question and the topic could be considered as ended. By opening a book on
astronomy, such as the *Astrophysical Quantities* [3], one may find the value

$$R_\odot = (6.955\,08\ \pm\ 0.000\,26) \times 10^8\ \text{m}$$

which appears as the best measure up to date of the solar radius. This estimate could be even considered as "definitive" and is thus widely used. However,

This chapter has been written by Jean-Pierre Rozelot with the help of the International Space Science Institute team, namely A. Bianda, C. Damiani, N. Kilifarska,
A. Kosovichev, J. Kuhn, S. Lefebvre and S. Turck-Chièze.

Rozelot, J.-P.: *What Is Coming: Issues Raised from Observation of the Shape of the Sun.*
Lect. Notes Phys. **765**, 15–43 (2009)
DOI 10.1007/978-3-540-87831-5_2   © Springer-Verlag Berlin Heidelberg 2009

looking carefully at this question, it is not so obvious. First, an absolute value is not yet determined. Just as an example, a discussion of measurements of the solar diameter made during the nineteenth century by Wittmann [89] yields $R_\odot$ = 696 265 ± 65 km (without any temporal trend), whereas measurements made by these authors at Izaña during the years 1990–2000 yield 960.63 ± 0.02 (arc second, at 1 AU), always without any significant cycle-dependence variations in excess of about 400 km [90, 91].[1] Such values are different from that adopted by Allen.[2] But giving a value of the solar diameter requires a definition, as the Sun is not a spherical solid. Several expressions can be given. The most commonly accepted is the diameter defined as the distance taken between the two opposite inflection points of the limb intensity profile, at a given wavelength. But other definitions can be used. For instance, an equipotential level of gravity (to a constant) perfectly defines the outer shape. Second, the Sun is a fluid body in rotation. It follows to first order an oblateness of the whole figure, and to other orders, deviations to sphericity. The diameter D under consideration must be thus identified. The semi-diameter R (radius, more frequently used) is referred to as equatorial, R_{eq}, or polar, R_{pol}, for which values are as follow [61]:

(a) if the Sun can be considered (for instance in stellar structure models) as a body rotating at a uniform rotation speed rate

$$R_{eq} = 6.95991756 \times 10^8 \text{ m and } R_{pol} = 6.95985961 \times 10^8 \text{ m (uniform rotation)}$$

and

(b) under a (surface) differential rotation speed rate

$$R_{eq} = 6.95991756 \times 10^8 \text{ m and } R_{pol} = 6.9598438 \times 10^8 \text{m (non-uniform rotation)}$$

At last, on a pure physical point of view, as the distribution of matter is not uniform inside the Sun (from the core to the surface), as well as the distribution of the velocity rates, the outer shape shows distortions which are linked to the successive gravitational moments. Hence, the solar radius, $R(\theta)$, must be a function of the latitude (θ). As a consequence, all layers that constitute the Sun are not spherical (Fig. 1). This has been already recognized for instance for the tachocline [30], which is prolate.

The knowledge of the value of the solar radius, once the definition is stated, is a key parameter not only in stellar physics but also in solar models. Indeed, the solar radius is a function of time. On very long-term evolution (several millenia), this has been recognized as the paradox of the faint young Sun.[3] On shorter term

---

[1] An attempt of solving discrepancies has been made by Habereitter et al. (ApJ., **675**, L53–L56, 2008) while this paper was in press.

[2] See also Table I in paper [41].

[3] The "faint young Sun paradox" was first pointed out by Carl Sagan and George Mullen [14]: it postulates that stars similar to the Sun should gradually brighten over their life time (excluding a very bright phase just after formation). This prediction is supported by the observation of lower brightness in young stars of solar type. It is generally acknowledged that the early Sun (the faint young Sun) had only 70% of the energy output that it has today. This would mean that the Earth would have been entirely frozen (no liquid water) in its early history, in contradiction with geological

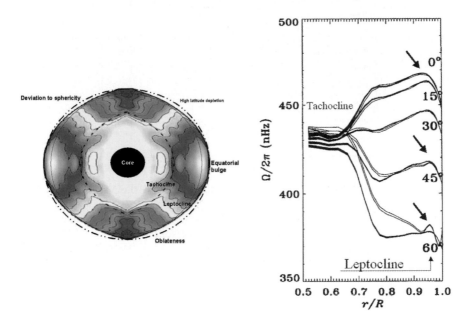

Fig. 1. *Left*: Due to the non-uniform distribution of mass and velocity rates inside the Sun (isocontours are shown), the resulting outer shape is not spherical and shows deviations to sphericity (exaggerated size here). However, the global shape remains oblate. The *inner dot line* (− −) shows the prolate *tachocline* and the *thin line* (—), the *leptocline*. *Right*: Several profiles of the rotation rate are plotted according to different heliographic latitudes. The changes indicated by an *arrow* show the seat of the *leptocline*

(since around 1600 up to now), it has been shown that the solar radius may also evolve with time (see Fig. 2 in [55], Fig. 3 in [67], or Figs. 2 and 3 in [68], all upgraded from [82]), likely on a very large periodic modulation, of about 110–120 years (from one extrema to the other one), the amplitude being not yet accurately determined.[4] On shorter periods of time (ranging over some solar cycles), the temporal variability has remained unclear for a long time, but it has been shown recently that it is in antiphase with the solar cycle for layers lying

---

observations of sedimentary rocks, which required the presence of flowing liquid water to form. (The case for Mars is even more extreme due to its greater distance from the Sun.) Does this paradox – between the icehouse that one would expect based on stellar evolution models and the geologic evidence for copious amounts of liquid water – indicate a problem with our stellar evolution models? Or is there another way around this conundrum?

[4] A significant 110 yr period in solar activity is not fully recognized. However Damon and Jirikovic [18, 19] identified two powerful harmonics considered as fundamental at 211.5 and 88.1 yr – the Suess and the Gleissberg cycles – which modulate the Schwabe 11 yr period and produce periods of maxima and minima in solar activity: the around 110–120 yr period could be a sub-harmonic.

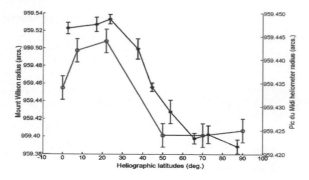

Fig. 2. Comparison between results obtained deduced from the Mount Wilson data, over 30 years of analysis (left scale) [43, 44], and those of the Pic du Midi, obtained on September, 1–4, 2001, where exceptional conditions of seeing were encountered (right scale) [72]. The observed solar limb contour does not follow an ellipsoidal shape and shows deviations to sphericity, as theory states [46]. However, the excess of asphericity found in the Mount Wilson data (120 mas around 20° of heliographic latitude with respect to 70° latitude) can be interpreted by the spectral domain; a contribution of the chromosphere can be suspected, as it was found that the chromosphere maybe oblate (Auchère et al., 36, L57 – L60, 1998)

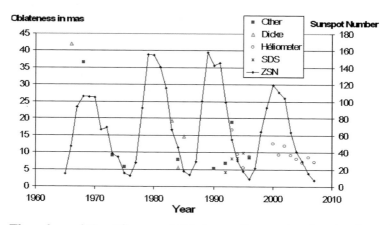

Fig. 3. The values of the difference ΔR between the equatorial and polar radii, according to several authors (see a list of estimates in [32] and in this chapter), plotted versus the international sunspot number. The oblateness seems to be in phase with the solar activity index. However the mean radius and the radius of the equatorial regions are out of phase. This result can be interpreted by means of the combined effects of the two first multipolar gravitational moments: the quadrupolar J_2 term is prevailing during lower activity periods of time, albeit the contribution of the hexadecapolar J_4 term is predominant during higher solar activity (and J_2 weaker)

at the very near surface of the Sun, and in phase for layers seated most deeper inside [45, 48, 49] (see Sect. 5.3).

The relevance of precise measurements of the Sun's shape can be summarized as follows:

- If $R(t)$ is known over a long period of time, ranging over several centuries, and even on undecennial cycles, then luminosity variations can be tackled (solar luminosity has increased over the life time of the Sun). By contrast, we do not know yet how radius variations on time scales ranging from seconds to hours, if they exist, may play a role in the luminosity variations. Hence, determination, in real time, of the so-called asphericity-luminosity parameter [44]

$$w = \partial ln(R)/\partial ln(L)$$

  is required. A table summarizing the estimated values of w is given in [25, 26]. Note also that the knowledge of this parameter is of high importance for the study of the Earth's upper atmosphere.
- If $R(\theta)$ is known, then asphericities coefficients c_n can be deduced, leading in principle to a determination of the solar gravitational moments J_n. The knowledge of these parameters is relevant to celestial mechanics and is required to set up precise ephemeris (due to the relation between J_n and the inclination of the orbits of planets, i.e., spin–orbit couplings) in a general relativistic description [59, 61].
- If $R(t)$ and $R(\theta)$ are known, then the solar core dynamics can be inferred. This can be achieved either through ground-based observations where the Fried parameter is larger than 15–20 cm or through dedicated space missions, such as SDO (Solar Dynamics Observatory) [39] or DynaMICCS/GOLF-NG (Dynamics and Magnetism from the Inner Core to the Corona of the Sun) [83, 84], expected to be launched by the end of 2008 (SDO) and 2012–2015 (DynaMICCS).

2 Observations of the Solar Shape

The solar shape is very difficult to observe and hence very difficult to measure because it requires an astrometric accuracy. If Dicke [21] can be considered as a pioneer in this task, his first attempts at Princeton were not convincing. Several other measurements, made between 1974 and 1994 (see a review in [59] or [65]), lead to more reliable results. Up to the 1990s, only the oblateness was searched for. To summarize, it was shown that, if the Sun were rotating at a uniform velocity rate, the oblateness is[5]

$$\Delta R = (R_{eq} - R_{pol}) = 6187 \quad \text{m} \quad \text{or} \quad 8.53 \quad \text{mas.} \tag{1}$$

---

[5] "*mas*" stands for milliarcsecond.

Note that ΔR (Eq. (2)) is upper bounded by 10.54 ± 0.25 mas as a maximum and 6.39 ± 1.31 mas as a minimum, according to the value adopted for the velocity rate (at the surface). See Sect. 4.

However, taking the differential rotation into account, the oblateness becomes

$$\Delta R = (R_{eq} - R_{pol}) = 7370 \quad \text{m} \quad \text{or} \quad 10.15 \quad \text{mas.} \tag{2}$$

It must be noted that the differential rotation increases the oblateness, in apparent contradiction with the theory of rotating stars. This can be explained by a change in the radial velocity rate near 45° latitude: $(d\omega/dr) = 0$ at this latitude, $(d\omega/dr) > 0$ at higher latitudes and $(d\omega/dr) < 0$ at lower latitudes.

Today, the best results concerning estimates of ΔR are given through three main different techniques. The results of the first one, deduced through balloon flights (in limited number) and the so-called SDS experiment, can be found in [79]. The second one, still into operation, has been developed at Mount Wilson Observatory (USA) [87]. Observations are based on a spectrographic analysis of the neutral iron line Fe I at 525 nm. Measurements have been recently re-analyzed by Lefebvre et al. [43, 46]. The third one is developed at the Pic du Midi Observatory by means of the scanning heliometer, initially conceived by J. Rösch. A full description of the apparatus can be found in [62] and the observational dependence of the solar radius with heliographic latitude is presented in [72]. A comparison of the results obtained by these two last techniques is given here in Fig. 2, and an analysis can be found in [43, 46]. Departures from a pure sphere are clearly seen: a bulge extends at the equator, up to around (30–40°), followed by a depression, the polar shape remaining oblate. Lastly, as measurements extend in time, it has been possible to study the temporal shape variations. SDS experiments yield an oblateness in phase opposition with the solar cycle, in contradiction with all other results: ground-based observations, (heliometer) [62] or space observations, either through SOHO-MDI [24] analysis or through recent RHESSI [28, 37] space experiments. Other results of the solar oblateness (mainly performed by Dicke and collaborators) can be found in [62] and are summarized in Table II given by Pireaux [59].

The variability of the solar oblateness with time can be briefly interpreted by the contribution of the two terms J_2 and J_4. Emilio et al. [24] reported a solar shape distortion using the Michelson Doppler Imager (MDI) aboard the Solar and Heliospheric Observatory (SOHO) satellite after correcting measurements for bright contamination. It was found that the shape distortion is nearly a pure oblateness term in 2001, while 1997 has a significant hexadecapolar (J_4) shape contribution. However, due to the fact that the hexadecapolar term might be of the same order of magnitude than the quadrupolar term, but obviously in opposite sign, it results that the equatorial radius at the surface is in antiphase with the solar cycle, which is consistent with the results deduced from the f-mode analysis.

Dicke and Rösch can be considered as precursors in the field of the solar shape. The first one has undeniably set the basis for the underlying physics of the oblateness (see also [67]). Even if his papers were often examined critically, they triggered a great amount of ideas which have moved astrophysics forward. The second one carefully examined the conditions of solar diameter observations, such as blurring effects or displacement of the inflection point toward the inner part of the disk (see also Hill and Oleson [36]). He defined also the *helioid* as the whole outer solar shape, in an analogy with the Earth's *geoid*.

Finally, a recent analysis of the data obtained at the Pic du Midi Observatory shows that the maximum value of the departures of the solar shape from a sphere does not exceed 20 mas; the oblateness varies slowly in time, in phase with the solar cycle (see figures in [62, 73, 74]). Figure 3 shows the variations with time of the oblateness as deduced from our own measurements (circles) and compared with those of Dicke (triangles), SDS experiments (cross) and other measurements (squares) found in the literature (see [32]).

One of the first attempts to understand theoretically solar surface distortions was made by Lefebvre [44] who showed that the thermal wind effect is one of the contributors at the solar surface. Note that the thermal wind (which is not the solar wind) is due to the difference in temperature between the pole and the equator and is the equivalent to the geostrophic effect, well studied by (Earth) meteorologists.

3 How Large Are the Temporal Variations of the Solar Diameter?

On physical grounds, temporal variability of the solar diameter cannot exceed 10 mas peak to peak in amplitude. Callebaut et al. [12] were certainly the first to point out that changes in solar gravitational energy, in the upper layers, necessarily involve limited variations in the size of the envelope. The mechanism can be described as follows, assuming hydrostatic equilibrium. Bearing in mind the definition of the energy $E_g = -\int Gm/r \, dm$, a thin shell of radius dr (or dm) in equilibrium under the gravitational force and the pressure gradient will expand or contract if any perturbation to these forces occurred. In Fazel et al. [25, 27], the authors improved the method and showed that any variations of the size of the solar envelope must be less than some 12 km of amplitude over a solar cycle, a value in agreement with those deduced from inversion of the helioseismic modes, or from space observations through the SOHO-MDI data analysis [40, 41].

Any other larger values are not consistent with astrophysical observations of other solar phenomena. For example, observed temporal irradiance changes, which are observed at a level of $\approx 1/00$ over the solar cycle, could be explained by a ≈ 200 mas changes in the solar diameter, if this mechanism ought to play a unique role. Such a large value is not realistic, as it would automatically cancel all other physical explanations and among them, the magnetism of the surface, which is known to explain most (but not all) of the irradiance variations (unless unknown physical mechanisms play a role at the extreme border of the limb).

As another example, consider the multipolar gravitational moments of the Sun. The injection of larger values of $\Delta R(t)$ in models which are tested in other respects (such as for the inclination of planetary orbits, theory of lunar motion, general relativity) would lead to major impossibilities. In the theory of lunar motion case, the inclusion of J_n estimates in a spin-orbital motion theory can be accurately confronted with observed lunar physical librations. As these librations are known to a few milliarcseconds of precision, it results that $\Delta R(t)$ is inevitably upper bounded [10, 66] by some 10–15 mas.

The next question the reader may ask is why solar astrolabes, distributed around the Earth (in France, Chile, Brazil and Turkey), are still measuring a diameter variability over the solar cycle of about 100–300 mas (sometimes more). A recent careful analysis, based on a statistical variographic analysis [8, 17], showed that measurements made by astrolabes may represent the fluctuations of the upper Earth atmosphere, i.e., the UTLS (Upper Troposphere–Lower Stratosphere) region, and *not*, as it is often claimed, the fluctuations of the lower atmosphere alone (the turbulence). As the UTLS region is modulated by solar activity [16], it results that astrolabes measure a part of the solar signal, but only a small part of it, as the singular spectrum analysis (SSA) shows and as it was suggested earlier ([55, 77]). Figure 4 shows the results obtained in the case of the French solar astrolabe data, but they are the same for other astrolabes. Only two eingenvalues are detected (i.e., the trend and a cyclic modulation), the remaining being noise. In fact, one can say that astrolabes are powerful instruments to measure the stratospheric variability (an amplification of the solar signal may also be produced in this zone): this point is not so trivial.

The last issue is the *phase*: Is the weak solar diameter variability in phase or not with magnetic activity? This question would deserve to be more widely debated, but let us jump to our conclusion, based on papers dealing with three different approaches: the original papers by Godier and Rozelot [31, 32], the papers by Fazel et al. [25, 27] and the papers by Lefebvre et al. [45, 49].

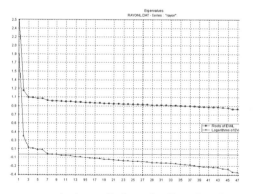

Fig. 4. Singular spectrum analysis applied to the French solar astrolabe data. It can be clearly seen that only two eigcnvalues are detected (*upper curve*; the *lower* one is in log unit), the main signal being noise [55, 77]. However, variographic analysis of the same data shows a clear correlation with the stratospheric signal coming from the UTLS region ([8, 17])

– The first quoted papers describe asphericities in the subsurface layers (through the so-called *Theory of Figures*): one asphericity is located around $0.7R_{\odot}$ (which is identified as the tachocline), and another one is located between 0.982 and 0.993 $R_{\odot}$, with two dips, at 0.986 $R_{\odot}$ and at 0.992 $R_{\odot}$; this last layer constitutes the *leptocline* (Fig. 2 in [33]).

– The second set of papers is based on the assumption that the effective temperature of the surface is nearly immutable, as suggested from observations made by Livingston at Kitt Peak [50, 51]. It is shown that to model the remaining part of the irradiance variations (i.e., the part which is not coming from surface magnetism phenomena), there may exist a phase shift in the $[dT, dR]$ plane, with a $dT(dR)$ curve separating solar variations in antiphase (for temperature values below 0.08 K), and in phase (for temperature values greater than 0.08 K) with solar irradiance variations [27]. It must be pointed out that in this case, the nonvariability of dT over the solar cycle could be explained by the flux tubes passing between the granules, without interaction; due to the magnetic pressure, one would expect a change in the mean size of the granules that would be thus shifted toward the smaller sizes, as magnetic activity is increasing. Such observations were already made at the Pic du Midi Observatory since 1997. As a consequence, the whole size of the Sun would decrease (anticorrelation with solar magnetism).

– The third set of papers reports changes of the Sun's subsurface stratification inferred from helioseismic inversions (see Sect. 5.3), for which a clear phase changing with depth is shown.

4 Solar Shape and Rotation

The study of the rotation of stars is not trivial. In theory, the problem is exceedingly simple and can be formulated as follows. Let us consider a single star that rotates along a fixed direction in space, with an angular velocity ω, and first assume that, for $\omega = 0$, the star is a gaseous body in gravitational equilibrium. The problem is to determine the outer shape of the star when the initial sphere is set rotating at an angular velocity ω. Such studies were conducted for the first time by Milne [54], then fully achieved by Chandrasekhar [15]. If the body is in uniform rotation ω and the density of the form $\rho = r^{-n}$ then the oblateness is given by

$$\varepsilon = (0.5 + 0.856\rho_c/\rho_m)\omega^2 R/g,$$

where ρ_m is the mean density of the star, ρ_c the density of the core, R the radius of the initial sphere and g the gravity at the surface. Applied to the Sun and to first order, this nice formula gives ($g = 2.7{\times}10^4 cm/s^2$, $\rho_c/\rho_m = 108.3$)[6]

$$\varepsilon = 1.10 \times 10^{-5} \pm 2.62 \times 10^{-7},$$

---

[6] According to the different values the ratio of central to mean density may take, ε lies between 0.504 and 0.513 $\omega^2 R/g$.

for a constant equatorial velocity rate (of $14°34 \pm 0°17/day$); ε decreases to $9.03 \times 10^{-6} \pm 4.26 \times 10^{-7}$ for a Sun rotating at a constant rate which is the $45°$ latitude velocity rate ($13°00 \pm 0°31/day$). At the pole, the uncertainty is a bit higher: ε is $6.7 \times 10^{-6} \pm 1.4 \times 10^{-6}$ (mean latitude velocity rate: $11°10 \pm 1°2/day$). Such values lead to a mean difference between the equatorial and the polar radii of respectively 10.54 ± 0.25 mas (equator), 8.67 ± 0.41 mas ($45°$) and 6.39 ± 1.31 mas ($90°$). In a metric scale, this amounts to a difference of respectively 7.6 ± 0.2 km, 6.3 ± 0.3 km and 4.6 ± 0.9 km. Note that in this case, the difference between R_{eq} and R_{sp} is 3.5 mas ($R_{eq} = 959.63''$). A ponderous process (the best estimate weighted by each error) yields

$$f = 8.33 \times 10^{-6} \pm 1.87 \times 10^{-6}.$$

The second point is to understand what happens if ω is not constant, not only in latitude (differential rotation) but also throughout the body, from the surface to the core. We are faced today with such problems, not only in the solar case but also for stars. With the advent of sophisticated techniques such as interferometry, one is now able to accurately determine the geometrical shape of the free boundary of stars, such as Altair or Achernar for which observations of the geometrical envelope have been made for the first time, respectively, by van Belle et al. [88] and Domiciano de Souza et al. [22].[7] But it would be of little or no interest to observe the geometric shape of a star – or that of the Sun – if one would not be able to infer some information on stellar – solar – physics. With such an approach, the purpose of theoreticians is to enumerate all the possible angular velocity distributions (from the center to the surface) that are compatible with the observed stellar – solar – surface. For stars, Maeder [53] examined the effects of rotation and among them, he described the equation of the surface with a rotation law which is differential, but only in the surface layer.

In other words, the knowledge of the angular velocity distribution from the core to the surface, together with the knowledge of the density function (related to the pressure function), completely determines the outer shape of the stars. Different techniques exist to observe such a figure. Once accurately determined, one would be able to go back to the physical properties of the body. This approach is called *Theory of Figures*.

This theory has been widely used in geophysics and is still used in specific cases, such as for the planet Mars, with an incredible accuracy ($J_{2\ Mars} = 1.860718 \times 10^{-3}$ according to [92], from a 75th degree and order model). Curiously, nothing or very little was done in the solar case, until 2001 by Rozelot et al. [67, 71]. The complexity of the rotation profile (Fig. 1) will highly infer on the photospheric shape: the outer figure is highly sensitive to the interior structure. Thus, in principle, accurate measurements of the limb shape distortions,

---

[7] Outer shape of other stars has been determined since then; see [45].

which are called "asphericities" (i.e., departures from the "helioid", the reference equilibrium surface of the Sun), combined with an accurate determination of the solar rotation provide useful constraints on the internal layers of the Sun (density, shear zones, surface circulation of the plasma, etc.). Figure 5 shows such asphericities that can be seen at a given spatial resolution. Alternatively, theoretical upper bounds could be inferred for the flattening which may exclude incorrect/biased observations. Lastly, another approach consists in analyzing the effects of rotation on stellar/solar p-mode frequencies, as the rotation of the bodies affects their oscillation frequencies [35].

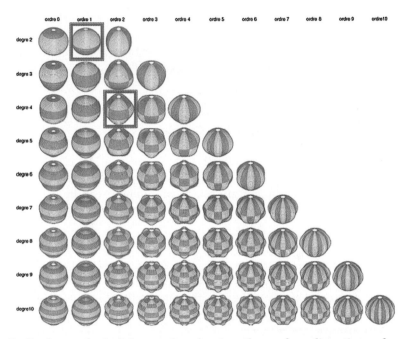

Fig. 5. Laplace spherical harmonics showing the surface distortions of a rotating fluid body. The solar oblateness ($n = 1$, $l = 2$) and the quadrupole moment ($n - 2$, $l = 4$) are illustrated in the two boxes (after R. Biancale, – personal communication)

5 The Solar Shape and Fundamental Physics

Of all the fundamental parameters of the Sun (diameter, mass, temperature, etc.), the successive gravitational moments that determine the solar moments of inertia are still poorly known. However, these moments have a physical meaning: they tell us how much the Sun's material contents deviate from a purely spherical distribution and how much the velocity rate differs from a uniform distribution. Thus, their precise determinations give indications on the solar internal structure.

The dynamic study of the gravitational moments until now is mainly based on solar observations (mainly through helioseismology but also through astronometric observations of the solar diameter) and solar models of rotation and density. Various methods (stellar structure equations coupled to a model of differential rotation, theory of the Figure of the Sun, helioseismology) lead to different estimates of J_n, which, if they agree on the order of magnitude, still diverge for their precise values [60, 73, 75].

5.1 Solar Asphericities

Asphericities, as defined before, can be computed according to the degree l and order n in the development of Laplace spherical harmonics in the general case of a rotating fluid body. Let ρ be the density (function of the radius r) and denote with a subscript 0, the lowest order l, spherically symmetric. Asphericities, described as [7]

$$c_l = -\rho_l/d\rho^{(0)}/dr \quad \text{(density)}, \tag{3}$$

$$s_l = -p_l/dp^{(0)}/dr \quad \text{(pressure)}, \tag{4}$$

measure the perturbation (nonspherically symmetric) and are usually expressed in terms of the normalized potential defined by $J_l = K\phi_l$, where $K = R_\odot/GM_\odot$ at the solar surface. The different gravitational moments can be written as

$$J_{2l} = \frac{R_\odot}{GM_\odot} \; \phi_{2l}(R_\odot), \tag{5}$$

where $\phi_{2l} = 0$ at the surface $r = R_\odot$. The function ϕ_{2l} is the solution to a differential equation requiring the knowledge of $\rho(r)$ and $\omega(r, \theta)$, where θ is the colatitude. A complete expression of ϕ_2 and ϕ_4 was provided by Armstrong and

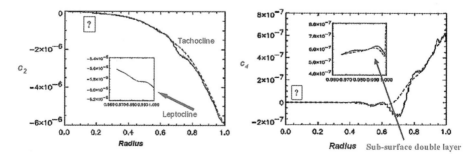

Fig. 6. Asphericity coefficient of degree 2 (*left*) and 4 (*right*), with respect to solar fractional radius ($r/R_\odot$): *solid* and *dashed lines*, respectively, density – Eq. (3) – and pressure asphericities – Eq. (4), after [7]. The dip in the curve at 0.7 $R_\odot$ locates the tachocline and indicates that this layer is prolate (instead of oblate for a bump). The near surface anomaly locates a new double subsurface layer, the "leptocline", that does not seem to be noticed before. See also [33]

Kuhn [7], using the solar standard rotation law $[\omega(\theta) = \omega_0 + \omega_2 \cos^2(\theta)]$, which permits to deduce c_l:

$$\frac{d^2\Phi_2}{dr^2} + \frac{2}{r}\frac{d\Phi_2}{dr} - 6\frac{\Phi_2}{r^2} = \frac{4\pi r^2}{M_r}\left[\Phi_2\frac{d\rho_0}{dr} - \frac{8}{21}\frac{2\omega_2}{\omega_0}r\rho_0\omega_0^2 - \frac{r^2}{3}\frac{d}{dr}\left(\rho_0\omega_0^2 + \frac{3}{7}\rho_0\frac{2\omega_2}{\omega_0}\omega_0^2\right)\right],$$

(6)

$$\frac{d^2\Phi_4}{dr^2} + \frac{2}{r}\frac{d\Phi_4}{dr} - \frac{20}{r^2}\Phi_4 = \frac{4\pi r^2}{M_r}\left[\Phi_4\frac{d\rho_0}{dr} + \frac{4}{35}\frac{2\omega_2}{\omega_0}r\rho_0\omega_0^2 - \frac{2}{35}r^2\frac{d}{dr}\left(\frac{2\omega_2}{\omega_0}\rho_0\omega_0^2\right)\right].$$

(7)

Results are shown in Fig. 6 for $l = 2$ and $l = 4$ (shape coefficients are expressed in units of the solar radius): a clear signature of the tachocline appears, at $r = 0.7\ R_\odot$, which is prolate (a dip of c_2 plotted as a function of the fractional radius $r/R_\odot$ and a bump of c_4 plotted as a function of $r/R_\odot$). However, the authors seem not to have noticed another aspherity in the c_n curves near the surface, i.e., an oblate layer, determined by a bump of c_2 and a dip of c_4, which is the signature at $r = 0.99\ R_\odot$ of another distorted shell, of different physical properties, the *leptocline* (Sect. 5.3)

5.2 Solar Gravitational Multipole Moments

Observations allow to constrain analytical rotation models, in colatitude θ and depth r. The first attempt to derive an analytical rotation law from helioseismic data has been made by Kosovichev [38]. Using such a law, several authors computed the gravitational moments (see Table 1), but discrepancies appeared, mainly for J_4. The discrepancy between the values obtained through different methods and authors can be explained by the use of different density models and by the way the differential equation Eq. (5) is integrated. It can be seen that the method using helioseimic data leads to multipole moment values lower than those obtained by other methods. However, the octopole moment, J_4, is much more sensitive than the quadrupole moment, J_2, to the presence of latitudinal and radial rotations in the convective zone. Ajabshirizadeh et al. [1] showed that the surface magnetism may reconcile the different approaches, between the Theory of Figures and numerical integration of Eq. (5): J_4 obtained by the first theory seems better matching observations. If we can adopt $(2.4 \pm 0.4)\times 10^{-7}$ as a good estimate for J_2, it remains that J_4 is very sensitive to the subsurface gradient of rotation: an estimate of $(4–7)\times 10^{-7}$ seems in better agreement with the observations. Finally, we can point out the formula linking J_2 and f in the presence of a magnetic field, as deduced by Ajabshrizadeh et al. ([1], see also Fig. 1):

$$J_2 = (2/3)f(1 - m') - (1/3)m$$

where m takes into account the velocity rate and m' is directly related to the magnetic moment of the rotating body. Kosovichev [38] noted that a subsurface shear layer results when the obtained helioseismically internal rotation is matched with the surface rotation. Hence, we suspected that the shallow layer near the surface may play an important role.

Table 1. Some solar gravitational multipole moments quoted from different authors and methods. The absolute order of magnitude of J_2 is 2.4×10^{-7}; J_4 is very sensitive to the subsurface gradient of rotation: an estimate of $(4-7) \times 10^{-7}$ seems better match the observations

References	Method	J_2	J_4	J_6	Others
Ulrich & Hawkins [84]	SSE + spots rotation law	$(10-15) \times 10^{-8}$	$(0.2-0.5) \times 10^{-8}$		
Gough [32]	First determination of helioseismic rot. rates	36×10^{-7}			
Campbell & Moffat [3]	Planetary orbits	$(5.5 \pm 1.3) \times 10^{-6}$			
Landgraf [40]	Astrometry of minor planets	$(0.6 \pm 5.8) \times 10^{-6}$			
Lydon & Sofia [50]	SDS experiment	1.84×10^{-7}	9.83×10^{-7}	4×10^{-8}	$J_8 = -4 \times 10^{-9}$ $J_{10} = -2 \times 10^{-10}$
Paternò et al. [54]	SSE + empirical rotation law and SDS	2.22×10^{-7}			
Pijpers [56]	SSE + GONG and SOI/MIDI data. Weighted value	$(2.14 \pm 0.09) \times 10^{-7}$ $(2.23 \pm 0.09) \times 10^{-7}$ $(2.18 \pm 0.06) \times 10^{-7}$			
Armstrong & Kuhn [7]	Vect. spher. harm. numerical error	-0.222×10^{-6} 0.002×10^{-6}	3.84×10^{-9} 0.4×10^{-9}		
Godier & Rozelot [29]	SSE + Kosovichev law	1.6×10^{-7}			
Roxburgh [61]	SSE + two models of rotation law	2.208×10^{-7} 2.206×10^{-7}	-4.46×10^{-9} -4.44×10^{-9}	-2.80×10^{-10} -2.79×10^{-10}	$J_8 = 1.49 \times 10^{-11}$ $J_8 = 1.48 \times 10^{-11}$
Rozelot et al. [67]	Theory of Figures	$-(6.13 \pm 2.52) \, 10^{-7}$ Note 3	3.4×10^{-7} Note 4		
Rozelot & Lefebvre [70]	Theory of Figures	-6.52×10^{-7}	4.20×10^{-7}	-9.46×10^{-9}	$J_8 = 2.94 \times 10^{-13}$

Table 1. (continued)

References	Method	J_2	J_4	J_6	Others
Rozelot et al. [71, 73, 74]	SSE + subsurface gradient of rotation (SGR)	$-2.28 \times 10^{-7} \pm 15\%$	Very sensitive to SGR Range: $\pm\,20\%$		
Ajabshirizadeh	With magnetic field	-2.613×10^{-7}	$+6.29 \times 10^{-7}$	-1.42×10^{-8}	$J_8 = +5.05 \times 10^{-13}$
et al. [2]	No magnetic field	-2.300×10^{-7}	$+6.29 \times 10^{-7}$	-1.42×10^{-8}	$J_8 = +5.05 \times 10^{-13}$

Note 1: SDS stands for solar disk sextant.

Note 2: SSE stands for stellar structure.

Note 3: The apparent large error comes from the fact that the value is a weighted average of several rotation rates.

Note 4: A mistake has been made in [33]: the second term of J_4 (i.e., mA_4) was incorrectly multiplied by "f" in the computations.

5.3 The Leptocline

Analyzing the temporal variation of f-mode frequencies for 1996–2004, Lefebvre and Kosovichev [45] have shown changes in the Sun's subsurface stratification. They have found a variability of the "helioseismic" radius in antiphase with the solar activity, the strongest variations of the stratification being just below the surface, around 0.995 $R_\odot$. On the other hand, the radius of the deeper layers of the Sun, between 0.975 and 0.99 $R_\odot$, changes in phase with the 11 year cycle (Fig. 7). A more careful analysis of these f-modes shows variations in the even-a coefficients, of nonnegligible amplitude, with both the frequency and the cycle, that imply the existence of asphericities in the subsurface layers [9, 47]. The conclusion is that this interface layer corresponding to the border between the interior of the Sun and its atmosphere is the seat of strong physical phenomena (in addition to non-homologous radius changes in time and depth), such as shearing, disturbance of the turbulent pressure, constraints upon the magnetic field, processes of ionization and, likely, inversion of the radial gradient of rotation and some tiny variations of the luminosity (Fig. 8).

Even if this layer is maybe more complex, involving another shell of some oblateness at the very near surface (unreachable for the moment to the f-modes), located at around 0.999 $R_\odot$, the proof is now made that the *leptocline* is a new and crucial zone that cannot be avoided in investigating the global properties of the Sun and its evolution on time scales of the order of months or years.

5.4 Solar Radius and Gravitational Energy Variations

As previously mentioned, the gravitational energy can be computed in the case of a nonspherical Sun, which leads to a variation of the solar luminosity L with

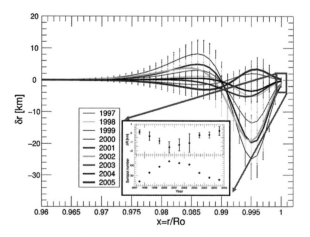

Fig. 7. Nonmonotonic variations of the radius in the most outer layers of the Sun. Below 0.99 $R_\odot$, the radius is varying in phase with the solar cycle with a maximal amplitude of about 10 km. Above 0.99 $R_\odot$ the radius is varying in antiphase with the solar cycle, which implies a compression with a maximal amplitude of about 30 km peak to peak amplitude and 2–3 km at the surface. After [45, 49]

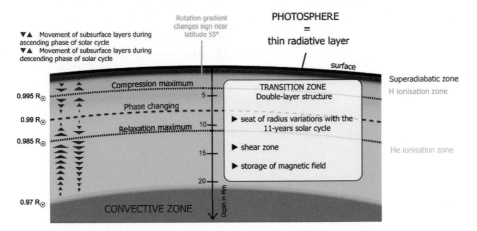

Fig. 8. A schematic view of the outer layers of the Sun shaping the leptocline. (After [47]; see also [33].) The most interesting feature is the changing phase near 0.99 $R_\odot$. This region is the seat of many phenomena: an oscillation phase of the seismic radius, together with a nonmonotonic expansion of this radius with depth, a change in the turbulent pressure, likely an inversion in the radial gradient of the rotation velocity rate at about $(45\text{--}50)°$ in latitude ($d\omega/dr$ is positive beyond these latitudes, negative below), opacities changes, superadiabicity, the cradle of hydrogen and helium ionization processes and probably the seat of in situ magnetic fields

dR according to a development of order n (see [26]). The authors made two computations, one with $n = 1$ (monotonic expansion with radius) and the other with $n = 2$ (non monotonic expansion).

Table 2 taken from [27], gives the results for two values of $\Delta L/L$. The first one is the usual adopted value, 0.0011, using TSI composite data from 1987 to 2001 [20]; mean value $L_\odot = 1366.495$ W/m². The second one is 0.00073, determined through a re-analysis of the composite TSI data over the period of time 1978–2004 [29]; mean value $L_\odot = 1365.993$ W/m². For $n = 2$ (the most likely case consistent with recent other results), the estimate of ΔR is smaller than the 8.9 km obtained in the case of a spherical Sun by Callebaut et al. [12]. However the $\Delta R/R$ agrees with that of [5], i.e., $\Delta R/R = 3\times10^{-6}$, that

Table 2. Variations of the solar radius computed in two cases, monotonic ($n = 1$) and non monotonic ($n = 2$) expansion, and for two mean values of $L_\odot$. The case $n = 2$ is the most likely. The sign $(-)$ indicates a shrinking, i.e., an anticorrelation of ΔR and L in this layer

$\Delta L/L = 0.0011$ [20]		$\Delta L/L = 0.00073$ [29]	
$\Delta R/R = -1.70 \times 10^{-5}$	$(n = 1)$	$\Delta R/R = -1.13 \times 10^{-5}$	$(n = 1)$
(or $\Delta R = 11.8$ km)		(or $\Delta R = 7.86$ km)	
$\Delta R/R = -8.38 \times 10^{-6}$	$(n = 2)$	$\Delta R/R = -5.56 \times 10^{-6}$	$(n = 2)$
(or $\Delta R = 5.83$ km)		(or $\Delta R = 3.87$ km)	

used f-mode frequency data sets from MDI (from May 1996 to August 2002) to estimate the solar seismic radius with an accuracy of about 0.6 km (see also among other authors [4, 78], for such a determination of the solar seismic radius to a high accuracy). It results that the asphericity-luminosity parameter w is -1.55×10^{-2} ($n = 1$) and -7.61×10^{-3} ($n = 2$).

Two points result from the analysis of the data. The first one concerns the "helioseismic radius" which does not coincide with the photospheric one, the photospheric estimate always being larger by about 300 km [11]. This point would require more specific attention in the future.

The second issue addresses the shrinking of the Sun with magnetic activity as pointed out by Dziembowski et al. [23], using f-mode data from the MDI instrument on board SOHO, from May 1996 to June 2000. They found a contraction of the Sun's outer layers during the rising phase of the solar cycle and inferred a total shrinkage of no more than 18 km. Using a larger database of 8 years and the same technique, Antia and Basu [6] set an upper limit of about 1 km on possible radius variations (using data sets from MDI, covering the period of May 1996 to March 2004). However, they demonstrated that the use of f-mode frequencies for $l < 120$ seems unreliable.

It results from the above discussion that the luminosity changes are likely produced in the outer shallow layer of the Sun. It thus appears that the leptocline might be the seat of the observed $1/_{oo}$ variations of the irradiance.

A recent application of the virial theorem to the radius changes in the Sun induced by magnetic variations [81] shows that the radius decreases around the time of maximum magnetic field strength. If we may have confidence in this result, by satisfying the conversion of total energy, it remains to explain why the observations show that the oblateness is in phase with the magnetic activity. This may be due to the reversal of the radial gradient of rotation near 45° of latitude, as explained before.

5.5 Results from Ground-Based Observations and Space Missions

Indeed, solar asphericities, encoded mainly by the first two coefficients c_2 and c_4, can be observed. An estimate of these two coefficients derived from SOHO-MDI space-based observations is at the surface [7]:

$$c_2 = (-5.27 \pm 0.38) \times 10^{-6} \quad \text{and} \quad c_4 = (+1.3 \pm 0.51) \times 10^{-6}. \tag{8}$$

These results were obtained by measuring small displacements of the solar limb darkening function (details are given in [40]), and the c_n coefficients are comparable to an isodensity surface level (see Eq. (3)). From Earth-based observations at the scanning heliometer of the Pic du Midi Observatory, we also obtained estimates of c_2 and c_4 coefficients. The mean ponderated values, computed over three years (values are given in [70]),[8] are

$$c_2 = (-6.56 \pm 0.18) \times 10^{-6} \quad \text{and} \quad c_4 = (+2.7 \pm 0.6) \times 10^{-6} \tag{9}$$

---

[8] A complete re-analysis of all the data, from 1996 to 2007, is under consideration.

that indicates (Sect. 2) a slight bulge extending from the equator to the edge of the royal zones (about $40°$ of latitude), with a depression beyond (at the pole itself, the ellipsoidal figure prevails).

Such a distorted shape, not exceeding some 20 mas of amplitude,[9] can be interpreted through the combination of the quadrupole and octopole terms, which, as shown previously, directly reflect the non-uniform velocity rate in surface (and depth). Moreover, this distribution implies a thermal wind effect, from the poles toward the equator [44]. The observed value of c_2, -6.6×10^{-6}, is not too far from the theoretical one, $\approx -(2/3)f \approx -5.9 \times 10^{-6}$ with $f = 8.9 \times 10^{-6}$ based on a solar model with a differential rotation law. It agrees also with the SOHO-MDI observations (see values given in (8)) and the theoretical estimate deduced from a vector harmonics solution [7], -5.87×10^{-6}. The coefficient c_4 remains difficult to match with the theory, which predicts $+(12/35)f^2$ for a uniform rotation law, and 0.616×10^{-6} for a differential one. The most likely explanation is that the shape coefficient c_4 is very sensitive to surface phenomena and differential rotation (as for J_4).

Only space-dedicated satellites will be able to definitively provide an answer to these questions.

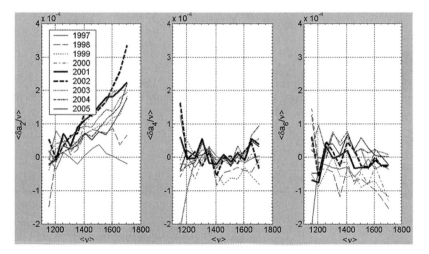

Fig. 9. Three first asphericities parameters γ_k, directly related to the even-a coefficients of f-modes. The first graph is significant and shows an antiphase correlation with the solar cycle (after [9])

---

[9] A first explanation (based on a temporal average) of such a result lies in the proposal made by Pecker [57]: "in the royal zones the existence of spots diminishes the solar brightness and thus the measured radius; but this effect is compensated and reinforced by the existence of faculae, which extends higher in latitudes. The measured radius is globally greater at the latitudes where faculae are statistically more numerous. At highest latitudes, no spots and faculae appear any longer, and the measured radius is consequently reduced".

5.6 Temporal Variations of the Asphericities Coefficients of the Solar Shape

If we are beginning to understand the significant physical character of the *lepto-cline*, we are far to know if the asphericities are variable with time. However, it can be reasonably thought that a temporal variability of such parameters might be due to the temporal variation of the internal structure. Lefebvre et al. [48] and Bedding et al. [9] reported first analysis concerning the temporal dependence of even-a coefficients. The available data (SOHO f-modes) permits to have access to the first 18 even-a coefficients, but for a sake of clarity, only the first three (γ_k) were computed (Fig. 9).

Each curve is an average difference over a specific year computed by reference to the minimum year of activity, 1996 (the error bars are not shown). The first graph on the left (γ_1) shows a negative trend, a frequency behavior almost flat and a clear behavior with the cycle in antiphase. The second graph dedicated to the variation of γ_2 shows a slight increase with the frequency, a change of sign at high frequency and no clear variations with the cycle. As far as the variation of γ_3 is concerned, the dispersion is too big to say anything.

In such a way, the outer shape would be time dependent, and this could explain also some tiny fluctuations of the irradiance. It is thus of interest to explore the whole chain, starting from the core up to the surface, to well understand the mechanisms of solar activity, then to get a better prediction, and to understand how the solar output may influence the atmosphere of our planet. One can judge such an investigation as ambitious, but we are today compelled to carefully examine all the sources of the solar variability to get a scientific opinion on the solar forcing – and even if it is to reject one of the processes.

5.7 Solar Shape and General Relativity

If, from a physical point of view, the multipolar moments lead to distortions of the solar surface (asphericities), they have also a dynamic role in the light deflection or in celestial mechanics. In the ephemerids computation, the determination of J_2 is strongly correlated with the determination of the Post-Newtonian Parameters (PPN) characterizing the relativistic theories of the gravitation. Lastly, the ignorance of J_2 is also a barrier to the determination of models of evolution of the solar system on the long term.

The relativistic aspects are crucial in the dynamic approach of the solar parameters and open interesting prospects for the future. In this context, it is interesting to obtain a dynamic constraint of J_2, independent of the solar models of rotation and density, being used thereafter to force the solar models. Such a study is relevant in the scope of space missions such as BeppiColombo (better determination of the PPN; possible measurement of the precession of the apside line of Mercury as a function of J_2), GAIA (better determination of the PPN, possible decorrelation PPN–J_2 thanks to the relativistic advance of the perihelion of planets and minor planets) and obviously *GOLF-NG* (precise determination of the rotation of the core where the quasi-totality of the mass is concentrated). Another key parameter of the solar models, which could also be constrained in a dynamic way is its spin, from the spin–orbit couplings which is introduced in celestial mechanics. From present solar system experiments (Lunar Laser Ranging, Cassini Doppler experiments, etc., see [59]), it turns out that general relativity is not excluded by those, as shown in the most up-to-date values in Fig. 10a. However, general relativity would be incompatible with the Mercury perihelion advance test if $J_2 = 0$ was assumed. But with a non zero J_2, general relativity agrees with this latter test, and there is still room for an alternative theory too (see Fig. 10b). *Space missions such as SDO or DynaMICCS-GOLF-NG should provide the necessary J_n measurements.*

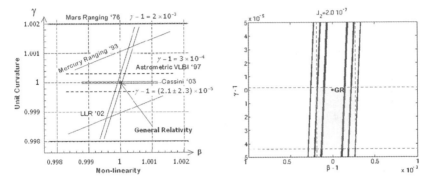

Fig. 10. *Left* (**a**) Thirty years of testing general relativity from space. *Right* (**b**) A given value of J_2 constitutes a test of the PN parameters β and γ. In the β and γ planes, the 1σ (the *smallest*), 2σ and 3σ (the *largest*) confidence level ellipses are plotted. Those are based on the values for the observed perihelion advance of Mercury, Δ_{wobs}, given in the literature and summarized in [59]. General relativity is still in the 3σ contours for the allowed theoretical values of J_2 quoted on the upper part of the chart

Regarding the solar core dynamics, the subject is of high priority for new investigations. Here again, space-dedicated missions, such as DynaMICCS/ GOLF-NG in a joint effort with SDO (Solar Dynamics Observatory), scheduled in a next future, should provide a new insight into the question [85].

6 Conclusion

From an historical point of view, the question of whether the diameter of the Sun evolves with time, or not, is very fertile. On time scales of the order of the millennium, the question of the solar luminosity can be tackled. On ranging time going from the medieval era up to now, the debate is not really closed. Wittmann [90] claimed, from Tobias measurements, that no secular solar diameter decrease can be inferred. We are more in favor of a long-term modulation, the Sun being bigger during periods of lower activity, such as during the Maunder minimum, and smaller in periods of more intense activity such as presently. On smaller time scales (over few cycles), models are still needed, but new input will come, paradoxically, from a better knowledge of the temporal evolution of the limb fluctuations (including solar diameter variations). On time scales of the order of months (or years), the variability is upper bounded by some 10 mas (and to 15 mas as an upper bound). Such estimates, deduced on physical grounds, are incompatible and irreconcilable with solar astrolabe measurements, which lead to one, even two orders or magnitude greater. The astrophysical consequences of such large variations would have been detected in indirect effects, such as lunar librations, which is not the case [65]. A possible explanation of the detected variations is through feedbacks mechanisms in the Earth UTLS zone [17]. From this point of view, it seems that the model proposed by Sofia et al. [80] (build to try to explain large variations of astrolabe data by the effect of magnetic field at the surface), which show an increase of the solar radius by a factor of approximately 1000 from a depth of 5 Mm to the solar surface, is not consistent (with the observations of the f-modes at the limb), in spite of its achieved formalism including magnetic field [49].

The question of the phase of the solar radius with activity depends on the depth, the diameter being in antiphase at the surface to progressively go to a phasing below 0.99 $R_\odot$ [45]. According to the analysis of f-mode frequencies, the Sun seems to be bigger in periods of lower activity (to a few km −30 as a maximum). Such an expansion could be due to magnetic fluxes passing through the gap between granules without interacting with them, the photospheric effective temperature playing a key role [51], confirmed by the gravitational energy variations in the upper layers [27]. Such an analysis leads to a phase shift of the solar luminosity with the solar cycle [25, 81].

The study of asphericities, directly linked with solar gravitational moments, is crucial not only for solar physics but also for astrometry (when computing light deflection in the vicinity of the Sun), celestial mechanics (relativistic precession of planets, planetary orbit inclination[10] and spin–orbit couplings) and future tests of alternative theories of gravitation (correlation of J_n with Post-Newtonian parameters): [59, 61].

Another issue is the solar changing shape. It has been shown that the outer solar shape significantly differs from a sphere, with a bulge at the equator, and a depressed zone at higher latitudes (the change being around 45°, due to the reversal of the radial velocity rate); the whole shape remains oblate at the pole. The Pic du Midi observations show *a variability of the whole oblate shape in phase with solar activity* [65] which *is not incompatible* with the above-mentioned long-term solar diameter modulation. Such in-phase dependence of the oblateness with the solar cycle has been confirmed through space by the RHESSI mission [37], at least an excess of the apparent oblateness with an equator to pole radius difference of 13.72 ± 0.44 mas (i.e., of the same order of magnitude that the mean value found at the Pic du Midi for the years 1993–1996: 11.5 ± 3.4 mas, see Table

---

[10] In the solar case, the potential expanded up to its quadrupole moment is given by $\Phi = -GM/r + J_2 GM R_{Eq}^2 (3 \sin^2 \eta - 1)$, where η is the angle relative to the equatorial plane.

1 in [66] or [64], the larger error coming from ground-based observations; note that the variations reflect temporal changes). Figure 11 shows the last results obtained at the Pic du Midi (F) Observatory (over a solar cycle), by means of the heliometer located in the so-called coupole J. Rösch. It is difficult to deny a phased variation with the cycle. If one wants to be very purist, one will be able to say that the flatness is 8.5 ± 3.5 mas. In such a case, the validity of the observational and analysis process is justified, as the theory gives 8.2 mas.

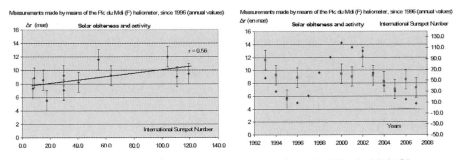

Fig. 11. Oblateness deduced from measurements made at the Pic du Midi Observatory by means of the heliometer since 1996

At last, due to the fact that the hexadecapolar term might be of the same order of magnitude than the quadrupolar one, but obviously in opposite sign, it results that the equatorial radius at the surface is in antiphase with the solar cycle, which is consistent with the results deduced from the f-mode analysis. We would like to emphasize again the key role of the leptocline in probing the sub-surface. To our mind, the inversion of the radial gradient of rotation at $50°$ contributes to solar asphericities, the whole shape remaining oblate. In period of lower activity, the equatorial diameter slightly increases, J_2 is predominant, J_4 has no influence and ε increases. In period of higher activity, the equatorial diameter slightly decreases under the influence of J_4 which is predominant, J_2 has no influence, so that ε decreases.

Accurate measurements from space observations are needed. They can be achieved by next generation of satellites, such as DynaMICCS/GOLF NG [83], SDO, or even balloon flights [79]. On a longer term, GAIA, which is expected to flight by the end of 2012, will allow to estimate the perihelion precession of Mercury and other small planets such as Icarus, Talos and Phaeton. In this case, it will be possible to separate the relativistic and the solar contributions in the perihelion advance, so that gravitational moments could be directly determined from dynamics, without the need of a solar model. Note also that presently dynamical estimates of J_2 are strongly correlated with the estimate of the Post-Newtonian parameter β, which, together with other PN parameters, characterizes relativistic theories of gravitation in observational tests. However, future PPN testing space missions, as well as non dedicated missions like GAIA, might help solve the problem.

According to the temporal variation of the f-mode frequencies, the very near solar surface is stratified in a thin double layer, interfacing the convective zone and the surface. This "leptocline" is the seat of many phenomena: an oscillation phase of the seismic radius, together with a nonmonotonic expansion of this radius with depth, a change in the turbulent pressure, likely an inversion in the radial gradient of the rotation velocity rate at about $45°$ in latitude, opacities changes, superadiabicity, the cradle of hydrogen and helium ionization processes and probably the seat of in situ magnetic fields [48]. Figure 8 shows a schematic view of the complex physics in this shear zone.

The last point could be an interrogation. Why helioseismology always leads to smaller values of the parameters under investigations? The rotation velocity rate at the surface is smaller than those obtained through other techniques (at the surface, see Table 1 given in [44]), as well as the quadrupole moment estimates and the radius itself. To disentangle all these points, we need to wait for space results: first, from the SDO (Solar Dynamics Observatory) mission, already accepted for a flight at the horizon of 2009–2010 with precise resolved velocity oscillation measures (HMI/SDO instrument); second, from GOLF-NG, the successor of GOLF/SOHO, as proposed in a future mission like DynaMICCS, for which the final aim is to reveal the complete 3D vision of the Sun[11].

The problem of determining the temporal diameter evolution of the Sun is still rich and fascinating. We hope to interest a broader community to deeply investigate this field.

Acknowledgments

This work was partly supported by the French Agency CNRS (UMR 6525). We thank also the International Space Science Institute (ISSI) in Berne, which helped us to hold scientific committees devoted to these questions (http://www.issi.unibe.ch/teams/SolDiam/).
The author thanks the whole team for valuable discussions.

References

1. Ajabshirizadeh, A., Rozelot, J.P., Fazel, Z.: Contribution on the solar magnetic field on gravitational moments., Sci. Iran. **15**(1), 144–149 (2008a)
2. Ajabshirizadeh, A. Rozelot, J.P., Fazel, N.: Sci. Iran., **15**, 144 (2008b)
3. Allen, C.W.: Astrophysical quantities, 4th edn., p. 340. Springer (2000)
4. Antia, H.M.: Estimate of solar radius from f-mode frequencies. Astron. Astrophys. **330**, 336 (1998)

---

[11] The microsatelitte Picard, scheduled to flight by 2009 is now conceived for other purposes than solar radius measurements –space wheather purposes and UV atmospheric images–, as the design –1 pixel per second of arc – cannot permit to achieved an astrometric precision.

5. Antia, H.M.: Does the Sun shrink with increasing magnetic activity? Astrophys. J. **590**, 567–572 (2003)
6. Antia, H.M., Basu, S.: ESA SP-559 **301** (2004)
7. Armstrong, J., Kuhn, J.R.: Interpreting the solar limb shape distortions'. Astrophys. J. **525**, 533–538 (1999)
8. Badache-Damiani, C., Rozelot, J.P.: Solar apparent radius variability: a new statistical approach to astrolabe multi-site observations. Mon. Not. R. Astron. Soc., **369**, 83–88 (2006)
9. Bedding, T., et al.: Solar dynamics, asphericities and gravitational moments: present state of the art. Joint Discussion 17. In: van der Hucht, K.A. (ed.), Highlights of Astronomy, Vol. 14, p. 1. XXVIth IAU General Assembly, Praha August (2006)
10. Bois, E., Girard, J.F.: Impact of the quadrupole moment of the sun on the dynamics of the earth-moon system. Celest. Mech. Dyn. Astron. **73**(1/4), 329–338 (1999)
11. Brown, T.M., Christensen-Dalsgaard, J.: Accurate determination of the solar photospheric radius. Astrophys. J. **500**, 195–198 (1998)
12. Callebaut, D.K., Makarov, V.I., Tlatov, A.G.: Gravitational energy, radius and solar cycle. ESA SP-477, 209–212 (2002)
13. Campbell, L., Moffat, J.W.: Astrophys. J., **275**, L77 (1983)
14. Carl Sagan, George Mullen: Earth and Mars: Evolution of atmospheres and surface temperatures. Science, **177**(4043), pp. 52–56 (1972)
15. Chandrasekhar, S.: The equilibrium of distorted polytropes. The rotational problem. Monthly Not. Roy. Astron. Soc. **93**, 390–406 (1933)
16. Coughlin, K.T., Tung, K.K.: 11-year cycle in the stratosphere extracted by the empirical mode decomposition method. Adv. Sp. Res. **34**, 323–329, (2004)
17. Damiani-Badache, C., Rozelot, J.P., Coughlin, K., Kilifarska, N.: Influence of the UTLS region on the astrolabes solar signal measurement. Mont. Not. Roy. Astr. **380**, 609–614 (2007)
18. Damon, P.E., Jirikovic, J.L.: The sun as a low-frequency harmonic oscillator. Radiocarbon **34**, 199–205 (1992)
19. Damon, P.E., Jirikovic, J.L.: Solar forcing of global climate change. In: Pap, J. (ed.) The Sun as a Variable Star, pp. 301–314. Cambrige University Press (1994)
20. Dewitte, S., Crommelynck, D., Mekaoui, S., Joukoff, A.: Non magnetic changes in the total solar irradiance. Solar Phys. **224**, 209 (2005)
21. Dicke, R.H., Goldenberg, H.M.: Solar oblateness and general relativity. Phys. Rev. Lett. **18**(9), 313 (1967)
22. Domiciano de Souza, A., Kervella, P., Jankov, S., Abe, L., Vakili, F., di Folco, E., Paresce, F.: The spinning-top Be star Achernar from VLTI-VINCI. Astron. Astrophys. **407**, L47–L50 (2003)
23. Dziembowski, W.A., Goode, P.R., Schou, J.: Does the Sun shrink with increasing magnetic activity? Astrophys. J. **553**, 897–904 (2001)
24. Emilio, M., Bush, R.I., Kuhn, J., Sherrer, P.: Astrophys. J., **660**, L161–L163 (2007)
25. Fazel, Z., Rozelot, J.P., Pireaux, S., Ajabszirizadeh, A., Lefebvre, S.: Solar irradiance, luminosity and photospheric effective temperature. In: Ermolli, I., Fox, P., Pap, J. (eds.) Mem. Soc. Astron. Ital. **76**, 961 (2005)
26. Fazel, Z.: Variabilité solaire: une approche globale tenant compte de l'hélioïde. Ph.D. Thesis, University of Nice-Sofia-Antipolis, 156p (2007)
27. Fazel, Z., Rozelot, J.P., Lefebvre, S., Ajabshirizadeh, A., Pireaux, S.: Solar gravitational energy and luminosity changes. New Astron. **13**(2), 65–72 (2008)

28. Fivian, M.D., Hudson, H.S., Lin, R.P., Zahid, H.J.: Solar shape measurements from RHESSI: a large excess oblateness. American Geophysical Union, Fall Meeting 2007, (abstract number SH53A-1076) (2007)
29. Frölich, C.: WRCPMODE, Annual report, 18 (2005)
30. Gilman, P. A., Dikpati, M.: Analysis of instability of latitudinal differential rotation and toroidal field in the solar tachocline using a magnetohydrodynamic shallow-water. Astrophys. J. **57**, 1031–1047 (2002)
31. Godier, S., Rozelot, J.P.: Solar parameters and shear regions: an unexpected dependent behavior. In: Garcia Lopez, R.J., Rebolo, R., Zapaterio Osorio, M.R. (eds.) Proceedings of the 11th Cambridge Workshop on Cool Star, Stellar Systems and the Sun. Tenerife (Spain), 4–10 October 1999. ASP Conference Series (San Francisco), Vol. 223, p. 649 (1999)
32. Godier, S., Rozelot, J.P.: The solar oblateness and its relationship with the structure of the tachocline and of the sun's subsurface. A & A **365**, 365–374 (2000)
33. Godier, S., Rozelot, J.P.: A new outlook on the differential theory of the solar quadrupole moment and of the oblateness. Solar Phys. **199**(2), 217–229 (2001)
34. Gough, D.O.: Nature, **298**, 334 (1982)
35. Goupil, M.: Effects of rotation on stellar p-mode frequencies. Lect. Notes Phys. **765** (2009)
36. Hill, H.A., Oleson, J.R.: The finite Fourier transform definition of an edge on the solar disk. Astrophy. J. **200**, 484–498 (1975)
37. Fivian, M.D., Hudson, H.S., Lin, R.P., Zahid, H.J.: A large excess in apparent solar oblateness due to surface magnetism, Science **322**, 5901 (2008)
38. Kosovichev, A.G.: Helioseismic constraints on the gradient of angular velocity at the base of the solar convection zone. Astrophys. J. **469**, L61 (1996)
39. Kosovichev, A.G., and HMI Science Team.: Helioseismology program for Solar Dynamics Observatory. Astron. Nachr. **328**, 339 (2007)
40. Kuhn, J.R., Bush, R., Scheick, X., Scherrer, P.: Nature, **392**, 155 (1998)
41. Kuhn, J.R., Bush, R.I., Emilio, M., Scherrer, P.H.: On the constancy of the solar diameter. Ap. J. **613**, 1241–1252 (2004)
42. Landgraf, W.: Solar Phys., **14**, 403 (1992)
43. Lefebvre, S., Bertello, L., Ulrich, R.K., Boyden, J.E., Rozelot, J.P.: Solar radius measurements at Mount Wilson. American Geophysical Union, Fall Meeting 2004, San Francisco (USA) (abstract SH53B-0314). Proceedings on http://adsabs.harvard.edu/abs/2004AGUFMSH53B0314L (2004)
44. Lefebvre, S., Rozelot, J.P.: Solar latitudinal distortions: from observations to theory. Astron. Astrophys., **419**, 1133–1140 (2004)
45. Lefebvre, S., Kosovichev, A.: Changes in the subsurface stratification of the sun with the 11-year activity cycle. Astrophys. J., **633**, L149–L152 (2005)
46. Lefebvre, S., Bertello, L., Ulrich, R.K., Boyden, J.E., Rozelot, J.P.: solar radius measurements at Mount Wilson observatory. Astrophys. J., **649**, 444–451 (2006)
47. Lefebvre, S., Kosovichev, A.G., Nghiem, P., Turck-Chièze, S., Rozelot, J.P.: Cyclic variability of the seismic solar radius from Soho/Midi and related physics. In: SOHO 18/GONG 2006/HELAS I – Conference, Shefield, U.K., August 7–11 2006, Beyond the Spherical Sun: A New Era of Helio- and Asteroseismology. ESA SP-624, CDROM, p. 9.1 (2006)
48. Lefebvre, S., Kosovichev, A.G., Rozelot, J.P.: Helioseismic measurements of solar radius changes from SOHO/MDI. In: SOHO-17: 10 Years of SOHO and Beyond. (Giardini-Naxos, I), ESA, SP-617, CD 43-1 (2006)

49. Lefebvre, S., Kosovichev, A., Rozelot, J.P.: Helioseismic test of nonhomologous solar radius changes with the 11-year activity cycle. Astrophys. J. **658**, L135–L138 (2007)
50. Livingston, W.C., Wallace, L.: The Sun's immutable basal quiet atmosphere. Solar Phys., **212**, 227 (2003)
51. Livingston, W.C., Gray, D., Wallace, L., White, O.R.: In: Sankarasubramanian, K., Penn, M., Pevstsop, A. (eds.) Large-Scale Structures and Their Role in Solar Activity, ASP Conference Series, Vol. 346, p. 353 (2005)
52. Lydon, T.J., Sofia, S.: Ap. J. Supp. S., **101**, 357–373 (1995)
53. Maeder, A.: Astron. & Astrophys. **347**, 185 (1999)
54. Milne, E.A.: The equilibrium of a rotating star. Monthly Not. Roy. Astron. Soc. **83**, 118–147 (1923)
55. Pap, J., Rozelot, J.P., Godier, S., Varadi, F.: On the relation between total irradiance and radius variations. A & A **372**, 1005–1018 (2001)
56. Paternò, L., Sofia, S., Di Mauro, M.P.: Astron. Astrophys., **314**, 940 (1996)
57. Pecker, J.C.: The changing sun, our star. Rev. Mex. Astron. Astrophys. **4**, 39 (1996)
58. Pijpers, F.P.: Monthly Notices Royal Astron. Soc., **297**, 76 (1998)
59. Pireaux, S., Rozelot, J.P.: Astron. Space Sci. **284**, 1159 (available also on www: astro-ph/0109032) (2003)
60. Pireaux, S., Lefebvre, S., Rozelot, J.-P.: Solar gravitational moments and solar core dynamics. In: Casoli, F., Contini, T., Hameury, J.M., Pagani, L.: (eds.) SF2A, 27th June–1st July 2005, EdP-Sciences Conference Series, pp. 121–122 (2005)
61. Pireaux, S., Rozelot, J.-P.: On the key role of a dynamical estimate of the Solar spin and gravitational multipole moments. In: Casoli, F., Contini, T., Hameury, J.M., Pagani, L. (eds.) SF2A, Section Astro Fondamentale: Géodésie spatiale, mécanique céleste, et systèmes de référence, 27th June–1st July 2005. EDP-Sciences Conference Series, p. 91 (2005)
62. Rösch, J., Rozelot, J.P., Deslandes, H., Desnoux, V.: A new estimate of the quadrupole moment of the Sun. Solar Phys. **165**, 1–11 (1996)
63. Roxburgh, I.W.: Astron. Astrophys., **377**, 688 (2001)
64. Rozelot, J.P.: Measurement of the Sun's changing sizes. In: Missions to the Sun, SPIE Proceedings, Vol. 2804, p. 241. Denver, Colo (USA) (1996)
65. Rozelot, J.P., Rösch, J.: Le Soleil change-t-il de forme? Comptes- Rendus Acad. Sciences Paris, **322**, série II, 637–644 (1996)
66. Rozelot, J.P., Bois, E.: New results concerning the solar oblateness and consequences on the solar interior. In: Balasubramaniam, K.S. (ed.) 18th NSO Workshop, Sacramento Peak, USA, Conf. of the Pacif. Astro. Soc., **140**, 75–82 (1998)
67. Rozelot, J.P.: Possible links between the solar radius variations and the Earth's climate evolution over the past four centuries, J. Atmos. Sol. Terr. Phys. **63**(4), 375–386 (2001a)
68. Rozelot, J.P.: Solar diameter variations and Earth's climate: the missing link. In: Schröder, W. (eds.) Solar Variability and Geomagnetism. IAGA-IASPEI Conference Meeting, Hanoi, August 26, 2001, Science Editions, Bremen (D), pp. 88–102 (2001b)
69. Rozelot, J.P., Godier, S., Lefebvre, S.: On the theory of the solar oblateness. Solar Phys. **198**, 223–240, Issue 2 and other figures in the attached CD-Rom (2001)
70. Rozelot, J.P.: The Sun's surface and subsurface, Investigating shape and irradiance. Lect. Notes Phys. **599**, 214p. Springer, Heidelberg (D) (2003)

71. Rozelot, J.P., Lefebvre, S.: The figure of the Sun, astrophysical consequences. In: The Sun Surface and Subsurface. Lecture Notes in Physics, Vol. 599, pp. 4–26. Springer ed. (2003)

72. Rozelot, J.P., Lefebvre, S., Desnoux, V.: High Angular Resolution Observations of the Solar limb shape at the Pic du Midi Observatory. In: Pevtsov, A.A., Uitenbroek, H. (eds.) Current Theoretical Models and High Resolution Solar Observations: Preparing for ATST. 21st NSO/SP workshop, Sunspot, New Mexico (USA), March 11–15, 2002. Astronomical Society of the Pacific, ASP Conference Series, Vol. 286, pp. 77–83. (2003)

73. Rozelot, J.P., Pireaux, S., Lefebvre, S., Ajabshirizadeh, A.: Solar rotation and gravitational moments: some astrophysical outcomes. SOHO 14/GONG 2004 Workshop, New Haven, ESA SP-559, 532 (2004a)

74. Rozelot, J.P., Lefebvre, S.: Gravitational distortions of the shape of the Sun: constraints on the models. In: Maeder, A., Eenens, P. (eds.) IAU Symposium No. 215: Session III: Rotation, Solar and Stellar Physics, Cancun (Mexique, 2002). 2004, 332 (2004b)

75. Rozelot, J-P., Lefebvre, S., Pireaux, S.: The Sun's Asphericities: Astrophysical Interest. Proceedings of the International Conference of Physics, 6–9 Janvier 2004, Amirkabir University of Technology, Teheran (Iran), p. 647 (2004c)

76. Rozelot, J.P., Pireaux, S., Lefebvre, S.: arXiv: astro-ph/0403382 v3 1 Apr. 2004d

77. Rozelot, J.P., Pireaux, S., Lefebvre, S., Fazel, Z.: Historical measurements of the Sun's diameter variations: some new comments. In: Schröder, W. (ed.) Historical Events and People in Aeronomy, Geomagnetism and Solar-Terrestrial Physics. Science Edition (2006), 10th Scientific Assembly of the International Association of Geomagnetism and Aeronomy (IAGA). Toulouse, F. (18–29 July), pp. 279–287. AKGGP/SHGCP, Bremen/Postdam (D) (2005)

78. Schou, J., Kosovichev, A., Goode, P.R., Dziembowski, W.A.: Determination of the Sun's seismic radius from the SoHO Michelson Doppler Imager. Astrophys. J. **489**, 197 (1997)

79. Sofia, S.: Mechanisms for global solar variability. In: Ermolli, I., Fox, P., Pap, J. (eds.) Mem. Soc. Astron. Ital. **76**, 768–772 (2005)

80. Sofia, S., Basu, S., Demarque, P., Li, L.: Astrophys. J., **632**, L147 (2005)

81. Stothers, R.B.: Astrophy. J., **653**, L73 (2006)

82. Toulmonde, M.: A & A **325**, 1174–1178 (1997)

83. Turck-Chièze, S., Rozelot, J.P. and other co-authors. The magnetism of the solar interior for a complete MHD solar vision, 2005 ESLAB Symposium, Trends in Space Science and Cosmic Vision 2020, 19–21 April 2005, ESTEC, Noordwijk, The Netherlands, ESA-SP 588, p. 193 (2005)

84. Turck-Chièze, S., Rozelot, J.P. and other co-authors: The DynaMICS perspective. SOHO 18/GONG 2006/HELAS I – Conference, Shefield, U.K., August 7–11 2006, Beyond the Spherical Sun: a new era of helio- and asteroseismology, ESA SP-624, CDROM (2006)

85. Turck-Chièze, S., Rozelot, J.P. and other co-authors: A mission for a complete and continuous view of the Sun to be published in Experimental Astronomy. (Dedicated to magnetism, space weather and space climate). (2008)

86. Ulrich, R.K., Hawkins, G.W.: Astrophys. J., **246**, 985 (1981) (and erratum, 1981, Astrophys. J., **249**, 831)

87. Ulrich, R.K., Bertello, L.: Nature, **377**, 214 (1995)

88. van Belle, G.T., Ciardi, D.R., Thompson, R.R., Akeson, R.L., Lada, E.A.: Altair's Oblateness and Rotation Velocity from Long-Baseline Interferometry. Astrophys. J. **559**(2), 1155–1164 (2001)
89. Wittman, A.D.: Tobias Mayer's observations of the Sun: evidence against a secular decrease of the solar diameter. Sol. Phys. **66**, 223 (1980)
90. Wittmann, A.D., Bonet Navaro, J.A., Wohl, H.: On the variability of the solar diameter. In: Cram, L., Thomas, J.H. (eds.), Physics of Sunspots, pp. 422–433. N.S.O. Publications (1981)
91. Wittmann, A.D., Bianda, M.: Drift-time measurements of the solar diameter 1990–2000: new limits on constancy. In: Wilson, A. (ed.) The Solar Cycle and Terrestrial Climate, Solar and Space Weather Euroconference (Santa Cruz de Tenerife, Tenerife, Spain). Proceedings of the 1st Solar and Space Weather Euroconference, September 25–29, 2000, Santa Cruz de Tenerife, Tenerife, Spain. Noordwijk, Netherlands. ESA SP, Vol. 463, p. 113 (2000)
92. Yuan, D.-N., Sjogren, W.L., Konopliv, A.S., Kucinskas, A.B.: Gravity field of Mars: a 75th degree and order model. J. Geophys. Res. **E10**(106), 23377–23401 (2001)

Effects of Rotation on Stellar p-Mode Frequencies

M.J. Goupil

Observatoire de Paris LESIA, 5 place Jules Janssen, 92190 Meudon principal cedex, France
mariejo.goupil@obspm.fr

Abstract Effects of stellar rotation on adiabatic oscillation frequencies of pressure modes are discussed. Methods to evaluate them are briefly exposed and some of their main results are presented.

1 Introduction

Many pulsating stars oscillate with pressure (or p-) normal modes for which pressure gradient is the main restoring force. A nonrotating consequently, frequencies of its free vibrational modes are $2\ell + 1$ degenerated when using a description by means of spherical harmonics, Y_ℓ^m (ℓ the degree, m the azimuthal indice) for the angular dependence of the variations. However, stars are rotating and their rotation affects their oscillation frequencies. This is an advantage since rotation breaks the spherical symmetry of the star; as a consequence, the $2\ell + 1$ degeneracy is lifted and gives access to information about the internal rotation. The information is relatively easy to extract if the star rotates slowly, provided observations are of high quality enough that they enable to achieve the necessary high accuracy of frequency measurements. On the other hand, when the star is rotating fast, one must properly take into account the consequences of the spheroidal shape of the star on the oscillation frequencies before extracting seismological information on the internal structure or the rotation profile of the star. Three methods have been used to investigate the effect of rotation on oscillation frequencies of stellar pressure modes: a variational principle, perturbation techniques and direct numerical integration of a 2D eigenvalue system. We briefly describe the three methods and their main results. All these methods assume the knowledge of the structure of the distorted star, which is obtained either by perturbation or by 2D calculations. This is also sketched when appropriate.

This lecture is primarily addressed to PhD students who possess some basic background in stellar seismology but wish to learn some details of how to handle the impacts of the rotation on the oscillation frequencies of a star. The

Goupil, M.: *Effects of Rotation on Stellar p-Mode Frequencies*. Lect. Notes Phys. **765**, 45–99 (2009)
DOI 10.1007/978-3-540-87831-5_3

present lecture cannot be exhaustive neither in the credit to authors nor to the numerous effects of rotation on stellar pulsation. Indeed, due to the importance of stellar rotation in a more general astrophysical context, a tremendous amount of work has been performed on stellar rotation and its interaction with stellar oscillations during the 1940s–1950s. For more details, we refer the reader to standard textbooks such as Ledoux and Walraven [83], Cox [35], Unno et al. [143] and to literature on the subject: Gough and Thompson [56], Gough [57], Christensen-Dalsgaard ([27], CD03) and references therein.

The organization of this contribution is as follows: as a start, some definitions and orders of magnitude are given in Sect. 2. Section 3 introduces the basic background in deriving the wave equation and eigenvalue problem modelling the adiabatic pulsations of a rotating star. Section 4 discusses the existence and consequence of a variational principle for rotating stars. Sections 5 and 6, respectively, concern calculation of the oscillation frequencies of slowly and moderately rotating stars by means of perturbation methods. Section 7 then turns to fast-rotating stars and the computation of their oscillation frequencies with nonperturbative techniques. Some comparisons of the results of perturbative and nonpertubative methods and comments on the range of validity of the first ones are given in the case of polytropic models. Section 8 briefly discusses observations of real stars and forward inferences on stellar internal rotation so far derived. Section 9 turns to inverse methods specifically applied to obtain the internal rotation profile. Finally results of inversions for the rotation profiles of some stars are presented in Sect. 9. To keep up with the allowed number of pages, the choice has been to show almost no figures and rather to refer in the text to plots in the literature. Some illustrations of the consequences of rotation on pulsations of stars can also be found in Goupil et al. [65].

2 Definitions and Orders of Magnitude

We will denote $\omega_{n,\ell}^{(0)}$ the pulsation of an oscillation mode with radial order n and degree ℓ for a nonrotating star and $\nu_{n,\ell}^{(0)} = \omega_{n,\ell}^{(0)}/2\pi$ the associated frequency in Hz. For a symmetric star of radius R and mass M and for any given eigenfrequency ω, we define the dimensionless frequency, σ:

$$\sigma = \frac{\omega}{(GM/R^3)^{1/2}} \tag{1}$$

Consider now a star which rotates with a rotational period P_{rot}. We denote $\Omega = 2\pi/P_{rot}$, its angular rotational velocity. One also defines the equatorial velocity $v_{eq} = R_{eq}\Omega$ where R_{eq} is the equatorial stellar radius; this is the velocity of the fluid at the equatorial level generated by rotation. One also uses the projected velocity $v \sin i$ where i is the angle between the line of sight and the rotation axis.

2.1 Why Are the Oscillation Frequencies of a Rotating Star Modified Compared to Those of a Nonrotating Star?

(i) A first effect is easy to understand as it is purely geometric. Let us consider a uniformly rotating star which oscillates with a pulsation frequency $\omega_{n,\ell}^{(0)}$ in the rotating frame. In an inertial frame, an observer will see oscillations with frequencies given by $\omega_{n,\ell,m} = \omega_{n,\ell}^{(0)} + m\Omega$ with $m \in [-\ell, \ell]$ with progade ($m > 0$) and retrograde ($m < 0$) modes. Each (n, ℓ) mode appears in a power spectrum as a frequency multiplet split by rotation composed of $2\ell + 1$ components.

(ii) At the same level (i.e. linear in the rotational angular velocity $O(\Omega)$), one must also take into account intrinsic effects of rotation in the rotating frame related to the Coriolis force which directly acts on the oscillating fluid velocity, deflecting the wave. Ledoux [82] has shown that the Coriolis acceleration modifies the pulsation such that in the observer frame it becomes

$$\omega_{n,\ell,m} = \omega_{n,\ell}^{(0)} + m\Omega(1 - C_{n,\ell}) + O(\Omega^2) \tag{2}$$

where $C_{n,\ell}$ is the Ledoux constant. The rotation rate then is simply given by

$$\Omega = (\omega_{n,\ell,m} - \omega_{n,\ell}^{(0)})/(m(1 - C_{n,\ell})) \tag{3}$$

where the Ledoux constant is assumed to be known from a nonrotating stellar model. However, the rotation of a star is not necessarily uniform: it can vary with latitude and depth; this again results in modifications of the frequencies. Actually these modifications are those one really wishes to identify. Indeed current open issues are the following: what are the regions inside the star where the rotation is uniform and where it is not? how strong are rotational gradients? what are the mechanisms which make the region uniform or in contrary generate differential rotation? how to model them or test the validity of their modelling?

The magnitude of the above effects is proportional to Ω. For a slowly rotating star indeed, effects of the centrifugal force, proportional to Ω^2, on oscillation frequencies are small and one usually ignores them. The *rotational splitting* of a mode with given (n, ℓ) then is defined as

$$\delta\omega_{n,\ell} = \frac{\omega_{n,\ell,m} - \omega_{n,\ell,0}}{m} \tag{4}$$

where $\omega_{n,\ell,0}$ can reasonably be identified with $\omega_{n,\ell}^{(0)}$ for a slow-rotating star.

(iii) For stars rotating faster, one must take into account the various effects due to the distortion of the star by the centrifugal force:

'mechanical' impact: the centrifugal force affects the wave propagation by deforming the resonant cavity where they propagate. Pioneer studies were due to Ledoux [81] and Cowling and Newing [34]. Ledoux [81] studied radial oscillations of a uniformaly rotating star by means of a variational principle and assuming uniform dilation and compression motions. He obtained the

squared frequency of the fundamental radial mode under the influence of the centrifugal force as

$$\omega^2 = -(3\Gamma_1 - 4)\frac{E_g}{I_0} + (5 - 3\Gamma_1)2T$$

where I_0, Γ_1, E_g and T, respectively, denote the moment of inertia, the adiabatic exponent (assumed constant throughout the star), gravitational potential and rotational kinetic energies. For a homogeneous spheroid in uniform rotation [129], $E_p/I_0 \sim 4\pi G\rho$ decreases whereas $T = \mathcal{M}/I_0 \sim \Omega^2/3$ increases ($\mathcal{M}$ is the total angular momentum) when Ω increases. Cowling and Newing [34] extended the study to nonradial oscillations of a rotating configuration and obtained numerical results for radial oscillations of a polytrope.

In order to obtain the rotational profile, one defines the **generalized rotational splitting** of a mode with given (n, ℓ) as

$$S_m = \frac{\omega_m - \omega_{-m}}{2m} \tag{5}$$

This second definition eliminates second-order perturbation effects (cf. Sect. 6.2) that eliminates the mechanical effects of the nonspherically symmetric distortion of the star on the oscillation frequencies.

'thermal' impact: Works by von Zeipel [145, 146], Eddington [48], Vogt [144] and Sweet [137] showed that the nonspherical shape of a rotating star causes a departure from thermal equilibrium which generates large-scale meridional currents. As already emphasized by Ledoux [81], this meridional circulation can influence the pulsations. Indeed it generates a differential rotation which causes hydrodynamical instabilities. The net effect results in transport of chemical elements and angular momentum in radiative regions and modifications of the thermodynamical state of the star. All these processes influence the star evolution and structure (Tassoul [141], Zahn [147]; see also Meynet and Maeder [97]; Mestel [96]; Zahn [148] for reviews), in particular the adiabatic sound speed and the rotation profiles, the two main quantities **accessible** with seismology. Hence it is important to note that even for a star which is assumed to be in near spherical symmetry and with a given rotation profile at the time of observation, there exists a nonzero difference between the zeroth order eigenfrequencies of a rotating model which has been evolved taking into account the effect of rotation and that of an 'equivalent' model at the same age evolved without rotation in which effect of rotation law is imposed afterwards or which has been evolved with rotation included only through the effective gravity. For a slow-rotating star, however, this difference can often be neglected.

2.2 How to Determine the Frequency Changes Due to Rotation?

The methods which must be used in order to determine precisely how and how much the oscillation frequencies are affected by rotation can be divided into two groups (apart from the variational technique): the *perturbative methods* on the one hand and the *direct numerical approaches or nonperturbative methods* on the other hand. The choice will depend whether the star is a rather slow or rather fast rotator. This classification can be loosely made according to values of dimensionless parameters. The ratio of the time scales related to Coriolis force and pulsation is roughly given:

$$\mu = \frac{P}{P_{rot}} = \frac{\Omega}{\omega} \tag{6}$$

Another important parameter is the ratio of rotational kinetic energy to the gravitational one. As an order of magnitude, it can be evaluated as

$$\epsilon_{eq} = \frac{\Omega^2 R_{eq}^2}{GM/R_{eq}} = \frac{\Omega^2}{(GM/R_{eq}^3)} = \mu^2 \, \sigma^2 \tag{7}$$

where M is the mass of the star; R_{eq} its radius at the equator; and G is the gravitational constant. One also uses the flatness, f, as defined by $f = 1 - R_{pol}/R_{eq}$, where R_{pol} is the radius at the pole.

In the following sections, we will then distinguish fast, moderate and slow rotators with respect to the pulsations of the stars according to these parameters. Stars for which these parameters obey ($\mu, \epsilon << 1$), ($\mu, \epsilon < 0.3$) or (μ or $\epsilon > 0.3$) are classified, respectively, as slow, moderate or fast rotators. For slow rotators, a first-order perturbation is enough; for a moderate rotator, a perturbation method can still be valid provided higher order contributions are included. Studying oscillations of fast rotators requires the use of 2D equilibrium models as well as solving a 2D eigenvalue system. However one must keep in mind that the above classification depends on the star as it depends on the frequency range of its excited modes. Perturbation methods cease indeed to be valid whenever the wavelength of the mode reaches the order of $\sim \epsilon R$ where R is the stellar radius, that is for p-modes with high enough frequencies.

In the solar case, $P_{rot,\odot}$ varies from $\sim$25 days at the equator up to $P_{rot,\odot} \sim 35$ days at the poles; then $\epsilon_{eq,\odot} = 2.1 \times 10^{-5}$ is very small. The splitting is of the order 0.456 µHz with a Coriolis strength $\mu = 1.4 \times 10^{-4}$. With $\mu, \epsilon << 1$, the Sun is considered as a slow rotator with respect to its pulsations. However, measurements of the rotational splittings are so accurate for the Sun – one is able to determine the internal rotation velocity of the Sun as a function of radius and latitude with unprecedented spatial resolution and accuracy – that one measures second-order effects in the rotational splittings which must be taken into account in order to obtain information on the solar magnetic field [46, 55] and to determine the gravitational quadrupole moment of the Sun with much higher accuracy [110].

For a δ Scuti-type variable star ($2M_\odot, 2R_\odot$), observations show that $v \sin i = 70-280$ km/s, that is $P_{rot} \sim 1, 45$ days to ~ 8 h, *Prot* is the period of rotation; this yields $\Omega = 5-20 \times 10^{-5}$ rad/s and a rotational splitting of the order of $\nu_{rot} = 8-32$ µHz. Typical values for the dimensionless parameters then are $\epsilon = 0.025-0.41$ and $\mu = 0.06-0.6$ for an oscillation frequency range of $200-500$ µHz.

Altair is an example of a rapidly rotating δ Scuti star with a rotation velocity between 190 and 250 km/s [119] which has been discovered to pulsate over at least 7 frequencies in the range 170–350 µHz [16]. With $M = 1.72 \pm 0.05 M_\odot$ and $R = 1.60 \pm 0.12$ for its mass and radius as found in the literature, one obtains $\epsilon \sim 0.305$ and $\mu \sim 0.09 - 0.2$ which definitely classes this star as a rapid rotator with respect to its pulsations.

A β Cepheid-type variable star $(9M_\odot, 5R_\odot)$ can reach $v \sin i = 300$ km/s, that is $P_{rot} < 20$ h. This yields $\Omega > 8.6 \ 10^{-5}$ rad/s and $\nu_{rot} > 14$ µHz. For the dimensionless parameters, one derives $\epsilon > 0.26$ and for $\nu \sim 500$ µHz, $\mu > 2.7 \times 10^{-2}$.

A much more evolved star such as a Cepheid is a radially pulsating star and a slow rotator. However period ratios of radial modes can be quite significantly affected by rotation as mentioned by Pamyatnykh ([103], Fig. 6) for a δ Scuti star and quantified by Suárez et al. [132, 135] for a Cepheid.

The above figures and classifications must be seen as illustrative and only a comparison between perturbative and nonperturbative approaches can give more precise delimitations between the different classes (Sect. 7.3).

3 A Wave Equation for a Rotating Star

A derivation of the wave equation for a rotating star can be found in Unno et al. [143]. The basic equations for a nonmagnetic, self-gravitating fluid are written as

$$\frac{\partial \mathbf{v}}{\partial t} + \mathbf{v} \cdot \nabla \mathbf{v} = -\frac{1}{\rho} \nabla p - \nabla \psi \tag{8}$$

$$\frac{\partial \rho}{\partial t} + \nabla \cdot (\rho \mathbf{v}) = 0$$

$$\rho T \frac{\partial S}{\partial t} + \rho T (\mathbf{v} \cdot \nabla) S = \rho \epsilon - \nabla \cdot \mathcal{F}$$

$$\nabla^2 \psi = 4\pi G \rho$$

where the quantities take their usual meaning and

$$\psi(\mathbf{r}, t) = G \int_V \frac{\rho(\mathbf{r}', t)}{|\mathbf{r} - \mathbf{r}'|} \, \boldsymbol{d}^3 r'$$

In an inertial frame, the effect of rotation appears through the inertial term $\mathbf{v} \cdot \nabla \mathbf{v}$. In the frame rotating with the star, the inertial term acts as two fictive forces: the Coriolis force $(-2\rho \boldsymbol{\Omega} \times \mathbf{v})$ which modifies the dynamics and insures the conservation of angular momentum and the centrifugal force $(-\rho \boldsymbol{\Omega} \times \boldsymbol{\Omega} \times \mathbf{r})$ which affects the structure of the equilibrium configuration of the star, i.e. the star is distorted and loses its spherical symmetry. The oscillations are described as a linear perturbation about an equilibrium or steady-state configuration which is assumed to be known.

3.1 Equilibrium Structure for a Rotating Star

We assume that in a steady-state configuration, i.e. all local time derivatives vanish, the quantities describing this equilibrium are given a subscript 0, for instance the steady-state velocity field is noted $\mathbf{v}_0$. The assumption of stationarity leads to

$$\frac{1}{\rho_0}\nabla p_0 + \nabla\psi_0 = -\mathbf{v_0}\cdot\nabla\mathbf{v_0} \qquad (9)$$

$$\nabla\cdot(\rho_0\mathbf{v}_0) = 0$$

$$\rho_0 T_0(\mathbf{v_0}\cdot\nabla)S_0 = \rho_0\epsilon_0 - \nabla\cdot\boldsymbol{\mathcal{F}}_0$$

$$\nabla^2\psi_0 = 4\pi G\rho_0$$

We consider an axisymmetric rotation $\boldsymbol{\Omega}$. In spherical coordinates, one has

$$\boldsymbol{\Omega}(r,\theta) = \Omega(r,\theta)\left(\cos\theta\ \mathbf{e_r} - \sin\theta\ \mathbf{e}_\theta\right)$$

The large-scale velocity field $\mathbf{v}_0 = v_\Omega + \boldsymbol{U}$ is composed of the rotational velocity and the meridional circulation. We consider here that the meridional circulation has no direct effect on the dynamics of the pulsation ($U << c_s$); hence one neglects it in the velocity field which therefore is only due to rotation and has the expression

$$\mathbf{v}_0(\mathbf{r}) = \boldsymbol{\Omega}\times\mathbf{r} = \Omega(r,\theta)\ r\sin\theta\ \mathbf{e}_\phi \qquad (10)$$

The steady-state configuration is then given by

$$-\frac{1}{\rho_0}\nabla p_0 - \nabla\psi_0 = \boldsymbol{\Omega}\times\boldsymbol{\Omega}\times\mathbf{r} = -r\Omega^2\sin\theta\ \mathbf{e_s} \qquad (11)$$

where $\mathbf{e_s} = \sin\theta\ \mathbf{e_r} + \cos\theta\ \mathbf{e}_\theta = \partial\mathbf{e}_\phi/\partial\phi$. Note that $-\boldsymbol{\Omega}\times\boldsymbol{\Omega}\times\mathbf{r} = r\Omega^2\mathbf{e_r}$ for $\theta = \pi/2$ is directed towards the $r > 0$ and we recognize the centrifugal acceleration.

Particular cases: barotropy and uniform rotation. When the centrifugal force can be considered as deriving from a potential (conservative rotation law), it can be included similarly to the gravity and the equilibrium equation becomes

$$-\frac{1}{\rho_0}\nabla p_0 - \nabla(\psi_0 - |\boldsymbol{\Omega}\times\mathbf{r}|^2/2) = 0$$

One sets $\psi_{eff}\equiv\psi_0 - |\boldsymbol{\Omega}\times\mathbf{r}|^2/2$ and obtains

$$\frac{dp_0}{dr} = -\rho_0\ g_{eff} \qquad\qquad \nabla\psi_{eff} = g_{eff}$$

The star keeps it spherical symmetry. This happens for a uniform rotation or for a cylindrical rotation $d\Omega/dz = 0$ in a cylindrical coordinate system [141]. The von Zeipel [145, 146] law states that for a barotrope and a conservative rotation law, isobars and isopycnics coincide with the level surfaces (constant potential):

$\rho = \rho(\psi_{eff}); \; T = T(\psi_{eff})$ with $\psi_{eff} = \psi + r\sin\theta\Omega$. Consequently the heat flux is proportional to local effective gravity $T_{eff}^4 \propto g_{eff}$ and poles are hotter than the equator (gravity darkening; see Tassoul [141] for details).

Shellular rotation $\Omega(r)$. Zahn [147] considered that in stellar radiative regions, highly anisotropic turbulence develops: this leads to an efficient homogenization in the horizontal plane but not in the vertical direction due to the stable stratification which inhibits vertical motions hence generating a shellular rotation. This was verified by Charbonneau [22] by means of 2D numerical simulations. The combined effects of rotation and turbulence induce a process of advection/diffusion of angular momentum and diffusion of chemical elements. The turbulence itself can arise from a dynamical shear instability generated by the differential rotation $\Omega(\theta)$ due to the advection of angular momentum by the meridional circulation (see for a review Talon [139]). For rotators with no angular momentum loss, a weak meridional circulation exists only to balance transport by differential rotation-induced turbulence. For rotators which lose angular momentum through wind, the meridional circulation must adjust so as to transport momentum towards the surface. Hence effects of a complex rotationally induced 2D process can be described in a 1D framework [89, 92, 147].

2D rotation: $\Omega(r,\theta)$. Several studies have developed numerical schema in order to build more and more realistic 2D rotating stellar models. As the full problem is quite difficult, simplifying assumptions have been made depending on the purpose of the study (Clement's series of papers from 1974 to 1998; Deupree from 1990 to 2001; see for a detailed historical review M. Rieutord [116], web site of the Brasil Corot meetings, CB3).

With the initial purpose of modelling the rapidly rotating, oblate Be star, Achernar, Jackson et al. [72, 73] neglected the evolution and dynamics of the star and focused their studies on a specified conservative rotation law. This allows them to solve a 2D partial differential equation for Poisson equation while dealing with only ordinary differential equations for the other equilibrium quantities. They were able to build stellar rotating models for a wide mass range ($2M_{\odot}-9M_{\odot}$) and for equatorial velocity up to 250 km/s for the most massive ones. Roxburgh [117] on the other hand computed 2D uniformaly rotating, barotrope zero age main sequence models again over a wide mass range and rotational velocities. The determination of the adiabatic nonradial oscillation frequencies of a rotating star only depends on the knowledge of $p_0(r,\theta)$, $\rho_0(r,\theta)$ and $\Gamma_1(r,\theta)$ and does not require that the thermal equilibrium be satisfied. Roxburgh [118] takes advantage of this property to build 2D acoustic models of rapidly rotating stars with a prescribed rotation profile $\Omega(r,\theta)$. The initial density profile at a given angle θ_m can be taken as the density profile of a 1D stellar model (which can include effects of transport and mixing due to rotation as described in the above section). Roxburgh [118] then solves iteratively the 2D hydrostatic and Poisson equations; the adiabatic index Γ_1 is then obtained through the equation of state and the knowledge of the hydrogen profile (again possibly derived from the 1D spherically averaged stellar model).

In all the above studies, the Ω profile is prescribed and stellar models are built at a given evolutionary stage with no inclusion of any feedback of rotation on evolution and structure. To proceed a step further with the purpose of including effects of evolution and internal dynamics, Rieutord (2006a, b), Espinosa Lara and Rieutord [52] worked at building more and more realistic rotating models using spectral methods for both directions r and θ. Espinosa Lara and Rieutord [52] succeeded in computing a fully radiative, baroclinic model in which the microphysics is treated in a simplified form and the star is assumed to be enclosed in a rigid sphere. Some interesting conclusions could nevertheless be drawn. For a stellar model rotating at 82% of the rotational break-up velocity, the baroclinic model is much less centrally condensed than a radiative polytrope. The steady state is characterized by poles hotter but rotating less rapidly than the equator. The authors found that the temperature contrast between poles and equators is less than that given by the von Zeipel model although this might be reduced with the use of a more realistic surface boundary condition. This is also found by Lovekin and Deupree [87] using the 2D rotating stellar code of Deupree. With increasing rotation, the equator cools down enough that a convective region seems to be able to develop. It is also found that the isothermals are more spherical than the isobars. As a consequence and in contrast with the Boussinesq case, the meridional currents circulate from equator to poles: the meridional velocity U_ϕ component decreases outward, because T decreases on an isobar from pole to equator. Worth to be noted also, differential rotation in latitude seems to keep the same general form for increasing Ω between 0.01 and 0.08 of the critical rotation rate (Fig. 10, Espinosa Lara and Rieutord [52]).

3.2 Oscillations: Linearization About the Equilibrium

For the oscillating, rotating fluid of interest here, the velocity field in each point of the space can be split into two components $\boldsymbol{v} = \boldsymbol{v}_0 + \boldsymbol{v}'$, where $\boldsymbol{v}'$ is the Eulerian perturbation of the velocity due to the oscillation. Similarly for any scalar quantity f (such as p, ρ, ψ), one writes $f = f_0 + f'$. The quantities with a prime represent the oscillations which are modelled as linear Eulerian perturbations about the equilibrium. One inserts these decompositions into the basic equations Eq. (9) and then linearizes with respect to the perturbations $\boldsymbol{v}', \rho'$, etc. about the equilibrium quantities. Only adiabatic oscillations are studied here; the system is then closed by using the linearized adiabatic relation:

$$\frac{\delta p}{p_0} = \Gamma_1 \left(\frac{\delta \rho}{\rho_0} \right) \quad \text{where} \quad \Gamma_1 \equiv \left(\frac{\partial p}{\partial \rho} \right)_S \tag{12}$$

since $\delta S = 0$ (S entropy) where δ denotes Lagrangian variations or in the Eulerian form

$$\frac{p'}{p_0} = \Gamma_1 \left(\frac{\rho'}{\rho_0} + \boldsymbol{\xi} \cdot A \right) \quad \text{where} \quad \mathbf{A} = \frac{1}{\Gamma_1} \frac{\nabla p_0}{p_0} - \frac{\nabla \rho_0}{\rho_0}$$

The system of linearized equations for adiabatic nonradial oscillations of a rotating star takes the form

$$\frac{\partial \mathbf{v'}}{\partial t} + \mathbf{v'} \cdot \boldsymbol{\nabla} \mathbf{v_0} + \mathbf{v_0} \cdot \boldsymbol{\nabla} \mathbf{v'} = -\frac{1}{\rho_0} \boldsymbol{\nabla} p' + \frac{\rho'}{\rho_0^2} \boldsymbol{\nabla} p_0 - \boldsymbol{\nabla} \psi'$$

$$\frac{\partial \rho'}{\partial t} + \boldsymbol{\nabla} \cdot (\rho' \mathbf{v_0} + \rho_0 \mathbf{v'}) = 0 \qquad (13)$$

$$\frac{p'}{p_0} = \Gamma_1 \left(\frac{\rho'}{\rho_0} - \boldsymbol{\xi} \cdot A \right)$$

$$\nabla^2 \psi' = 4\pi G \rho'$$

The displacement is related to the Lagrangian velocity by

$$\delta v = \frac{D\boldsymbol{\xi}}{Dt} = \frac{\partial \boldsymbol{\xi}}{\partial t} + \mathbf{v_0} \cdot \boldsymbol{\nabla} \boldsymbol{\xi}$$

where D/Dt is the Lagrangian derivative [78, 79]. Using the linearized relation between the Lagrangian and Eulerian velocities $\mathbf{v'} = \delta \mathbf{v} - (\boldsymbol{\xi} \cdot \boldsymbol{\nabla}) \mathbf{v_0}$, the Eulerian velocity perturbation $\mathbf{v'}$ is related to the displacement $\boldsymbol{\xi}$ by

$$\mathbf{v'} = \frac{\partial \boldsymbol{\xi}}{\partial t} + (\mathbf{v_0} \cdot \boldsymbol{\nabla}) \boldsymbol{\xi} - (\boldsymbol{\xi} \cdot \boldsymbol{\nabla}) \mathbf{v_0} \qquad (14)$$

3.3 Time and Azimuthal Dependences of the Oscillation

One adopts a spherical coordinate system with the polar axis coinciding with the star rotation axis. For an axisymmetric star, the system Eq. (13) is separable in ϕ and one can then seek quite generally for a solution of the form

$$\boldsymbol{\xi}(\mathbf{r}, t) \propto \boldsymbol{\xi}(r, \theta, t) \, e^{im\phi}$$

In a description based on spherical harmonics, an eigenfunction will be decomposed on the spherical harmonics with the same m. Because of the assumed existence of a steady-state configuration, one can represent all scalar perturbations by $f'(\mathbf{r}, t) = f'(r, \theta) \, e^{i(\omega t + m\phi)}$ and for the fluid displacement vector $\boldsymbol{\xi}(\mathbf{r}, t) = \tilde{\boldsymbol{\xi}}(r, \theta) \, e^{i(\omega t + m\phi)}$. The relation between $\mathbf{v'}$ and the displacement (Eq. (14)) becomes

$$\mathbf{v'} = i(\omega + m\Omega)\tilde{\boldsymbol{\xi}} - (\tilde{\boldsymbol{\xi}} \cdot \boldsymbol{\nabla} \Omega) r \sin \theta \, \mathbf{e}_\phi \qquad (15)$$

From now on, we drop the tilde for the displacement vector field.

3.4 The Eigenvalue System

One assumes that the configuration is axisymmetric and that the motion is purely rotational, i.e. one then has $B = \mathbf{v_0} \cdot \boldsymbol{\nabla} = \Omega \, \partial/\partial\phi$ and $Bf \equiv \mathbf{v_0} \cdot \boldsymbol{\nabla} f = \Omega \partial f/\partial\phi$ for a scalar f. Using Eqs. (13), (14), (15) and the equality

$$B\xi \equiv \boldsymbol{v_0} \cdot \boldsymbol{\nabla}\xi = im\Omega\xi + \boldsymbol{\Omega} \times \xi \tag{16}$$

for a vector field ξ, one obtains the wave equation:

$$\frac{1}{\rho_0}\mathcal{L}\xi - \hat{\omega}^2\xi + 2\hat{\omega}\, i\, \boldsymbol{\Omega} \times \xi - (\xi \cdot \boldsymbol{\nabla}\Omega^2)\, r\sin\theta\, \boldsymbol{e}_s = 0 \tag{17}$$

where we have defined

$$\mathcal{L}\xi = \boldsymbol{\nabla}p' - \frac{\rho'}{\rho_0}\boldsymbol{\nabla}p_0 + \rho_0\,\boldsymbol{\nabla}\psi' \tag{18}$$

and $\hat{\omega} = \omega + m\Omega(r,\theta)$. The continuity equation in Eq. (13) becomes

$$\rho' + \boldsymbol{\nabla} \cdot (\rho_0\xi) = 0 \tag{19}$$

The system is completed with the linearized adiabatic relation Eq. (12). The first two terms in Eq. (17) subsist in the absence of rotation and provide the eigenvalue problem for a nonrotating star: $\mathcal{L}\xi - \omega^2\rho_0\xi - 0$. The first additional term in Eq. (17) is due to the Coriolis force; the second additional term only exists in the presence of a nonuniform rotation. The centrifugal force comes into play through its effects on the structure, i.e. through the quantities $p_0, \rho_0, \dots$. The usual boundary conditions are the requirements that the solutions must keep a regular behaviour in the centre and at the surface. More specifically one usually asks that $\delta p = 0$ at the surface and the gravity potential goes to zero at infinity. At the centre the regularity conditions are expressed as $f' = O(r^\ell)$ for a scalar quantity for instance (see Unno et al. [143]).

Equations (17), (18), (19) together with the perturbed adiabatic and Poisson equations plus the boundary conditions give rise to an eigenvalue problem for the oscillation of a rotating star. One looks for ω and ξ, given Ω and the equilibrium structure. Although it is not necessary, the angular dependence of the eigenfunctions is most commonly looked under the form of a decomposition over the basis formed with the spherical harmonics (with a given m):

$$\xi_m(\mathbf{r}) = \sum_{\ell \geq |m|}^{+\infty} (\,\xi_{r,\ell}(r)\, Y_\ell^m(\theta,\phi)\,\mathbf{e_r} + \xi_{h,\ell}(r)\,\boldsymbol{\nabla}\,Y_\ell^m(\theta,\phi)$$

$$+ \gamma_\ell(r)\,\mathbf{e_r} \times \boldsymbol{\nabla}Y_\ell^m(\theta,\phi)\,) \tag{20}$$

$$f'_m(\mathbf{r}) = \sum_{\ell \geq |m|}^{+\infty} f'^m_\ell(r)\, Y_\ell^m(\theta,\phi) \tag{21}$$

The first two terms in $\xi_m(\mathbf{r})$ represent the spheroidal part of the oscillation whereas the third represents the toroidal part. If the star possesses the equatorial symmetry $\theta \implies \pi - \theta$ (with $\Omega(r, \pi - \theta) = \Omega(r,\theta)$), the eigenmodes can be classified into two groups: symmetric (or even) modes and antisymmetric (or odd) modes with respect to the equator. One then obtains two different systems of differential equations, one for each parity (see Unno et al. [143]; Hansen et al. [70]). Symmetry with respect to the equatorial plane induces the property

$$\omega_{n,\ell,-m}(\Omega) = -\omega_{n,\ell,m}(\Omega) \tag{22}$$

3.5 Symmetry Properties of the Operators and Orthogonality Relations

Symmetries of the operators in Eqs. (17), (18), (19) and their consequences were studied by Lynden-Bell and Ostriker [88]; Dyson and Schutz [45], Schutz [127, 128], Clement [33]; see also Unno et al. [143], Reese [112] and Reese et al. [113] for a polytrope in uniform rotation. One first introduces the inner product between two arbitrary vector fields $\boldsymbol{\eta}$ and $\boldsymbol{\xi}$:

$$< \boldsymbol{\eta}|\boldsymbol{\xi} > \equiv \int_V (\boldsymbol{\eta}^*(r) \cdot \boldsymbol{\xi}(\mathbf{r})) \, \mathrm{d}^3\mathbf{r} \tag{23}$$

where $\boldsymbol{\eta}^*$ is the complex conjugate of $\boldsymbol{\eta}$. We also recall that the adjoint operator $\mathcal{Q}^+$ of an operator $\mathcal{Q}$ is defined such that $< \mathcal{Q}^+(\boldsymbol{\eta})|\boldsymbol{\xi} > \equiv < \boldsymbol{\eta}|\mathcal{Q}(\boldsymbol{\xi}) >$ for any $\boldsymbol{\xi} \in$ domain of $\mathcal{Q}$. An operator $\mathcal{Q}$ is symmetric with respect to an inner product if for any nonsingular vector fields $\boldsymbol{\eta}$ and $\boldsymbol{\xi}$ defined in the unperturbed volume and having continuous first derivatives everywhere, one has

$$< \mathcal{Q}(\boldsymbol{\eta})|\boldsymbol{\xi} > = < \boldsymbol{\eta}|\mathcal{Q}(\boldsymbol{\xi}) > \tag{24}$$

This is equivalent to

$$\int \boldsymbol{\eta}^* \cdot \mathcal{Q}(\boldsymbol{\xi}) \mathrm{d}^3\mathbf{r} = \left(\int \mathcal{Q}(\boldsymbol{\eta}) \cdot \boldsymbol{\xi}^* \mathrm{d}^3\mathbf{r} \right)^*$$

Let start from the fluid motion equation

$$\frac{D\mathbf{v}}{Dt} = \left(\frac{\partial}{\partial t} + B \right) \mathbf{v} = -\frac{1}{\rho} \boldsymbol{\nabla} p + \boldsymbol{\nabla} \psi$$

with B defined in Eq. (16) above. If the configuration is assumed to be in steady state, B commutes with $\partial/\partial t$. Using this property, Lynden-Bell and Ostriker [88] have shown that the wave equation Eq. (17) can be cast under the form

$$\frac{D^2\boldsymbol{\xi}}{Dt^2} = \delta \left(-\frac{1}{\rho} \boldsymbol{\nabla} p + \boldsymbol{\nabla} \psi \right)$$

where $\boldsymbol{\xi}$ is the fluid Lagrangian displacement and δ represents a Lagrangian variation. As

$$\frac{D^2\boldsymbol{\xi}}{Dt^2} = \frac{\partial^2\boldsymbol{\xi}}{\partial t^2} + 2\,B\frac{\partial\boldsymbol{\xi}}{\partial t} + B^2\boldsymbol{\xi} \tag{25}$$

one must then solve

$$\frac{\partial^2\boldsymbol{\xi}}{\partial t^2} + 2\,B\frac{\partial\boldsymbol{\xi}}{\partial t} = C(\boldsymbol{\xi}) \tag{26}$$

where the Lagrangian expression for $C(\boldsymbol{\xi})$ can be found in Lynden-Bell and Ostriker [88] or Dyson and Schutz [45] for instance. It is also given by

$$\mathcal{C}(\boldsymbol{\xi}_m) = -B^2(\boldsymbol{\xi}_m) - \frac{1}{\rho_0}\mathcal{L}(\boldsymbol{\xi}_m) + K(\boldsymbol{\xi}_m)$$

where $\mathcal{L}(\boldsymbol{\xi})$ is defined by Eq. (18) and

$$K(\boldsymbol{\xi}) = \boldsymbol{\xi} \cdot \boldsymbol{\nabla}\left(-\frac{1}{\rho_0}\boldsymbol{\nabla}p_0 + \boldsymbol{\nabla}\psi_0\right) = \boldsymbol{\xi} \cdot \boldsymbol{\nabla}\left(-\Omega^2 r \sin\theta e_s\right)$$

where we have used Eq. (11). Equation (26) is valid for any axisymmetric rotation law for a steady-state rotating configuration. Assuming again a time dependence e^{iwt} for $\boldsymbol{\xi}$ and the scalar variables, the linear adiabatic perturbations of a differentially rotating, axisymmetric stellar model (Eqs. (17), (18)) are then cast into the form [45]

$$\mathcal{Q}(\boldsymbol{\xi}_m) \equiv -\omega^2\rho_0\boldsymbol{\xi}_m + i\,w\mathcal{B}(\boldsymbol{\xi}_m) + \mathcal{C}(\boldsymbol{\xi}_m) = 0 \tag{27}$$

where we have defined

$$\mathcal{B}(\boldsymbol{\xi}_m) = 2\rho_0\left(\boldsymbol{\Omega} \times \boldsymbol{\xi}_m + im\Omega\boldsymbol{\xi}_m\right) \tag{28}$$

Lynden-Bell and Ostriker [88] have shown that C is symmetric and B is antisymmetric, i.e. for any $\boldsymbol{\eta}$ and $\boldsymbol{\xi}$,

$$<\boldsymbol{\eta}|C(\boldsymbol{\xi})>=<C(\boldsymbol{\eta})|\boldsymbol{\xi}>;\quad <\boldsymbol{\eta}|\mathcal{B}(\boldsymbol{\xi})>= - <\mathcal{B}(\boldsymbol{\eta})|\boldsymbol{\xi}>$$

and actually C is real and B is purely imaginary. Dyson and Schutz [45] and Schutz [127, 128] studied the general properties of eigenfunctions of a rotating star. We focus here on a much simpler issue, for later use. Let $\boldsymbol{\xi}_1$ and $\boldsymbol{\xi}_2$ be two eigenfunctions associated with two eigenvalues λ_1 and λ_2; they verify the equalities

$$\mathcal{Q}(\boldsymbol{\xi}_1) = \lambda_1^2\rho_0\boldsymbol{\xi}_1 + \lambda_1\mathcal{B}(\boldsymbol{\xi}_1) + \mathcal{C}(\boldsymbol{\xi}_1) = 0 \tag{29}$$
$$\mathcal{Q}(\boldsymbol{\xi}_2) = \lambda_2^2\rho_0\boldsymbol{\xi}_2 + \lambda_2\mathcal{B}(\boldsymbol{\xi}_2) + \mathcal{C}(\boldsymbol{\xi}_2) = 0 \tag{30}$$

Taking the inner product of Eq. (29) with $<\boldsymbol{\xi}_2|$, one gets

$$\lambda_1^2 <\boldsymbol{\xi}_2|\rho_0\boldsymbol{\xi}_1> +\lambda_1 <\boldsymbol{\xi}_2|\mathcal{B}(\boldsymbol{\xi}_1)> + <\boldsymbol{\xi}_2|\mathcal{C}(\boldsymbol{\xi}_1)>= 0 \tag{31}$$

Similarly taking the inner product of Eq. (30) with $<\boldsymbol{\xi}_1|$, one obtains

$$\lambda_2^2 <\boldsymbol{\xi}_1\rho_0|\boldsymbol{\xi}_2> +\lambda_2 <\boldsymbol{\xi}_1|\mathcal{B}(\boldsymbol{\xi}_2)> + <\boldsymbol{\xi}_1|\mathcal{C}(\boldsymbol{\xi}_2)>= 0 \tag{32}$$

We now make use of the symmetry properties of C and B, i.e. $<\boldsymbol{\xi}_1|\mathcal{C}(\boldsymbol{\xi}_2)>=<\mathcal{C}(\boldsymbol{\xi}_1)|\boldsymbol{\xi}_2>$ and $<\boldsymbol{\xi}_1|\mathcal{B}(\boldsymbol{\xi}_2)>= - <\mathcal{B}(\boldsymbol{\xi}_1)|\boldsymbol{\xi}_2>$ which give for the last equality Eq. (32):

$$\lambda_2^2 <\boldsymbol{\xi}_1\rho_0|\boldsymbol{\xi}_2> -\lambda_2 <\mathcal{B}(\boldsymbol{\xi}_1)|\boldsymbol{\xi}_2> + <\mathcal{C}(\boldsymbol{\xi}_1)|\boldsymbol{\xi}_2>= 0$$

Taking the complex conjugate of this equation yields

$$\lambda_2^{*2} < \boldsymbol{\xi}_2|\rho_0\boldsymbol{\xi}_1 > -\lambda_2^* < \boldsymbol{\xi}_2|\mathcal{B}(\boldsymbol{\xi}_1) > + < \boldsymbol{\xi}_2|\mathcal{C}(\boldsymbol{\xi}_1) >= 0$$

which can be compared to Eq. (31). The difference between these two equations yields the relation

$$(\lambda_1^2 - \lambda_2^{*2}) < \boldsymbol{\xi}_2|\rho_0\boldsymbol{\xi}_1 > +(\lambda_1 + \lambda_2^*) < \boldsymbol{\xi}_2|\mathcal{B}(\boldsymbol{\xi}_1) >= 0 \qquad (33)$$

For a pure imaginary eigenfrequency $\lambda = i\omega$, this reduces to $(\omega_1^2 - \omega_2^2) < \boldsymbol{\xi}_2|\rho_0\boldsymbol{\xi}_1 > +i(\omega_1 - \omega_2) < \boldsymbol{\xi}_2|\mathcal{B}(\boldsymbol{\xi}_1) >= 0$. For $\omega_1 \neq \omega_2$, one obtains the orthogonality relation:

$$< \boldsymbol{\xi}_2|(\omega_1 + \omega_2)\rho_0 + i\mathcal{B}(\boldsymbol{\xi}_1) >= 0 \qquad (34)$$

which is valid for any rotating, axisymmetric star. A similar relation although in the context of a perturbative third-order approach was derived in Soufi et al. [130].

4 A Variational Principle

Chandrasekhar [20] has shown that nonradial oscillations of a nonrotating star can be treated as a variational problem. This was extended to uniformly rotating bodies by Chandrasekhar and Lebovitz [21] and nonuniform axisymmetric rotation for a steady-state configuration by Lebovitz [78, 79]. The symmetry properties of the operators in Eqs. (17), (18) indeed leads to the existence of a *variational principle* which states as follows: let a real number ζ and a function $\eta(r)$ be related by the relation

$$\zeta^2 < \eta|\rho_0\eta > +\zeta < \eta|B\eta > + < \eta|C\eta >= 0$$

where B and C are antisymmetric and symmetric operators. One assumes that $\eta(r)$ is a trial function close enough but not equal to the unknown true eigenfunction $\xi(r)$ and one sets $\eta = \xi + \delta\xi$. The value of ζ will be close but different from the true unknown eigenvalue λ with $\zeta = \lambda + \delta\lambda$. Inserting the expressions for ζ and η in the above equation and linearizing, one gets

$$2(\delta\lambda)\lambda < \xi|\rho\xi > +\lambda^2\delta(< \xi|\rho\xi >) =$$
$$(\delta\lambda) < \xi|B\xi > +\lambda(\delta < \xi|B\xi >) + \delta(< \xi|C\xi >) \qquad (35)$$

which is rewritten as

$$\delta\lambda = \frac{2\,\mathcal{R}(< \delta\xi|(-\lambda^2\rho\xi + \lambda B\xi + C\xi) >)}{(2\lambda < \xi|\rho\xi > - < \xi|B\xi >)} \qquad (36)$$

where $\mathcal{R}$ means real part. The numerator in the right-hand side of Eq. (36) vanishes since ξ is the eigenfunction associated to λ; one then has $\delta\lambda = 0$ (i.e. $\zeta = \lambda$) for any arbitrary $\delta\xi \neq 0$. The reciprocal is also true. Hence the eigenfrequencies are invariant to first order to a change in the function ξ. The existence of such a variational principle has the consequence that any trial function used instead of the true eigenfunction gives an estimation of the eigenvalue accurate enough: the error on λ is quadratic in the error on the eigenfunction.

Variational expressions for the eigenfrequencies. It is possible to use Eq. (27) to derive an integral expression for the eigenfrequency. Taking the inner product with ξ^* and integrating over the entire volume of the star, one gets $-\omega^2 + \omega b + c = 0$, where

$$b = \frac{i}{I} \int_0^R \boldsymbol{\xi}_m^* \cdot \mathcal{B}(\boldsymbol{\xi}_m) d^3\mathbf{r}; \qquad c = \frac{1}{I} \int_0^R \boldsymbol{\xi}_m^* \cdot C(\boldsymbol{\xi}_m) d^3\mathbf{r}$$

and

$$I = \int_0^R (\boldsymbol{\xi}_m^* \cdot \boldsymbol{\xi}_m) \, \rho_0 d^3\mathbf{r} \tag{37}$$

b and c are real since the operators are Hermitian. Then one obtains $\omega = \left(b \pm \sqrt{b^2 + 4c}\right)/2$. For $\Omega = 0$, one recovers $\omega^2 = c = 1/I \int_0^R \boldsymbol{\xi}_m^* \cdot \mathcal{L}(\boldsymbol{\xi}_m) d^3\mathbf{r} \equiv \omega_0^2$. Since c does not depend on m ($-\ell < m < \ell$), the eigenfrequencies of nonradial pulsations in a nonrotating star are $(2\ell + 1)$-fold degenerate.

Clement [32] applied the variational principle to compute eigenfrequencies of polytropes. Clement [33] again used a variational approach to compute the non-axisymmetric modes for a $15M_\odot$ uniformly rotating 2D model. For the p-mode eigenfunctions entering the integral expression, as they have a dominant contribution from the poloidal part, Clement chose trial functions in the form of the gradient of a longitudinal potential $\xi_\mathbf{p} \sim \nabla\phi_1$; the potential ϕ_1 is then expanded in powers of Legendre polynomials and the expansion coefficients are determined by the stationary condition on the eigenfrequency. Saio [123] neglected the perturbation of gravitational potential ψ' in Eq. (27) (which is justified in the case of high-frequency p-modes) and derived a simplified expression for the coefficients b and c. As the error of the frequencies is much smaller than that of the eigenfunctions, it is possible to use the numerical eigenfunctions (obtained by solving numerically the eigensystem Eq. (17)) in the integrals for the coefficients b and c defined above. The resulting eigenfrequencies ω_{var} can be compared to the numerical ones. If an error $\delta\xi$ exists for ξ in the numerical calculation, the resulting error on ω will be $\omega_{var} = \omega + O(\|\delta\xi\|^2)$. This property has been used for instance by Christensen-Dalsgaard and Mullan [25] to check the accuracy of their numerical computations of eigenfrequencies of nonrotating polytropes. Reese et al. [113] used it to check the numerical accuracy of the results of their 2D eigenvalue system for a rotating polytrope.

5 Slow Rotators

At the lowest order, $O(\Omega)$, only the Coriolis force plays a role and one can ignore the direct effect of the centrifugal oblateness on the oscillation frequencies for slow rotators like the Sun for example. From here on, we use the DG92 notations. The eigensystem Eqs. (17), (18) then reduces to

$$\mathcal{L}(\xi) - \hat{\omega}^2 \rho_0 \, \xi + 2\hat{\omega}\rho_0 \, i\boldsymbol{\Omega} \times \xi = 0$$

The scalar perturbations $f = (p', \rho', \psi')$ are expanded up to first order $f' = f_0' + f_1'$ and similarly for the fluid displacement $\xi = \xi_0 + \xi_1$; this yields $\mathcal{L} = \mathcal{L}_0 + \mathcal{L}_1$ where the operators $\mathcal{L}_0$ and $\mathcal{L}_1$ take the same form than $\mathcal{L}$ with the perturbed quantities subscribed with 0 and 1, respectively. Note that the equilibrium quantities entering $\mathcal{L}$ can come from a rotating stellar model; although strictly speaking, this would be somewhat inconsistent. The eigenfrequency is written as $\omega = \omega_0 + \omega_1$ and $\hat{\omega} = \omega + m\Omega = \omega_0 + \omega_1 + m\Omega$. The subscript 0 refers to solutions of the problem at the zeroth order which is specified by

$$\mathcal{L}_0 \, \boldsymbol{\xi}_0 - \omega_0^2 \rho_0 \, \boldsymbol{\xi}_0 = 0 \tag{38}$$

This yields the zeroth-order eigenvalues ω_0 and eigenfunctions $\boldsymbol{\xi}_0$ for a given (ℓ, n) set. We recall that the zeroth-order operator $\mathcal{L}_0$ is self-adjoint:

$$< \boldsymbol{\xi}_0 | \mathcal{L}_0 \, \boldsymbol{\xi}_0 > = < \mathcal{L}_0 \boldsymbol{\xi}_0 | \, \boldsymbol{\xi}_0 > \tag{39}$$

Consequently the right and left eigenfunctions are identical and orthogonal to one another:

$$< \boldsymbol{\xi}_{0i} | \boldsymbol{\xi}_{0j} > = \int \rho_0 (\boldsymbol{\xi}_{0i}^* \cdot \boldsymbol{\xi}_{0j}) \mathbf{d^3 r} = I \, \delta_{ij}$$

with I the oscillatory moment of inertia Eq. (37) and δ_{ij} is the Kroncckcr symbol, i, j label eigenfunctions and are shortcuts for the subscripts (n_i, ℓ_i) and (n_j, ℓ_j). Using the integral form, the zeroth-order solution is then given by

$$\omega_0^2 = \frac{1}{I} < \boldsymbol{\xi}_0 | \mathcal{L}_0 \, \boldsymbol{\xi}_0 > \tag{40}$$

In practice, ω_0^2 is obtained as the numerical solution of the zeroth-order eigensystem. At the first order in Ω, one must solve

$$\mathcal{L}_0 \, \boldsymbol{\xi}_1 - \omega_0^2 \, \rho_0 \, \boldsymbol{\xi}_1 = 2\omega_0 \, \rho_0 \, (\omega_1 + K) \, \boldsymbol{\xi}_0 \tag{41}$$

with $K = m\Omega - i\boldsymbol{\Omega}\times$. One seeks a solution for the correction ω_1. The integral solution is again obtained by projecting Eq. (41) onto the eigenfunction $\boldsymbol{\xi}_0$ using the inner product Eq. (23). $< \boldsymbol{\xi}_0 | \mathcal{L}_0 \boldsymbol{\xi}_1 > - \omega_0^2 < \boldsymbol{\xi}_0^* | \rho_0 \boldsymbol{\xi}_1 > = 2\omega_0 < \boldsymbol{\xi}_0 | \rho_0 (\omega_1 + K) \boldsymbol{\xi}_0 >$ which admits a solution provided $< \boldsymbol{\xi}_0 | (\omega_1 + K) \boldsymbol{\xi}_0 > = 0$ is satisfied, which yields

$$\omega_1 = -\frac{1}{I} < \boldsymbol{\xi}_0 | K \, \boldsymbol{\xi}_0 > \tag{42}$$

Due to the existence of the variational principle discussed in Sect. 4, if one is interested only in the eigenfrequency, one need not know the correction to the eigenfunction $\boldsymbol{\xi}_1$. Note however that one obtains the correction ω_1 numerically by integrating the $O(\Omega)$ eigensystem as shown by Hansen et al. [69]. Comparison of the variational expression and the numerical one can be used to check the results.

5.1 Rotational Splitting

The integral expression of Eq. (42) is

$$\omega_1 = \frac{1}{I} \int \rho_0 \left(m\Omega |\boldsymbol{\xi}_0|^2 - i\, \boldsymbol{\xi}_0^* \cdot (\boldsymbol{\Omega} \times \boldsymbol{\xi}_0) \right) \mathbf{d^3 r} \tag{43}$$

which requires the knowledge of the zeroth-order displacement eigenfunction $\boldsymbol{\xi}_0$. Each mode is usually determined as a sum of spherical harmonics Eq. (20). Solving the zeroth-order eigensystem Eq. (38) shows that actually the zeroth-order solution can be written with a single spherical harmonics:

$$\boldsymbol{\xi}_0(\boldsymbol{r}) = \xi_r(r) Y_\ell^m \,\mathbf{e_r} + \xi_h(r) \boldsymbol{\nabla}_h Y_\ell^m \tag{44}$$

where the gradient $\boldsymbol{\nabla}_h$ is defined as

$$\boldsymbol{\nabla}_h = \mathbf{e}_\theta \frac{\partial}{\partial \theta} + \mathbf{e}_\phi \frac{1}{\partial \sin\theta} \frac{\partial}{\partial \phi}$$

Using this expression, it is straightforward to show that

$$m\Omega |\boldsymbol{\xi}_0|^2 \; -i\, \boldsymbol{\xi}_0^* \cdot (\boldsymbol{\Omega} \times \boldsymbol{\xi}_0) =$$

$$m\Omega \left[\; (|\xi_r|^2 - (\xi_r^* \xi_h + cc)) |Y_\ell^m|^2 + \right.$$

$$\left. |\xi_h|^2 \left(\boldsymbol{\nabla}_h Y_{\ell,m}^* \cdot \boldsymbol{\nabla}_h Y_\ell^m - \frac{\cos\theta}{\sin\theta} \frac{\partial |Y_\ell^m|^2}{\partial\theta} \right) \right]$$

Inserting this expression into Eq. (43) yields the rotation splitting, Eq. (4), in the compact form:

$$\delta\omega_m = \frac{\omega_{1,j}}{m} = \int_0^R \int_0^\pi K_{n,\ell,m}(r,\theta)\; \Omega(r,\theta)\; r d\theta dr \tag{45}$$

where $K_{n\ell}$ is called *rotational kernel*; its expression can be found in Schou et al. [124, 125], Pijpers [109] or CD03. For instance, the sectoral modes ($|m| = \ell$) for increasing ℓ get increasingly confined towards the equator. Hence the measure of the splittings of sectoral modes provides a measure of the equatorial velocity.

Shellular rotation $\Omega(r)$. When the rotation can be assumed independent of θ, the expression of the rotational splitting reduces to

$$\delta\omega_{n,\ell} = \int_0^R K_{n,\ell}(r)\; \Omega(r)\; dr \tag{46}$$

with

$$K_{n\ell}(r) = \frac{1}{I} \left(\xi_r^2 + \Lambda \xi_h^2 - 2\xi_r \xi_h + \xi_h^2 \right) \rho_0\, r^2$$

and

$$I = \int_0^R \rho_0 r^2 \left(\xi_r^2 + \Lambda \xi_h^2 \right) dr$$

and $\Lambda = \ell(\ell + 1)$ and R the stellar radius. One also finds in the literature

$$\delta\omega_{n,\ell} = m\bar{\Omega}(1 - C_{n,\ell} - J_{n,\ell}) \tag{47}$$

in the observer frame, with $\bar{\Omega}$ an averaged or the surface rotation rate. The Ledoux constant is

$$C_{n,\ell} = \frac{1}{I} \int_0^R \rho_0 r^2 \, (2\xi_r\xi_h + \xi_h^2) dr$$

and

$$J_{n,\ell} = \frac{1}{I} \int_0^R \rho_0 r^2 \, (\Omega(r) - \bar{\Omega}) \, (\xi_r^2 + \Lambda\xi_h^2 - 2\xi_r\xi_h - \xi_h^2) dr$$

A particular case is that of *solid-body rotation* for which one derives $\delta\omega_{n,\ell} = m\Omega(1 - C_{n\ell})$. It is worth noting that for $|\xi_h| << |\xi_r|$, $C_{n,\ell} \to 0$. As

$$\frac{\xi_h}{|\xi_r|} \propto \frac{1}{\sigma^2}$$

where σ is the normalized frequency, $C_{n\ell} \sim 0$ for high-frequency p-modes, and the measure of the rotational splitting $\delta\omega_{n,\ell,m} = \Omega$ then is a quasi-direct measure of the rotational angular velocity.

Forward techniques compute the splittings with Eqs. (45), (46) by assuming a rotation profile. The rotational kernels $K_{n\ell}$ are assumed to be known and calculated for the appropriate stellar model. The results are compared with the observed splittings. The integral relations Eqs. (45), (46) can be inverted to provide Ω. The observed splittings for several different modes (n, ℓ) constitute the data set (see Sect. 9).

Latitudinal dependence. It is convenient to assume a rotation of the type

$$\Omega(r, \theta) = \sum_{s=0}^{s_{max}} \Omega_{2s}(r) \, (\cos\theta)^{2s} \tag{48}$$

The expression for the rotational splitting then becomes (see, for instance, Hansen et al. [70]; DG92 [47]; CD03 [28])

$$\delta\omega_m = \sum_{s=0}^{s_{max}} \int_0^R \Omega_{2s}(r) \, K_{n,\ell,m,s}(r) \, \rho_0 r^2 dr$$

with

$$K_{n,\ell,m,s}(r) = |\xi_r|^2 - (\xi_r^*\xi_h + cc)\mathcal{S}_s + |\xi_h|^2((2s-1)\mathcal{S}_{s-1} - \Lambda\mathcal{S}_s)$$

where

$$\mathcal{S}_s = \int_0^\pi d\theta \sin\theta(\cos\theta)^{2s} \, |Y_\ell^m(\theta, \phi)|^2$$

and $\mathcal{S}_{-1} = 0$; $\mathcal{S}_0 = 1$. For later purpose, we also give

$$\mathcal{S}_1 = \frac{1}{4\Lambda - 3}(-2m^2 + 2\Lambda - 1) \tag{49}$$

$$\mathcal{S}_2 = \frac{1}{4\Lambda - 15}\left[\mathcal{S}_1(-2m^2 + 2\Lambda - 9) + 1\right]$$

More generally $\mathcal{S}_s$ for any s is given by a recurrent relation (Eq. (31) in DG92). Results of 2D inversion in the solar case are discussed in Sect. 9.3.

6 Moderately Rotating Pulsating Stars

For moderate rotators, one needs to include second-order corrections to the frequency ω_2 (not to say higher order contributions, see Sect. 6.2). At the second-order level, several contributions must be included.

1. The eigenfunction is modified by the Coriolis acceleration and becomes $\xi = \xi_0 + \xi_1$, with the eigenfunction correction including a spheroidal (ξ_{1p}) and a toroidal (ξ_{1t}) component, respectively: $\xi_1 = \xi_{1p} + \xi_{1t}$ which results in frequency corrections labelled ω_{2P} et ω_{2T}.
2. Oscillatory inertia is also modified $I = I_0 + I_2$ where $I_2 = <\xi_1|\xi_1>$. We recall that $<\xi_0|\xi_1> = 0 = <\xi_1|\xi_0>$ and $<\xi_0|\xi_2> = 0 = <\xi_2|\xi_0>$. This causes a second-order frequency correction ω_2^I.
3. The centrifugal force generates changes in the structure of the star:
   (i) Geometrical distortion which can be split into two components:
      – *spherically symmetric* distortion due to the latitudinally averaged centrifugal force which generates an effective gravity

$$g_{eff}(r) = g(r) - (2/3)r\Omega^2 \tag{50}$$

   Its principal effect is to decrease slightly the central density and to modify the radius of the star. The corresponding frequency change is proportional to a quantity which we denote as Z_1,

      – *nonspherically symmetric* distortion. This is the dominant second-order effect and is denoted as ω_{2D} as in DG92.

   (ii) Rotationally induced transport and mixing modify the internal stratification and influence the evolution of the steady configuration. This effect is responsible for a frequency change which is proportional to a quantity denoted as Z_2. The frequency correction induced by modifications of the steady configuration is then proportional to $Z_1 + Z_2$. Finally the (second-order) frequency correction is given by

$$\omega_{2,n,\ell,m} = \frac{\Omega^2}{\omega^{(0)}}(Z_1 + Z_2) + \omega_{2I} + \omega_{2P} + \omega_{2T} + \omega_{2D} + \frac{\omega_1^2}{\omega_0} \tag{51}$$

In order to obtain explicit expressions and quantitative estimates for these corrections, one starts again with Eq. (17). The expansions are carried out up to second order $f' = f'_0 + f'_1 + f'_2$ for any scalar perturbed quantity and for the displacement vector field $\xi = \xi_0 + \xi_1 + \xi_2$. Assuming that the zeroth-order and first-order systems have been solved, it remains to solve the second-order system of equations. The knowledge of the first-order correction to the *eigenfunction* is indeed required to compute ω_2. One also needs to include the effects of the centrifugal force on the equilibrium structure. For moderate rotation, it is enough to use a perturbative technique to compute the centrifugal distortion. Two approaches have been used: a mapping technique and a direct perturbative method.

6.1 Mapping Technique

Because the shape of the star is distorted, one must in principle work in a *spheroidal* coordinate system and take into account the oblateness of the surface in the boundary conditions. It is possible however to work in a *spherical* coordinate system with simple (classical) boundary conditions by defining a mapping between the coordinate system (r, θ, ϕ) in the spheroidal volume of the star and the corresponding one (ζ, θ, ϕ) in a spherical volume. This mapping then consists in defining a transformation $(r, \theta, \phi) \Longrightarrow (\zeta, \theta, \phi)$ where r and ζ are related through the transformation

$$r(\zeta, \theta) = \zeta \left(1 - h_2(\zeta, \Omega) P_2(\cos\theta)\right) \tag{52}$$

where $P_2(\theta)$ is the second-order Legendre polynomial. The function h_2 is chosen so that surfaces of constant ζ are surfaces of constant pressure, in particular the surface of the star is given by $\zeta = R$. This gives for the oblateness of the star $r_{eq} - r_{pole} = (3/2)h_2(\zeta, \Omega) \zeta$ between the equatorial and polar radii.

This approach was carried out by Simon [129] and Lebovitz [78, 79]. Goughu and Thompson [55] developed the formalism under the Cowling approximation – which consists in neglecting the Eulerian perturbation to the gravitational potential – for a general stellar model and a prescribed rotation law $\Omega(r, \theta)$. The authors considered the wave equation Eq. (17) from Lynden-Bell and Ostriker [88] which in our notation is (see also Sect. 4)

$$\mathcal{L}(\xi) - \rho_0 \omega^2(\xi) = i\omega\mathcal{B}(\xi) + \mathcal{N}(\xi)$$

where

$$\mathcal{L}(\xi) = \nabla \left((p_0 - \rho_0 c_{s0}^2)\nabla \cdot \xi \right. \tag{53}$$
$$\left. - \xi \cdot \nabla p_0 \right) - p_0 \nabla(\nabla \cdot \xi) + \xi \cdot \nabla(ln\rho_0)\nabla p_0$$

with $\mathcal{N}\xi = \rho_0 \left[\xi_0 \cdot \nabla(v_0 \cdot \nabla v_0) - (v_0 \cdot \nabla)^2\xi\right]$ and $\mathcal{B}(\xi)$ is defined in Eq. (28). GT90 actually were interested in the effects of a magnetic field on adiabatic oscillation frequencies and included the Lorenz force contribution to the above equation which we disregard here. The equilibrium quantities p_0, ρ_0 and the adiabatic sound speed c_{s0}^2 can be expanded about their values in the spherical volume, so, for example,

$$p_0(\mathbf{r}) = p_{00}(\zeta) + \epsilon^2 p_2(\zeta)\ P_2(\cos\theta) \tag{54}$$

This procedure has the advantage of retaining simplicity in the boundary conditions which are $\delta P = 0$ at surface and $\psi' = 0$ matching the solution of Laplace equation away outward from the surface. GT90 found that for slow rotation, these boundary conditions are far enough. However, they also discussed in detail the boundary conditions obtained by matching the interior with an isothermal atmosphere in hydrostatic equilibrium.

The price to pay for the change of coordinate system Eq. (52) is additional terms coming from the decomposition of the gradients such that $\nabla = \nabla_0 + \nabla_\Omega$ where

$$\nabla_0 = \mathbf{e_r}\frac{\partial}{\partial\zeta} + \mathbf{e}_\theta\frac{1}{\zeta}\frac{\partial}{\partial\theta} + \mathbf{e}_\phi\frac{1}{\zeta\sin\theta}\frac{\partial}{\partial\phi} \tag{55}$$

$$\nabla_\Omega = \mathbf{e_r}\,\frac{d(\zeta h_2)}{d\zeta}\,P_2(\theta)\frac{\partial}{\partial\zeta} + \mathbf{e}_\theta\left(h_2\frac{dP_2(\theta)}{d\theta}\frac{\partial}{\partial\zeta} + \frac{1}{\zeta}h_2 P_2(\theta)\frac{\partial}{\partial\theta}\right)$$

$$+\mathbf{e}_\phi h_2 P_2(\theta)\frac{1}{\zeta\sin\theta}\frac{\partial}{\partial\phi} \tag{56}$$

so that $\mathcal{L} = \mathcal{L}_0 + \mathcal{L}_\Omega$. An inhomogeneous differential equation and boundary conditions for $h_2(\zeta)$ are obtained by matching the gravitational potential with the vacuum potential in $\zeta > R$ and requiring that the transformation be regular at $\zeta = 0$.

Simon [129] carried the expansion up to second order. He derived an integral expression for the second-order eigenfrequency correction for nonradial p-modes for a prescribed rotation law $\Omega(r,\theta)$. He then studied the particular case of a uniformly rotating homogeneous spheroid and computed the frequency for the radial fundamental mode which was found to be of the form $\sigma = \sigma_0 + \beta\ \bar{\Omega}^2$, $\bar{\Omega}$ is the dimensionless rotation rate $\Omega/(GM/R^3)^{1/2}$, retrieving results similar to those of Ledoux [81] and Chandrasekhar and Lebovitz [17–19]. Chlebowski [24] derived the expressions for the correcting coefficients for nonradial modes and wrote the frequencies under the form

$$\sigma = \sigma_0 - m(1 - C_{n,\ell})\ \hat{\Omega} + \frac{1}{2\sigma_0}(P_{n,\ell} - m^2\ Q_{n,\ell})\ \hat{\Omega}^2 \tag{57}$$

Numerical results were obtained for g-modes of white dwarfs. Later Saio [122] developed an equivalent procedure and applied it to the study of a polytrope in uniform rotation. He put the scaled eigenfrequency under the convenient form

$$\sigma_{n,\ell,m} = \sigma_{0,n,\ell} + m\ \hat{\Omega}\ (1 - C_{nl}) + \frac{\hat{\Omega}^2}{\sigma_0}((Z + X_1 + X_2)$$

$$+m^2(Y_1 + Y_2)\ ) \tag{58}$$

in the observed frame. σ_0 is the dimensionless frequency of a nonrotating polytrope having the same central pressure and density than the rotating one. Saio found a significant departure from equidistance due to both spherical Z and nonspherical X2, Y2 distortion. GT90 applied this formalism to a shellular rotation for a solar model. Burke and Thompson [15] used the same approach and correct some errors in GT90. They did not include the poloidal correction of the first-order correction to eigenfunction ξ_{1p}. They computed p-mode eigenfrequencies for stellar models with a variety of masses and ages in the Cowling approximation.

6.2 Perturbative Approach

This procedure has been first developed by Hansen et al. [69] and DG92.

Perturbative approach for the structure. Equation (11) is decomposed as

$$-\frac{1}{\rho_0}\frac{\partial p_0}{\partial r} = \frac{\partial \psi_0}{\partial r} + \frac{2}{3}r\Omega^2(r,\theta)\left(1 - P_2(\cos\theta)\right) \tag{59}$$

$$-\frac{1}{\rho_0}\frac{\partial p_0}{\partial \theta} = \frac{\partial \psi_0}{\partial \theta} + \frac{1}{3}r^2\Omega^2(r,\theta)\frac{dP_2(\cos\theta)}{d\theta} \tag{60}$$

where the acceleration term $-r\Omega^2\sin\theta\,\boldsymbol{e}_s$ has been written in terms of Legendre polynomials. Pressure, density and gravitational potential are then expanded in terms of even Legendre polynomials due to the equatorial symmetry. One keeps only the first two terms, i.e. for the pressure, for instance one sets

$$p_0(r,\theta) = p_{00}(r) + p_2(r)\ P_2(\cos\theta) \tag{61}$$

Inserting this expansion into Eqs. (59),(60), one gets on the one hand one equation for the spherically symmetric perturbed part of the stellar model $p_{00}, \rho_{00}, \psi_{00}$ and on the other hand equations for the nonspherical part of the stellar model p_2, ρ_2, ψ_2.

Spherically symmetric distortion. For a shellular rotation $\Omega(r)$, Eq. (61) and Eq. (59) yield:

$$-\frac{1}{\rho_{00}}\frac{dp_{00}}{dr} = \frac{d\psi_{00}}{dr} + \frac{2}{3}r\Omega^2 \tag{62}$$

This equation replaces the hydrostatic equation for a nonrotating star. The horizontally averaged centrifugal force modifies the effect of gravity and the equilibrium pressure must adjust in order to balance an 'effective' gravity g_{eff}, (Eq. (50)). This effect can easily be implemented in an evolutionary code by solving numerically for a classical spherically symmetric (nonrotating) stellar model but using the effective gravity (Kippenhahn and Weigert [77]).

Nonspherically symmetric distortion of the equilibrium structure. For a shellular rotation $\Omega(r)$, Eqs. (61), (59), (60) yield two equations:

$$\frac{dp_2}{dr} = -\rho_{00}\frac{d\psi_2}{dr} - \rho_2\frac{d\psi_{00}}{dr} - \frac{2}{3}r\rho_{00}\ \Omega^2 \tag{63}$$

$$p_2(r) = -\rho_{00}(\psi_2 + \frac{1}{3}r^2\Omega^2) \tag{64}$$

The perturbed part of the Poisson equation is

$$\frac{1}{r^2}\frac{d}{dr}\left(r^2\frac{d\psi_2}{dr}\right) - \frac{6}{r^2}\psi_2 = 4\pi G\rho_2 \tag{65}$$

Details for the integration of the system can be found in DG92 [46], Soufi et al. [130]. Generalization to the case of $\Omega(r,0)$ is treated in DG92 and GT90.

Second-order effects from uniform rotation: centrifugal force and uniform rotation. For pedagogical reasons, the effects of the Coriolis force are ignored in this section. Including only the frequency corrections due to the centrifugal force remains a good approximation for high-frequency p-modes. Indeed the Coriolis force becomes negligible for large μ (Eq. (6)). We also restrict the study to uniform rotation. In this simplified case, the eigenfrequency and eigenfunction are expanded as $\omega_0 + \omega_2$ and $\xi_0 + \xi_2$, respectively. One starts again with Eqs. (17), (18):

$$\mathcal{L}_0(\xi_0 + \xi_2) + \mathcal{L}_2(\xi_0) - \hat{\omega}^2(\rho_{00} + \rho_2)\,(\xi_0 + \xi_2) = 0 \tag{66}$$

where

$$\mathcal{L}_0\xi = \nabla p' - \frac{\rho'}{\rho_{00}}\nabla p_{00} + \rho_{00}\nabla\psi' \tag{67}$$

$$\mathcal{L}_2\xi = \rho_2\left(\frac{\rho'}{\rho_{00}}\nabla p_{00} + \nabla\psi'\right) - \frac{\rho'}{\rho_{00}}\nabla p_2 \tag{68}$$

6.2.1 Effect of Uniform Rotation on ω_0

From Eq. (66), one first gets the zeroth-order system Eq. (38) with the zeroth-order solution given by Eqs. (40), (44). However, now, second-order effects are indirectly inscribed in ω_0 as the equilibrium quantities $p_{00}, \rho_{00}, \psi_{00}$ are modified by the effective gravity Eq. (50). This effect is taken into account for instance in Soufi et al. [130], Goupil et al. [61], Daszynska-Daszkiewicz et al. [37, 38] and Suarez et al. [132, 133]. It is quite small in general: it slightly changes the radius of the star for given mass and age compared to a nonrotating model. Therefore the tracks of a nonrotating equilibrium model and a model with same mass but including spherical distortion do not coincide. Figure 1 of Goupil et al. [61] shows evolutionary tracks of main sequence $1.80M_\odot$ stellar models where gravity is modified by rotation for initial rotational velocities 50, 100 and 150 km/s. Local conservation of angular momentum is assumed for the evolution of the rotation with age. The evolutionary track for a spherically distorted model is shifted to lower luminosities than that of nonrotating model with same mass. It therefore corresponds to a track of nonrotating models with smaller masses; the larger the initial rotation velocity, the smaller the mass of the nonrotating model. The rotation modification of gravity corresponds to a change of mass smaller than $0.02M_\odot$ for initial velocities and masses in the above ranges.

This has the consequence that a comparison of the oscillation frequencies of a nonrotating model and a model including rotation depends on the choice of the nonrotating model. Several choices of nonrotating models are possible. Indeed, for given physical input, chemical composition and rotation rate, two additional parameters must be specified to define a model. The simplest and meaningful choice is that these parameters be fixed at the same value for both models. Saio [122] compared frequencies of polytropes with and without rotation keeping the central pressure and density constant (by keeping the polytropic index constant). For realistic stellar models, other possible choices are the mass and the radius; the mass and the central hydrogen content for main sequence stars; and the effective temperature and the luminosity. The frequency difference for a given mode between a nonrotating model and a rotating model depends on the choice of these two constants. Christensen-Dalsgaard and Thompson [26] kept the luminosity and the effective temperature constant and showed that the results for radial modes for a polytrope are quite similar to those obtained for realistic models. They also found that the frequency difference obtained between nonrotating and rotating polytropes keeping constant mass and radius as they chose can simply be modelled by adding a constant value to the difference obtained by keeping constant central density and pressure (as in Saio [122]) at least for asymptotic, i.e. high n radial modes: $\delta\sigma/\sigma = \hat{\Omega}^2\,(X_1 + Z)/\sigma^2 + 0.33$ where Z is interpolated from Saio's table.

Behaviour of ω_2. The second-order correction to the eigenfrequency, ω_2, is obtained from Eq. (66) written under the form

$$\mathcal{L}_0\boldsymbol{\xi}_2 - \rho_{00}\,\omega_0^2\boldsymbol{\xi}_2 = 2\rho_{00}\,\omega_0\,\omega_2\,\boldsymbol{\xi}_0 - \left(\mathcal{L}_2 - \omega_0^2\rho_2\right)\boldsymbol{\xi}_0 \tag{69}$$

together with the second-order perturbed continuity and Poisson equations. Projection of $\boldsymbol{\xi}_0$ onto this equation (recalling that $< \xi_0|\mathcal{L}_0\xi_2 >=< \mathcal{L}_0\xi_0|\xi_2 >=< \rho_0\hat{\omega}_0^2\xi_0|\xi_2 >$) yields

$$\omega_2 = \frac{1}{2\omega_0 I}\int_{V0}\xi_0^*\cdot\left(\mathcal{L}_2 - \omega_0^2\rho_2\right)\xi_0\,d^3r \tag{70}$$

Some algebraic manipulations leads to

$$\omega_{2,n\ell,m} = \frac{\Omega^2}{\omega_{0,n,\ell}}\left(D_{1,n,\ell} + m^2 D_{2,n,\ell}\right)$$

for each mode (n, ℓ, m) (DG92). The m, Ω dependence of Eq. (70) can be understood as a consequence of symmetry properties.

6.2.2 Consequences of Symmetry Propertics

When only the centrifugal force is considered, the eigenfrequencies obtained by perturbation are of the form $\omega_{n,\ell,m} = w_{n,\ell,m}^{(0)} + \Omega^2\,w_{n,\ell,m}^{(2)} + \,....$ More generally, following Lignières et al. [85] and Reese et al. [113], we write the frequency as an expansion in power of Ω

$$\omega_{n,\ell,m}(\Omega) = \sum_{j \geq 0} w^{(j)}_{n,\ell,m} \, \Omega^j$$

where j is an integer. The symmetry property $\omega_{n,\ell,m}(\Omega) = \omega_{n,\ell,m}(-\Omega)$ which exists in the absence of the Coriolis force imposes $w^{(2j-1)}_{n,\ell,m} = 0$. Hence the perturbative expansion of the frequency in powers of Ω only involves even powers in Ω. Furthermore the reflexion symmetry also imposes $\omega_{n,\ell,-m}(-\Omega) = \omega_{n,\ell,m}(\Omega)$, that is $w^{(2j)}_{n,\ell,m} = w^{(2j)}_{n,\ell,-m}$. Hence the even-order coefficients are even functions of m. As a consequence, the generalized rotational splitting vanishes $S_m = (\omega_{n,\ell,m} - \omega_{n,\ell,-m})/m = 0$ in absence of the Coriolis force.

6.2.3 Near-Degeneracy

Chandrasekhar and Lebovitz [17–19] discussed the effect of near-degeneracy in the stellar oscillation context, that is the existence of small denominators which is a classical problem of the perturbation methods in quantum mechanics for instance. Indeed the nth order correction of an eigenfunction usually involves small denominators of the form $1/(\omega_{0j} - \omega_{0k})$ which can make this correction as large as the $(n-1)$th order eigenfunction correction, thereby invalidating the perturbation expansion. Hence when two modes, labelled say j and k, are such that $(\omega_{0j} - \omega_{0k}) \sim 0$, one must consider them as degenerate or 'coupled'. This leads to an additional correction to the eigenfrequencies. Several works in the context of stellar pulsation have included this correction: Simon [129], DG92, Suarez et al. [133, 134], Soufi et al. [130], Karami et al. [75]. The eigenfunction correction, $\boldsymbol{\xi}_{2k}$ for the mode labelled k, is obtained by assuming that it can be written in the zeroth-order normal mode basis as

$$\boldsymbol{\xi}_{2k} = \sum_{j \neq k} \alpha_j \boldsymbol{\xi}_{0j} \tag{71}$$

where the unknowns now are the α_j coefficients. Inserting Eq. (71) into Eq. (69), one obtains

$$\sum_{j \neq k} \alpha_j \left(\mathcal{L}_0 - \rho_{00} \, \hat{\omega}_{0k}^2 \right) \boldsymbol{\xi}_{0j} \tag{72}$$

$$= 2\rho_{00} \, \omega_{0k} \, \omega_{2k} \, \boldsymbol{\xi}_{0k} - \left(\mathcal{L}_2 \boldsymbol{\xi}_{0k} - \omega_{0k}^2 \rho_2 \boldsymbol{\xi}_{0k} \right)$$

Taking the inner product Eq. (23) with $\boldsymbol{\xi}_{0k}$ yields the solution for ω_{2k}.

$$\alpha_j = -\frac{1}{\left(\omega_{0j}^2 - \omega_{0k}^2 \right) I_j} \int_0^R \boldsymbol{\xi}_{0j}^* \cdot \left(\mathcal{L}_2 - \omega_{0k}^2 \rho_2 \right) \boldsymbol{\xi}_{0k} \, \mathbf{d}^3 \mathbf{r} \tag{73}$$

Hence the correction to the eigenfunction of mode k is given by

$$\boldsymbol{\xi}_{2k} = -\sum_{j \neq k} \frac{D_{jk}}{\left(\omega_{0k}^2 - \omega_{0j}^2 \right)} \tag{74}$$

where

$$D_{jk} = \frac{1}{I_j} \int_V \boldsymbol{\xi}_{0j}^* \cdot \left(\mathcal{L}_2 - \omega_{0k}^2 \rho_2\right) \boldsymbol{\xi}_{0k} \, \mathbf{d^3r} \tag{75}$$

Note that $D_{jj} = 2\omega_{0j}\omega_{2j}$, the second-order (nondegenerate) frequency correction for mode j. Let two degenerate modes such that $\omega_a \sim \omega_b$ with $\omega_a = \omega_{0a} + \omega_{2a}$ and $\omega_b = \omega_{0b} + \omega_{2b}$ where ω_{0a} and ω_{2a} be given by Eqs. (40), (70). One seeks for an eigenfunction of the form $\boldsymbol{\xi}_{0ab} = c_a \, \boldsymbol{\xi}_{0a} + c_b \, \boldsymbol{\xi}_{0b}$ associated with an eigenvalue ω. Let us define

$$\bar{\omega} = \frac{\omega_{0a} + \omega_{0b}}{2}; \qquad \delta\omega = \omega_{0a} - \omega_{0b}$$

Inserting these expressions into Eq. (69) and taking the inner product with ξ_{0a} and ξ_{0b}, respectively, yield the following system of linear equations:

$$c_a \left(\omega_{0a}^2 + 2\omega_{0a} \, \omega_{2a} - \omega^2\right) + c_b \, D_{ab} = 0$$
$$c_a \, D_{ba} + c_b \left(2\omega_{0b} \, \omega_{2b} + \omega_{0b}^2 - \omega^2\right) = 0 \tag{76}$$

where D_{jk} is given by Eq. (75) above. Up to second order in Ω, one can write $\omega_{0a}^2 + 2\omega_{0a} \, \omega_{2a} \sim \omega_a^2$ so that

$$c_a \left(\omega_a^2 - \omega^2\right) + c_b D_{ab} = 0 \ ; \ c_a D_{ba} + c_b \left(\omega_b^2 - \omega^2\right) = 0$$

The solutions must then verify

$$\omega^4 - \omega^2(\omega_a^2 + \omega_b^2) + \omega_a^2\omega_b^2 - D_{ab} \, D_{ba} = 0$$

Accordingly, the frequencies of modified degenerate modes are

$$\omega_\pm^2 = \frac{1}{2}(\omega_a^2 + \omega_b^2) \pm \frac{1}{2}\sqrt{(\omega_a^2 - \omega_b^2)^2 + 4D_{ab}D_{ba}} \tag{77}$$

The corresponding eigenfunctions $\xi_\pm$ are composed of contributions from both zeroth-order modes a and b in respective parts determined by the ratio c_b/c_a. For vanishing coupling coefficients or negligible in front of the frequency difference, i.e. $4\sqrt{D_{ab}D_{ba}} << |\omega_a^2 - \omega_b^2|$, one recovers the uncoupled second-order eigenfrequencies $\omega_\pm = \omega_{a,b} = \omega_{0a,b} + \omega_{2a,b}$.

When the effect of the Coriolis force is taken into account up to third order and for nonuniform rotation, the calculation is slightly more complicated but the procedure remains in the same spirit [130].

Full second order and beyond and shellular rotation. In this section, we consider both the Coriolis and the centrifugal forces and assume a shellular rotation $\Omega(r)$, unless otherwise stated. At the end of this section, we also briefly discuss the extension of the calculation up to third order included. Let us then return to the eigenvalue system Eqs. (17), (18) which includes both forces and a differential rotation. When compared to Eq. (66), additional terms are present which are due to Coriolis; another additional term is proportional to the gradient of Ω, i.e.

$$\mathbf{f} \equiv \mathcal{L}_0\xi - \hat{\omega}^2\rho_{00} \, \xi + 2\hat{\omega}\rho_{00} \, i\boldsymbol{\Omega} \times \xi - \rho_{00} \left(\xi \cdot \nabla\Omega^2\right) r\sin\theta \mathbf{e_s}$$
$$+ \left(\mathcal{L}_2\xi - \hat{\omega}^2\rho_2\xi\right) + 2i\hat{\omega}\rho_2\Omega \times \xi = 0 \tag{78}$$

where the operators $\mathcal{L}_0, \mathcal{L}_2$ have been defined in Eqs. (67), (68).

6.2.4 Effects of a Shellular Rotation on ω_0

When solving the zeroth-order eigensystem (Eq. (38)), the quantities p_{00}, ρ_{00} can be provided by solving a 1D spherically symmetric stellar model which includes rotationally induced mixing of chemical elements and transport of angular momentum. which can significantly change the evolution of a rotating star compared to one which is not rotating. Examples of comparison of evolutionary tracks of models with and without rotationally induced transport can be found in the literature: evolutionary tracks for a $9M_\odot$ stellar model evolved with an initial rotational velocity of $v = 100$ and 300 km/s [138]; for massive stars [97]; for a $1.85M_\odot$ stellar model evolved with an initial rotational velocity of $v = 70$ or 100 km/s [62]; for low-mass stars [102]. The main sequence lasts longer for the rotating model as mixing fuel-fresh H to the burning core. This also causes for the rotational model a larger increase in its luminosity with time. The evolution of the rotating star can then be quite different from that of a nonrotating one. For an intermediate mass main sequence star with a convective core, the evolution is closer to that of a nonrotating model with overshoot. The inner structure in the vicinity of the core is therefore quite different. Hence, at zeroth order, the eigenfrequency ω_0 differs from the eigenfrequency $\omega^{(0)}$ of a nonrotating star or that of a uniformaly rotating star because one must take into account the rotationally induced transport which is likely to occur in the presence of a differential rotation. At the same location in the HR diagram, the Brunt–Väissälä frequencies in the central regions are similar for the rotating and nonrotating stars (both with no overshoot) but the Brunt–Väissälä frequency for the overshoot model is quite different (see Fig. 4. Goupil and Talon [62]). Hence, one expects large frequency differences for low-frequency and mixed modes which are sensitive to these layers between a rotating (no overshoot) model and a nonrotating model with overshoot one.

Consequences of mixing on solar-like oscillations have been studied for several specific stars. An exemple is a 1.5 $M_\odot$ stellar model with an initial velocity of 150 km/s (Eggenberger, PhD; Mathis et al., [93]). At the same location in the HR diagram, the evolutionary stages, respectively, are $X_c = 0.33$ and $X_c = 0.443$ for the central hydrogen relative mass content. The effect remains small on the averaged large separation as mass and radius are similar $(< \omega_{n,\ell} - \omega_{n-1,\ell} >_n /(2\pi) \sim 70.40\,\mu\text{Hz}$ without rotation against $69.94\,\mu\text{Hz}$ with rotation). The difference for the averaged small separation between $\ell = 0, 2$ $(< \omega_{n,\ell} - \omega_{n-1,\ell+2} >_n /(2\pi) \sim 5.07\,\mu\text{Hz}$ without rotation against $5.76\,\mu\text{Hz}$ with rotation) is large enough to be detected with the space seismic experiment CoRoT [7].

For lower mass stars, undergoing angular momentum losses, the effects might be more subtil. β Vir is a solar-like star with a mass 1.21–1.28 $M_\odot$ and an effective temperature $\sim 6130\,\text{K}$. It was selected as one best candidate target for the unfortunate seismic space mission EVRIS [59, 98]. Solar-like oscillations for this star have later been detected from ground with Harps: 31 frequencies between 0.7 and 2.4 μHz. This star is a slow rotator with a $v \sin i$ between 3 and 7 km/s. Eggenberger and Carrier [49] have modelled this star with the evolutionary

Geneva code with the assumption of rotational mixing of type I [89, 91, 147]. The computed frequencies show that one cannot reproduce simultaneously the large and small separations. The authors stress that a large dispersion of the large separation for the nonradial modes exists which could be attributed to nonresolved splittings. Three models with initial velocity $v = 4$, 6.8 and 8.2 km/s (i.e. with magnetic breaking) have been studied (Fig. 8 of Eggenberger and Carrier [49]). As the rotation is so slow, its effect on the structure, hence on ω_0 and the small separation, is very small. This small separation however is found larger than for a corresponding nonrotating model. The rotationally induced transport is not efficient enough to impose a uniform rotation in the radiative zone. The models give $\Omega_c/\Omega_s \sim 3.12$. Such a Ω gradient ought to be detectable with the rotational splittings, provided they could be detected. The mean value of the splitting is found smaller when the rotation is uniform (0.6 instead of 0.8–0.9 nμHz). Hence if future observations indicate that this 1.3 $M_\odot$ star, like the Sun, is uniformaly rotating, one will have to call for an additional mechanism to transport angular momentum as in the solar case.

A precise knowledge of ω_0 can also serve as a test for the transport efficiency of the horizontal turbulence as reported by Mathis et al. [93]. The investigated case is that of a calibrated solar model. The initial rotational velocity is taken to be 0, 10, 30 or 100 km/s. The seismic properties are compared for three different prescriptions for the horizontal turbulence transport coefficient, D_h given, respectively, by Zahn [147], Maeder [90] and Mathis et al. [91]. The more recent prescriptions lead to increased transport and mixing and therefore to a larger effect on the eigenfrequencies compared to nonrotating model. Increasing the rotation leads to an increase in the value of the small separation but its variation with frequency remains similar. Going from Zahn's transport coefficient to Maeder's coefficient results in the increase with Ω to be larger when ω is increased.

6.2.5 Solving for the First-Order Eigenfunction Correction ξ_1

The resolution for the first-order frequency correction has been discussed in Sect. 5 above. However solving for the second-order frequency correction, ω_2, requires the knowledge of the first-order eigenfunction

$$\xi_1 = \xi_1^P + \xi_1^T \tag{79}$$

which is composed of a poloidal part ξ_1^P and a toroidal part ξ_1^T. DG92 provide a detailed procedure for calculating ξ_1 and ω_2 and the nonspherical distortion of the star for a differential rotation $\Omega(r, \theta)$ for a prescribed rotation law Eq. (48). The toroidal part is obtained by taking the radial curl of Eq. (41). One can obtain the poloidal part by two possible methods: the first one consists in expanding the poloidal part in terms of unperturbed eigenvectors; one then gets the standard expression as Eq. (74). However as stressed by DG92, this method involves an infinite sum which in practice must be truncated at some level. An alternative approach is to solve Eq. (41) directly following Hansen et al. [69] and Saio [122] in deriving equations for the radial eigenfunctions corresponding to ξ_1^P. DG92 generalized it so as to include latitudinally differential rotation. For a shellular rotation, the poloidal and toroidal parts are looked under the respective form:

$$\xi_{1,\ell,m}^P = \xi_{1r}(r)\, Y_\ell^m\, \mathbf{e_r} + \xi_{1h}(r)\boldsymbol{\nabla}_h Y_\ell^m$$

$$\xi_{1,\ell,m}^T = \sum_{\ell,m} \tau_1(r)\, \mathbf{e_r} \times \boldsymbol{\nabla}_h Y_\ell^m$$

For the vector field $\boldsymbol{f}$ defined in Eq. (78) to vanish, one must impose

$$\int Y_\ell^{m*}\, (\mathbf{e_r} \cdot \boldsymbol{f})\, d\underline{\Omega} = 0 \rightarrow (\xi_1^P)_{r,\ell}$$

$$\int Y_\ell^{m*}\, (\boldsymbol{\nabla}_h \cdot \boldsymbol{f})\, d\underline{\Omega} = 0 \rightarrow (\xi_1^P)_{h,\ell} \qquad (80)$$

$$\int Y_\ell^{m*}\, (\mathbf{e_r} \cdot \boldsymbol{\nabla} \times \boldsymbol{f})\, d\underline{\Omega} = 0 \rightarrow \tau_{1,\ell+1}(r), \tau_{1,\ell-1}(r)$$

with $d\underline{\Omega} = r^2 \sin\theta d\theta d\phi$. These conditions provide differential equations for the two components $\xi_{1r}, \xi_{1h}(r)$ and analytical expressions for $\tau_1(r)$. The components $\xi_{1r}, \xi_{1h}(r)$ of the poloidal part are obtained numerically; the numerical resolution of this system also provides the first-order correction ω_1 which can be compared to the integral value Eq. (43) (Hansen et al. [69], DG92 [46]). For near-degenerate modes a and b, the solution is

$$\omega_{1\pm} = \bar{\omega}_1 \pm \sqrt{(\Delta\omega_1)^2 + 4\omega_{1,ab}^2}$$

where

$$\bar{\omega}_1 = \frac{\omega_{1a} + \omega_{1b}}{2}; \quad \Delta\omega_1 = \omega_{1a} - \omega_{1b}$$

and for the coupling coefficient

$$\omega_{1,ab} = - < \xi_{0a}|(m\Omega - i\boldsymbol{\Omega}\times)\xi_{0b} >$$

i.e. for a shellular rotation (see also Suárez et al. [132, 133]),

$$\omega_{1,ab} = m\frac{1}{I}\delta_{\ell_a,\ell_b}\delta_{m_a,m_b} \int_0^R dr \rho_0 r^2\, \Omega(r)$$
$$(\xi_{ra}\xi_{rb} + \Lambda\xi_{ha}\xi_{hb} - \xi_{ra}\xi_{hb} - \xi_{ha}\xi_{rb} - \xi_{ha}\xi_{hb}) \qquad (81)$$

where δ_{ℓ_a,ℓ_b} and δ_{m_a,m_b} are Kroenecker symbols and $\Lambda = \ell_a(\ell_a + 1)$.

6.2.6 Solving for the Second-Order Frequency Correction ω_2

Once the first-order system is fully solved, the second-order system can be solved with the same procedure described in Sect. 6.2 above:

$$\mathcal{L}_0 \boldsymbol{\xi}_2 - \omega_0^2 \rho_{00} \boldsymbol{\xi}_2 = (2\omega_0 \omega_2 + 2m\Omega\omega_1 + m^2\Omega^2 + \omega_1^2)\rho_{00}\ \boldsymbol{\xi}_0$$
$$+ 2\omega_0(m\Omega + \omega_1)\rho_{00}\ \boldsymbol{\xi}_1 - 2\omega_0\rho_{00}i\boldsymbol{\Omega} \times \boldsymbol{\xi}_1$$
$$+ \rho_{00}(\boldsymbol{\xi}_0 \cdot \nabla\Omega^2)\ r\sin\theta\mathbf{e_s} - (\mathcal{L}_2\boldsymbol{\xi}_0 - \omega_0^2\rho_2\boldsymbol{\xi}_0) \qquad (82)$$

where the operators $\mathcal{L}_0, \mathcal{L}_2$ have been defined in Eqs. (67), (68). DG92 finally obtained $\omega_2 = \omega_2^T + \omega_2^P + \omega_2^I + \omega_2^D + \omega_1^2/2\omega_0$. DG92 established the integral expressions for each of the above contributions to ω_2 in the general case of a rotation law Eq. (48) and showed that for a given (n, ℓ, m) mode, $\omega_{2,n,\ell,m}$ is a polynomial in m^2. Following the notations of Eqs. (57), (58), the frequency computed up to second order for a shellular rotation can be cast under the form

$$\omega_{n,\ell,m} = \omega_{n,\ell}^{(0)} + m\bar{\Omega}(1 - C_{n\ell} - J_{n,\ell}) + \Omega^2\ (D_{1,n,\ell,m} + m^2 D_{2,n,\ell,m}) \qquad (83)$$

Suárez et al. [133] provided the equivalence between Saio's and DG92 notations for a shellular rotation included effect of degeneracy up to second order.

6.2.7 Including Third-Order Effects

In order to compute the third-order frequency correction, a classical perturbation procedure requires the knowledge of the second-order eigenfunction $\boldsymbol{\xi_2}$. However it is possible to build a pseudo-zeroth-order system which avoids the lengthy computations of eigenfunctions at two successive orders including degeneracy. The procedure is developed in Soufi et al. [130]; see also Karami et al. [75]. The wave equation Eq. (17) is written as $\mathcal{F}_0(\xi, \omega) + \mathcal{F}_c(\xi, \omega) = 0$. Part of the Coriolis force is included in the pseudo-zeroth-order $\mathcal{F}_0(\xi, \omega)$. As a consequence, the first-order frequency and eigenfunction corrections are implicitly included in the pseudo-zeroth-order solution; first-order degeneracy is implicitly included as well. One seeks for an eigenfrequency of the form $\tilde{\omega}_0 + \omega_c$ where $\tilde{\omega}_0$ is solution of the pseudo-zeroth-order eigensystem and ω_c is a frequency correction. The solution is then expanded as $\xi = \tilde{\xi}_0 + \xi_c + O(\Omega^4)$ where $\tilde{\xi}_0$ takes the form

$$\tilde{\xi}_0(\boldsymbol{r}) = \xi_{r,m}(r)Y_\ell^m \mathbf{e_r} + \xi_{h,m}(r)\nabla_h Y_\ell^m$$
$$+ \tau_{\ell+1,m}(r)\boldsymbol{e}_r \times \boldsymbol{\nabla}_h Y_{\ell+1}^m + \tau_{\ell-1,m}(r)\boldsymbol{e}_r \times \boldsymbol{\nabla}_h Y_{\ell-1}^m \qquad (84)$$

The pseudo-zeroth-order system is $\mathcal{F}_0(\tilde{\xi}_0, \tilde{\omega}_0) = 0$ and $\tilde{\xi}_0$ is then solution of a differential equation system which must be numerically solved and provides the pseudo-zeroth-order eigenfrequency $\tilde{\omega}_0$. ξ_c is a correction to the eigenvector field $\tilde{\xi}_0$ and is solution of the system $\mathcal{F}_0(\xi, \omega) + \mathcal{F}_c(\tilde{\xi}_0, \tilde{\omega}_0) - \mathcal{F}_0(\tilde{\xi}_0, \tilde{\omega}_0) = 0$ arising from $O(\Omega^2)$ contributions. The solvability condition for this equation yields the frequency correction ω_c. Part of the third-order contribution due to Coriolis force is implicitly included in pseudo-zeroth order. The price to pay is that (1) the pseudo-zeroth-order numerical eigensystem is now m-dependent and (2) the eigenfunctions are no longer orthogonal with respect to the inner product Eq. (23). However they are orthogonal with respect to Eq. (34).

For a given nondegenerate (n, ℓ, m) mode, the frequency is then obtained as

$$
\begin{aligned}
\omega_{n,\ell,m} = \tilde{\omega}_{0,n,\ell,m} &+ \bar{\Omega}^2 \left(D_{1,n,\ell,m} + m^2 \, D_{2,n,\ell,m} \right) \\
&+ \bar{\Omega}^3 \, m \left(T_{1,n,\ell,m} + m^2 \, T_{2,n,\ell,m} \right)
\end{aligned} \tag{85}
$$

where $\bar{\Omega}$ is a constant rotation (for instance a depth average or the surface value). For convenience we define a frequency $\omega_{0,n,\ell,m}$ such that

$$
\omega_{0,n,\ell,m} = \tilde{\omega}_{0,n,\ell,m} - m\bar{\Omega}(1 - C_{n,\ell,m} - J_{n,\ell,m}) \tag{86}
$$

so that one writes the eigenfrequency in a more familiar form

$$
\begin{aligned}
\omega_{n,\ell,m} = \omega_{0,n,\ell,m} &+ m\bar{\Omega}(1 - C_{n,\ell,m} - J_{n,\ell,m}) \\
&+ \bar{\Omega}^2 \left(D_{1,n,\ell,m} + m^2 \, D_{2,n,\ell,m} \right) \\
&+ \bar{\Omega}^3 \, m \left(T_{1,n,\ell,m} + m^2 \, T_{2,n,\ell,m} \right)
\end{aligned} \tag{87}
$$

To a good approximation, when cubic-order effects are not too large, one has $\omega_{0,n,\ell,m} \sim \omega_{0,n,\ell}$ where $\omega_{0,n,\ell}$ includes only the $O(\Omega^2)$ effects of spherically symmetric distortion. Because of the symmetry property

$$
\omega_{n,\ell,m}(\Omega) = \omega_{n,\ell,-m}(-\Omega)
$$

the coefficients in Eq. (87) verify

$$
\tilde{\omega}_{0,n,\ell,m} = \tilde{\omega}_{0,n,\ell,-m} \; ; \; D_{j,n,\ell,m} = D_{j,n,\ell,-m} \; ; \; T_{j,n,\ell,m} = T_{j,n,\ell,-m} \tag{88}
$$

for $j = 1, 2$. It is also convenient to cast the result under the following form:

$$
\begin{aligned}
\omega_{n,\ell,m} = \omega_{0,n,\ell,m} &+ m\bar{\Omega}(1 - C_{n,\ell,m} - J_{n,\ell,m}) \\
&+ \frac{\bar{\Omega}^2}{\tilde{\omega}_0} \left((X_1 + m^2 Y_1) + (X_2 + m^2 Y_2) \right) \\
&+ \frac{\bar{\Omega}^3}{\tilde{\omega}_0^2} \, m \left(\mathcal{S}_1 + m^2 \, \mathcal{S}_2 \right)
\end{aligned} \tag{89}
$$

where notations similar to Saio81's are used but generalized to shellular rotation [133]. When modes are degenerate, one uses the same procedure as described in the above paragraph and for a two-mode degenerate coupling, the frequencies are then given by [37, 130]

$$
\omega_{\pm} = \bar{\omega}_{n,\ell,m} \pm h_{n,\ell,m} \tag{90}
$$

where

$$
\bar{\omega}_{n,\ell,m} = \frac{1}{2}(\omega_{n,\ell,m} + \omega_{n-1,\ell+2,m})
$$

$$
h_{n,\ell,m} = \frac{1}{2}\sqrt{d_{n,\ell,m}^2 + 4H_{n,\ell,m}^2}; \quad d_{n,\ell,m} = \omega_{n,\ell,m} - \omega_{n-1,\ell+2,m}
$$

This generalizes to three-mode coupling which becomes quite common when rotation is large.

6.2.8 Rotation Splitting

Using Eq. (87) and symmetry properties, one derives for the rotational splittings Eq. (5) the following expression:

$$S_m = \bar{\Omega} \left(1 - C_{n,\ell,m} - J_{n,\ell,m}\right) + \bar{\Omega}^3 \left(T_{1,n,\ell,m} + m^2 T_{2,n,\ell,m}\right) \tag{91}$$

which is free of second-order nonspherically symmetric distortion effect. For degenerate modes, the expression for the splittings S_m is more complicated and can be derived from Eq. (90). Asymmetry of the split multiplets, or departure from equal splittings for nondegenerate modes, is measured by

$$\Delta\omega_{n,\ell,m} = \omega_{n,\ell,m=0} - \frac{1}{2}(\omega_{n,\ell,m} + \omega_{n,\ell,-m}) \tag{92}$$

Using Eq. (87), its expression becomes

$$\Delta\omega_{n,\ell,m} = (\omega_{0,n,\ell,0} - \omega_{0,n,\ell,m})$$
$$+ \bar{\Omega}^2 \left((D_{1,n,\ell,0} - D_{1,n,\ell,m}) - m^2 D_{2,n,\ell,m}\right)$$

where we have used the symmetry properties Eq. (88). When cubic-order effects on $\omega_{0,n,\ell,m}$ are small, $D_{2,n,\ell,m}$ dominates over the third-order differences $(\omega_{0,n,\ell,m} - \omega_{0,n,\ell,0})$ and $(D_{1,n,\ell,0} - D_{1,n,\ell,m})$ so that one can most often consider

$$\Delta\omega_{n,\ell,m} \sim -m^2 \, \bar{\Omega}^2 \, D_{2,n,\ell,m} \tag{93}$$

6.3 Some Theoretical Results: Case of a Polytrope

As mentionned in Sect. 6.1, Simon [129] built a mapping between a spheroidal coordinate system and a spherical one and computed the second-order effects for radial modes of a polytrope of index $n_{polyt} = 3$ and specific heat coefficient $\Gamma_1 = 5/3$. He found for the dimensionless squared frequency (in units of $4\pi G\rho_c$, ρ_c being the central density): $\sigma^2 = \sigma_0^2 + \beta\Omega^2$ with $\sigma_0^2 = 0.057, \beta = -3.858$ for the fundamental radial mode which includes the effect of the first-order toroidal contribution $8/3\Omega^2$ ($m = 0$, $X_1 = 8/3$, $X_2 = 0$ in Eq. (89)) to σ and the approximate effect of distortion. These results agreed with the earlier work by Cowling and Newing [34], but in a somewhat disagreement with results of other previous works such as Chandrasekhar and Lebovitz [17–19]. Clement [29, 30] using a variational principle computed the frequency for an axisymmetric $\ell = 2$ mode; he found $\sigma^2 = 8.014 - 0.255\hat{\Omega}^2$.

Saio [122] studied the effect of uniform rotation upon nonradial oscillation frequencies up to second order $O(\Omega^2)$ and computed the frequency corrections for a polytrope of index $n_{polyt} = 3$ and $\gamma = 5/3$ and for $\ell = 0, 1, 2, 3$ modes with radial order n up to $n = 6$ for p-modes. His numerical results were in agreement with those of Simon [129], Clement [29, 30] and Chlebowski [24]. For the radial fundamental mode, he found $\sigma^2 = 9.252 - 3.79 \, \hat{\Omega}^2$ to be compared to Simon's result in the same units $\sigma = 9.249 - 3.86 \, \hat{\Omega}^2$. He wrote the corrected frequency under the convenient form Eq. (58). The quantity $Z + X_2 + m^2 Y_2$ arises from the distortion of the equilibrium model. The first-order eigenfunctions were computed by solving directly the appropriate system of equations which he gave in appendix, generalizing the approach derived by Hansen et al. [69] in the Cowling approximation. Perturbations of the structure were obtained from the tabulated results of Chandrashekhar and Lebovitz [17–19]. Equation (58) shows that second-order effects break the symmetry of the rotational splitting which exists at first order. For nonradial p-modes, he found that the effect of distortion of the equilibrium model dominates the frequency correction: Z is large and negative and dominates over X_2 which is also large and positive; Y_2 is negative and much larger than Y_1 which is positive. $|Z|, X_2$ and $|Y_2|$ increase with radial order of the p-modes.

Table 1 gives values of the coefficients in Eq. (89) for a polytrope with $n = 3$ and $\gamma = 5/3$ computed using Soufi et al. [130] approach. Columns S_1 and S_2 are the cubic-order coefficients appearing explicitly in Eq. (89); they are

Table 1. Coefficients of Eq. (87) assuming a uniform rotation for a polytrope with polytropic index 3 and adiabatic index $\gamma = 5/3$. The squared frequency σ_0^2 is the dimensionless squared frequency $\omega^2/(GM/R^3)$. Spherical distortion of the polytrope has not been included

$\ell = 1$

n	σ_0^2	$C_{n,\ell}$	X_1	Y_1	X_2	Y_2	S_1	S_2	$<W_2>$
1	11.400	0.028	0.776	0.980	2.898	−4.347	0.694	−0.243	0.127
2	21.540	0.034	0.773	0.919	5.829	−8.743	0.345	0.231	0.135
3	34.896	0.033	0.773	0.872	9.677	−14.515	−0.038	0.767	0.139
4	51.467	0.031	0.776	0.839	14.427	−21.641	−0.436	1.334	0.140
5	71.234	0.027	0.778	0.815	20.069	−30.103	−0.835	1.906	0.141
6	94.177	0.024	0.781	0.798	26.593	−39.890	−1.233	2.483	0.141
7	120.280	0.021	0.783	0.785	33.995	−50.992	−1.634	3.067	0.141
8	149.529	0.019	0.785	0.775	42.269	−63.404	−2.031	3.647	0.141
9	181.911	0.017	0.787	0.768	51.413	−77.119	−2.426	4.226	0.141
10	217.417	0.015	0.788	0.761	61.422	−92.134	−2.818	4.803	0.141
11	256.040	0.013	0.789	0.756	72.296	−108.445	−3.207	5.377	0.141
12	297.770	0.012	0.790	0.752	84.033	−126.050	−3.593	5.948	0.141
13	342.604	0.011	0.791	0.748	96.631	−144.947	−3.976	6.515	0.141
14	390.535	0.010	0.792	0.744	110.090	−165.136	−4.355	7.077	0.141
15	441.560	0.009	0.793	0.741	124.410	−186.615	−4.730	7.633	0.141
16	495.676	0.008	0.793	0.739	139.590	−209.386	−5.101	8.184	0.141
17	552.879	0.008	0.794	0.736	155.631	−233.447	−5.466	8.727	0.141
18	613.167	0.007	0.794	0.734	172.533	−258.800	−5.825	9.262	0.141
19	676.539	0.006	0.795	0.731	190.296	−285.445	−6.178	9.787	0.141
20	742.992	0.006	0.795	0.729	208.922	−313.383	−6.524	10.302	0.141
21	812.525	0.006	0.796	0.727	228.411	−342.616	−6.861	10.805	0.141
22	885.139	0.005	0.796	0.725	248.764	−373.145	−7.190	11.295	0.141
23	960.830	0.005	0.796	0.723	269.982	−404.972	−7.509	11.772	0.140

found to increase steadily with the radial order as for the second-order distortion coefficients but they are smaller and increase more slowly with the frequency. The last column lists the asymptotic coefficient $< W_2 >$ defined in Eq. (94).

6.4 Some Theoretical Results: Realistic Stellar Models

Frequency comparisons between polytropic and realistic stellar models. Table 2 of DG92 compares results from a polytropic and a realistic stellar models for the first-order splitting coefficient $C_{n,\ell}$ and the second-order coefficient $D_L = Y_1 + Y_2$ appearing in Eq. (89). The authors computed D_L assuming rigid rotation for a 2 $M_\odot$ at two evolutionary stages one with a central hydrogen content $X_c = 0.699$ (ZAMS) and a more evolved model with $X_c = 0.313$. They compared with the results for a $n_{polyt} = 3$ polytrope. Apart for mixed modes, the polytropic and realistic values for D_L are comparable $D_L < 0, |D_L|$ increases with n. Differences were found larger for C_L coefficients than for D_L ones. The largest differences arise for modes in avoided crossing on the C_l coefficients.

The solar case and solar-like pulsators. GT90 computed the high p-mode frequencies up to second order for a solar model assuming uniform rotation as well as several shellular rotation laws. They found that centrifugal distortion is the dominant second-order effect, of the order of a few dozen nHz for a surface rotation frequency ν_s of about 0.5 μHz and either a uniform rotation or a rotation profile with a core rotation of $\sim 2\nu_s$ and a first-order splitting of ~ 440 nHz. DG92 investigated these effects for a realistic solar model and a differential rotation $\Omega(r, \theta)$. The authors studied the second-order effects on the splittings δ_m (Eq. (45)) for the Sun and found that of the nonspherical distortion Y_2 in Eq. (89) dominates (hence Eq. (93)) with values as $\Omega^2/\omega_0 \sim 0.1$ nHz which must be mulitplied by $\sigma^2 \sim 100$–1000 for solar p-mode, in agreement with GT90. Because the nonspherically symmetric distortion dominates for high-frequency nondegenerate modes, one can write for these modes

$$\omega_{2,n,\ell,m} \sim \omega_2^D \ (DG92) = \frac{\bar{\Omega}^2}{\omega_0}(X_2 + m^2 Y_2) \ (Saio81)$$

It is also convenient to write the quantity $X_2 + m^2 Y_2$ as

$$X_2 + m^2 Y_2 \sim \mathcal{Q}_{2\ell,m} \ \mathcal{D}_{asymp} \quad \text{with} \quad \mathcal{Q}_{2,\ell,m} = \frac{\Lambda - 3m^2}{4\Lambda - 3}$$

For high radial order, an asymptotic analysis indeed shows that $\mathcal{D}_{asymp} \sim \sigma_0^2 < W_2 >$ where $< W_2 >$ is an integral over the distorted structure quantities which depend on the nonspherically rotational perturbation of the gravitational potential (DG92, Fig. 8 of Goupil et al. [61]). This explains the linear increase with the frequency, ω_0, of the second-order correction (Saio81 [122], GT90 [55], DG92 [46], Goupil et al. [63]) which in dimensionless form behaves as

$$\sigma_{2m} = \hat{\Omega}^2 \ \mathcal{Q}_{2,\ell,m} \ \sigma_0 \ < W_2 > \tag{94}$$

The asymptotic quantity $< \mathcal{W}_2 >$ is listed in Table 1 for a polytrope with $n_{polyt} = 3$ and $\Gamma_1 = 5/3$. Burke and Thompson (2006) computed the second-order effects for a 1 $M_\odot$ evolving along the main sequence and for a 1.5 ZAMS model. They also find that the second-order dimensionless coefficients vary little with age for 1 $M_\odot$ and vary in a homologous way for different masses along the ZAMS.

A linear increase with radial order is also the case for the degenerate second-order coupling coefficient (Suárez et al. [133]): for two modes a and b ($\omega_{0a} \sim \omega_{0b}$):

$$H_{ab} \sim \omega_{2D,ab} \sim \frac{\Omega^2}{\omega_0} \, \mathcal{Q}_{2,\ell,m} \, \mathcal{D}_{ab,asymp}$$

with $\mathcal{D}_{ab,asymp} \sim \sigma_0 \; < \mathcal{W}_2 >$ as it is also dominated by nonspherically symmetric distortion.

6.4.1 Near-Degeneracy and Small Separation

P-mode frequency small separations, defined as $\omega_{n-1,\ell,0} - \omega_{n,\ell+2,0}$, are of the order of a few dozen µHz for solar-like main sequence low and intermediate mass stars. When the star is rotating fast enough (F,G,K main sequence stars have surface projected rotational velocities between 10 and 40 km/s), the frequencies are modified by an amount which can be significant, particularly when they are degenerate. Close frequencies as those involved in small separations favour the occurrence of degeneracy-induced modifications. As a consequence, the small separation can be quite affected by rotation. This was stressed by Soufi et al. [130]. Quantitative estimates have been obtained by Dziembowski and Goupil [47] for 1 $M_\odot$ and $v = 10$ and 20 km/s and Goupil et al. [63] for a 1.4 $M_\odot$ and a 1.54 $M_\odot$ rotating stellar models with a surface rotational velocity of 20, 30 and 35 km/s. Changes in the small separations are of the order of $0.1 - 0.2$ µHz (corresponding to a change of 0.1–0.2 Gyr for the age of the star) and increase with the frequency. Provided enough components of the rotational splittings are available, it is possible to remove most of these contamination effects in order to recover a 'classical' small separation (Dziembowski and Goupil [47] and Fig. 4 in Goupil et al. [63]). As rotation-induced distortion of the equilibrium significantly affects the small separation of high-order p-modes, it also affects the shape of the ridges in an echelle diagram. The effect is larger in the upper part of the diagram, i.e. at high frequency (Goupil and Dupret [66], Fig. 6) but as one can decontaminate the small separation, it is also possible to recover an echelle diagram free of rotationally induced pollution effects (Goupil et al. [65]; Lochard et al. 2008, in prep.).

Delta Scuti stars. DG92 computed the second-order frequencies for 2 $M_\odot$ with a uniform rotation velocity of 100 km/s and discussed the departure from equal rotational splittings induced by distortion ω_{2D} for low-frequency modes in the range of the fundamental radial mode. The distortion, hence the departure from equal splitting, is larger for the trapped mode (or mixed mode) in this low-frequency regime where all the other modes have a predominantly g-mode character. Goupil et al. [61] investigated the effects of moderate rotation (initial radial velocity of 100 km/s) on rotational splittings of δ Scuti stars using the third-order perturbative approach of Soufi et al. [130]. A $1.8\,M_\odot$ stellar model half way on the main sequence was studied. Effects of successive perturbation order contributions on the oscillation frequencies are shown for modes $\ell = 0$ and $\ell = 2$ modes (Fig. 3). Changes in the frequency pattern appearance in a power spectrum are mainly due to centrifugal distortion and are shown in the particular case of the δ Scuti star FG Vir for three values of the initial rotation velocity 10, 46 and 92 km/s. The first-order equidistant pattern of the rotational splittings is totally lost at 92 km/s. The third-order effects in the generalized rotational splittings (Eq. (91)) computed for a uniform rotation are found relatively small. Although the rotation is uniform, S_m show strong variations of the order of µHz with the frequency due to the presence of mixed modes (Fig. 6). The true (uniform) rotation rate can however be recovered when combining well-chosen components of the multiplets. Departure from equal splittings for nondegenerate modes as measured by Eq. (92) is again found to be dominated by the nonspherically symmetric centrifugal distortion contribution. The splitting asymmetry then becomes

$$\Delta\omega_m \sim m^2\,\omega_0\,\frac{3}{4\Lambda - 3}\,\frac{\Omega^2}{(GM/R^3)}\, <W_2>$$

Pamyatnykh [103] give quantitative estimates of the effect of mode near-degenerate coupling on nonradial p-mode frequencies (Fig. 5) and on period ratios of radial modes (Fig. 6) for a 1.8 $M_\odot$ main sequence model with a surface rotational velocity of 92 km/s. The induced modifications can be quite significant. Suárez et al. [132, 133] studied second-order effects for a 1.5 $M_\odot$ mass star, representative of a delta Scuti star and for a prescribed shellular rotation law assuming a surface velocity of 100 km/s. Comparing frequencies for models assuming a uniform rotation on the one hand and a shellular rotation law on the other hand, they found differences of about 1–3 µHz for high-frequency p-modes and larger for lower frequencies. The authors investigated consequences of degeneracy due to rotation which are also illustrated in Goupil et al. [65]. Burke and Thompson (2006) computed the second-order frequencies for a 1.98 $M_\odot$ mid-main sequence star representative of a δ Scuti star. Their Fig. 3 shows that the coefficients vary little with radial order n except for mixed modes.

7 Fast Rotators: Nonperturbative Approaches

Several types of fast-rotating pulsating stars are known to exist. One good example is the nearby A-type star, Altair. This star is rotating fast with a $v \sim 227$ km/s and is flattened with a ratio $R_{eq}/R_{pole} \sim 1.23 - 1.28$; $\mu = 0.08 - 0.2$. Interferometric observations have revealed a gravity-darkening effect in accordance with von Zeipel theory for this star with the equatorial layers cooler ($\sim$6800 K instead of $\sim$8700 K) and 60–70% darker than the poles [42, 100]. Altair is a delta Scuti star and its power spectrum shows seven frequencies from WIRE observations [16]. Modelling of these pulsations has been attempted by Suárez et al. [131]. For such a star, it is likely that a perturbative approach is no longer valid neither for the equilibrium model nor for the computation of the oscillation properties.

7.1 Formalisms

Oscillation properties are computed for a stellar model which is considered in static equilibrium. The rotating model is no longer perturbative but is assumed to keep the axisymmetry and must then be described as a 2D configuration. Hence, the equations are separable in ϕ but no longer in (r, θ) variables in a spherical coordinate system. In order to study the structure of a rotating star, several works have developed various techniques with the goal of building 2D rotating equilibrium configurations as mentionned in Sect. 3.1.

Eigenvalue problem. Once the equilibrium configuration is built, the goal is to calculate the adiabatic oscillations of a given model defined by the quantities $\rho(r, \theta), p((r, \theta)$. etc. When the star is rotating fast, as the latitudinal and radial dependences (in θ, r) are no longer separable, one here again deals with a 2D computation. Linearization of the equations, ϕ variable separation and the hypothesis of a steady-state configuration allow to write the displacement eigenfunction as $\xi \propto e^{i(\omega t - m\phi)}$ where m is an integer. Solving the associated eigenvalue problem has led to a series of different studies starting with Clement [31]; see Reese [112] for a detailed bibliography. One of the techniques is to expand the solution as a series of spherical harmonics for the angular dependence:

$$\boldsymbol{\xi}_m = e^{i\omega t} \ r \ \sum_{\ell \geq |m|}^{\infty} (\ S_\ell(r) Y_\ell^m(\theta, \phi) \boldsymbol{e}_r + H_\ell(r) \boldsymbol{\nabla}_h Y_\ell^m(\theta, \phi)$$

$$+ T_\ell(r) \ \boldsymbol{e}_r \times \boldsymbol{\nabla}_h Y_\ell^m(\theta, \phi) \ ) \tag{95}$$

and $f' = \sum_{\ell \geq |m|}^{\infty} f'_\ell(r) Y_\ell^m(\theta, \phi) e^{i\omega t}$. One obtains an infinite set of coupled differential equations for the depth dependence of the eigenfunctions. The properties of axisymmetry and symmetry with respect to the equator ($\theta \to \pi - \theta$) cause a decoupling of the problem into two independent eigenvalue systems [84, 143]. For any integer $j \geq 0$:

Even modes (i.e. sym/equator) are $\ell = |m| + 2j + 2, \ell' = \ell + 1$, i.e. ($m = 0, \ell = 0, 2, 4 \ldots$); ($m \pm 1, \ell = 1, 3, 5 \ldots$);
Odd modes (i.e. antisym/equator) are $\ell = |m| + 2j - 1, \ell' = \ell - 1$, i.e. ($m = 0, \ell = 1, 3, 5, \ldots$); ($m \pm 1, \ell = 2, 4, 5 \ldots$).

Lee and Saio [84] used this technique to study the g-modes of a 10 $M_\odot$ stellar model; the frequencies were computed by keeping only the first two harmonics in the series Eq. (95). Note that the equilibrium model was obtained by means of perturbation as developed by Kippenhan et al. [76].

More recently, Espinosa [50, 51] considered the effect of the centrifugal force only, neglecting the Coriolis force, assumed the Cowling approximation and neglected the Brünt–Väissälä frequency in the adiabatic oscillation equation ($N^2 \ll \omega^2$). These above assumptions are valid for high-frequency p-modes. He also assumed a uniform rotation. The numerical resolution was based on a finite difference method. Espinosa [50, 51] studied first a model with a uniform density and then turned to a realistic model built by Claret with ϵ^2 between 0 and 0.3. An alternative approach has been developed by Reese et al. [113] who studied a polytrope in uniform rotation, that is the structure is built according to

$$p_0 = K\rho_0^\gamma; \quad \Delta\psi_0 = 4\pi G\rho_0$$

$$0 = -\nabla p_0 - \rho_0 \nabla \left(\psi_0 - \frac{1}{2}\Omega^2 s^2 \right) \tag{96}$$

The computation of adiabatic oscillation frequencies of p-modes is based on a 2D approach which uses spectral methods in both dimensions r, θ with expansions in spherical harmonics for the angular part and in Chebitchev polynomials for the radial dependence. For any function, f,

$$f(r, \theta, \phi) = \sum_{\ell=0}^{\infty} \sum_{m=|\ell|}^{\infty} f_m^\ell(r) Y_\ell^m(\theta, \phi)$$

Each of the radial functions f_m^ℓ is written as $f_m^\ell(r) = \Sigma_{j=0}^\infty a_j^{\ell,m} T_j(2r-1)$. For the velocity field,

$$v(r, \theta, \phi) = \Sigma_{\ell=0}^\infty \left(\Sigma_{m=|\ell|}^\infty u_m^\ell(r) \boldsymbol{R}_\ell^m + v_m^\ell(r) \boldsymbol{S}_\ell^m + w_m^\ell(r) \boldsymbol{T}_\ell^m \right)$$

where $\boldsymbol{R}_\ell^m, \boldsymbol{S}_\ell^m$ and $\boldsymbol{T}_\ell^m$ constitute a basis which becomes the usual spherical basis $Y_\ell^m \boldsymbol{e}_r, \nabla Y_\ell^m, \boldsymbol{e}_r \times \nabla Y_\ell^m$ (as in Eq. (95)) when $\Omega \to 0$. According to the authors, 120 points with a spectral method correspond to 5000 points with a finite difference method at least when the structure is smooth as is that of a polytrope. Furthermore, the authors chose a coordinate system which is better adapted to the oblate geometry (ζ, θ, ϕ) as proposed earlier in another context by Bonazzola et al. [10]. The relation between the star radius in spherical (r) and oblate (ζ) coordinate systems is given by

$$r(\zeta, \phi) = \zeta \left(1 - \epsilon + \frac{5\zeta^2 - 3\zeta^4}{2}(R_s(\theta) - 1 + \epsilon) \right)$$

where $R_s(\theta)$ is the surface radius and ϵ is the oblateness parameter. The surface boundary condition is taken as $\psi' = 0$ at large distance of the surface of the star. The numerical computation solves a full matrix system.

7.2 Some Results and Conclusions

Espinosa [50, 51] and Lignières et al. [85] studied the effect of the centrifugal force. The first study chose an oblateness parameter in the range $\epsilon = 0-0.3$ while the second one investigated the range $\epsilon = 0 - 0.15; \Omega/\Omega_k = 0$ up to 0.59, that is a rotational velocity up to 150 km/s for a δ Scuti where $\Omega_k = (GM/R_{eq}^3)^{1/2}$ with M the stellar mass and R_{eq} the equatorial radius. The effects of the centrifugal force increase with increasing ω (Fig. 2 of Lignières et al. [85]). Frequencies of even and odd modes behave differently: frequencies of even modes tend to increase whereas those of odd modes decrease. The effect of the centrifugal force on p-modes is to contract the eigenfrequency spectrum. These effects are more important for high n and low ℓ (see Fig. 3 of Lignière et al. [85]). Reese et al. [113] computed frequencies of a polytrope with a polytropic index $n = 3$, using 70–80 ℓ degree spherical harmonics and 60–80 points for the radial grid in the spectral decompositions. Both the centrifugal and Coriolis forces are included. The Coriolis effects are found to decrease with increasing ω. For high-frequency modes, the effects due to the centrifugal force dominate. Visualization of oscillations of a distorted rotating star can be seen on the web site of D. Reese or Fig. 4 of Espinosa [50, 51].

Mode classification. In general, the oscillation modes of a fast-rotating star look like modes for a spherically symmetric star and a mode classification remains possible: the dominant ℓ degree most often corresponds to the degree of the mode for a zero rotation. This is shown by following the mode starting from a zero rotation and progressively increasing the rotation rate. However following the mode with increasing rotation rate is made more complicated by the presence of mixed modes. For the highest rotational velocities, the dominant degree ℓ is no longer the degree of the mode with zero rotation. The dominant ℓ can be as large as $\ell_0 + 6$ where ℓ_0 is the degree of the mode for the nonrotating model. In that case, one can wonder whether the mode would still be visible in disk-integrated light from ground. Would that contribute to explain the complicated patterns seen in power spectra of fast-rotating δ Scuti stars where no frequency is detected in large domains in between frequency ranges where many frequencies are detected for the same star?

Mode pairing . Espinosa [50, 51] found that the numerical frequencies computed assuming a uniform rotation and with the aforementioned approximation obey the relation $\omega_{n,\ell,m} = \bar{\omega}_{n,\ell,m} + m\Omega - f_{n,\ell}(m^2, \Omega) - (-)^{\ell+m}g_{n,\ell}(m^2, \Omega)$ where f and g are positive definite and monotonically increasing functions with m^2. The combined effect of the third and last terms produces compressions of the frequencies within the multiplets which generate pairing of modes with different types of symmetry. Indeed for increasing rotation, frequencies tend to assemble by pair of modes with different values of m (Fig. 1 of Espinosa [50, 51], Lignières et al. [85], Reese et al. [113]). This effect starts to appear for larger ℓ when the rotation is increased. The origin of such a pairing is attributed to the nonspherical distortion of the structure. This effect might have already been observed in the Fourier spectra of δ Scuti star light curves.

New regularities in frequency spectra. Figure 7.3 in PhD thesis of Reese [113] shows that when the radial order n increases, the large separation still tends towards a constant value but the small separation no longer decreases to 0. Figure 7.2 also shows that the small separation between $\ell = 0, n$ and $\ell = 2, n-1$ modes for instance increases for increasing rotation as the frequency of the $\ell = 2$

mode decreases whereas that of the radial one remains almost constant until some rotation is reached beyond which the small separation is now between the $\ell = 2, n - 1$ and the radial mode below, i.e. $\ell = 0, n - 1$. This behaviour is reminiscent of the avoided crossing behaviour of the same modes when the (nonrotating) stellar model evolves, i.e. with decreasing effective temperature or central hydrogen content. Reese et al. [113] modelled the behaviour of the numerical frequencies as $\omega_{n,\ell,m} = \Delta_n n + \Delta_\ell \ell - m\Omega + \Delta_m |m| + \alpha_\pm$ where the Δ coefficients are constant depending on Ω but not on n, ℓ, m, $\alpha_\pm$ depends on Ω and on the $(\ell + m)$ parity of the mode.

Concentration of the amplitude towards the equator. The centrifugal distortion is found to cause a high concentration of the mode amplitude towards the equator (Fig. 8 of Lignière et al. [85]).

From a rotating polytrope to a more realistic stellar model. Most of the above effects obtained for a polytrope appear to be also verified when one turns to the oscillation frequencies of a more realistic stellar model (Espinosa [50, 51], PhD thesis). However, the effect of mode couplings seems to be stronger in the realistic case. This perhaps might be expected as the structure of a polytrope is smoother than that of a more realistic stellar model.

7.3 Validity of the Results from Perturbation Techniques

From the previous chapter and what has been described in this chapter, one concludes that some results of perturbative and nonpertubative methods are qualitatively similar. On the other hand, some new types of behaviour seem to appear when the perturbative methods are no longer valid. For concrete studies, one then needs to quantify the domain of validity of perturbation methods. In order to do so, one must compare the results coming from the perturbation methods under the form Eq. (87), with the numerical frequencies computed with a nonperturbative approach. Reese et al. [113] determined the dependence of the frequency ω in a function of the rotation rate Ω by means of a polynomial fit of the form

$$\omega_{fit}(\Omega) = A_0 + A_1 \, \Omega + A_2 \, \Omega^2 + A_3 \, \Omega^3 + O(\Omega^4)$$

The A_j coefficients are determined by fitting the numerical frequencies computed with the nonperturbative approach. The authors then compared the numerical frequencies to their counterpart at a given level of approximation. The comparison is carried out with the differences $\delta_j = \omega - \omega_{app,j}$ where $\omega_{app,j} = \sum_1^j A_j \Omega^j$. When only the centrifugal force in included, the error between second-order perturbative and nonperturbative frequencies is $\sim 11\mu$Hz for $\Omega/\Omega_K = 0.24$ for a typical delta Scuti star with a rotational velocity of 100 km/s (Fig. 12, Lignières et al. [85]). The relative error reaches 3% at $\Omega/\Omega_K = 0.3$. Reese et al. [113] considered more appropriate quantities for the comparison:

$$\frac{\omega_{n,\ell,m} + \omega_{n,\ell,-m}}{2} = \omega^0_{n,\ell,m} + \omega^2_{n,\ell,m}\Omega^2 + \omega^4_{n,\ell,m}\Omega^4 \tag{97}$$

$$\frac{\omega_{n,\ell,m} - \omega_{n,\ell,-m}}{2} = \omega^1_{n,\ell,m}\Omega + \omega^3_{n,\ell,m}\Omega^3 + \omega^5_{n,\ell,m}\Omega^5 \tag{98}$$

Frequency differences between perturbative and nonperturbative computations are the same for two modes with opposite m and same ℓ, n. Reese et al. [113] found that the error which is introduced between perturbative frequencies and nonperturbative ones is larger than the accuracy of frequency measurement (0.4 µHz) for 150 days of observation for a solar-like oscillating star (0.08 µHz) (Fig. 4 of Reese et al. [113]). In the frequency range considered and with COROT's accuracy, it is found that complete calculations are required beyond $vsini = 50$ km/s for a $R = 2.3\,R_\odot, M = 1.9\,M_\odot$ polytropic star. Furthermore, it is shown that the main differences between complete and perturbative calculations come essentially from the centrifugal distortion. A comparison between the results of nonperturbative calculations and those from perturbative computations as described in the previous section is also necessary and is currently being carried out.

8 Observed Rotational Splittings and Forward Inferences

8.1 Solar-Like Pulsators

Several low-to-intermediate mass main sequence stars are known to oscillate with solar-like oscillations. However, they are rotating too slow and in all but one instance and apart for the Sun, their splittings have not been detected yet.

α **Cen A**. Fifty days of space photometric observations of this star with Wire led to the detection of 18 frequencies in the range 1700–2650 µHz which were measured with an accuracy of 1–2 µHz [53]. Detected modes have $\ell = 0$ to 2 and radial order $n = 14$–18. These observations allowed to determine an average value for the rotational splitting of 0.54 ± 0.22 µHz.

Procyon. Solar-like oscillations for Procyon have been detected in velocity measurements ([28, 80], Hekker et al. 2007). Some controversy exists as to whether these oscillations can be detected in photometry even from space. As a prototype for a 1.5 $M_\odot$ main sequence star, Bonanno et al. [9] have modelled the latitudinally differential rotation in the outer convective layers for this star. Their models give a flatness of order 0.1–0.45 which is comparable with the solar case 0.25. The authors computed a latitudinal averaged $\Omega(r)$. The splittings differ by a few nHz depending on the latitudinal shear obtained in the models generating the differential rotation. Such a small value could be detected with Corot after averaging over several multiplets.

HD 49933. The space experiment Corot, which was successfully launched in December 2006, seems to be able to keep its promises and makes possible to detect mean splitting values as well as individual splittings at least in some favourable cases. As the first solar-like target for this mission, HD49933 was observed from ground with the spectrograph HARPS [101]. These observations confirmed the expectation that this star is a solar-like oscillator. The star shows a high degree of stellar activity and the rotation appears to be quite rapid with a period of the order of 4–8 days. The star was the object of a hare and hound exercise in order to see whether splittings can indeed be detected and which accuracy can be expected. This exercise indicated that for an assumed rotation of 1.4–2.9 µHz, many splittings could be detected. Only a few correct ones however were measured within 0.5 µHz from the correct value and with error bars less than 0.5 µHz [4].

8.2 δ Scuti Stars

These stars are nonradial multiperiodic variables with excited modes in the vicinity of the fundamental. The main problem for these stars is mode identification. As it is not yet possible to assign n, l, m values with full confidence to the observed frequencies, it is difficult to get information on the star. As far as rotational splitting is concerned, a few detections have been claimed.

GX Peg belongs to a spectroscopic binary system. Knowing the surface rotation from the binarity, the identification of a splitting led to the conclusion that the core is rotating faster than the outer layers [58]. The rotationally split multiplet was found asymmetric which could be attributed to distortion of the star by the centrifugal force.

FG Vir is one of the best studied delta Scuti star. Its variability has been extensively observed and analysed and the results have been the subject of many observations and theoretical studies over the past 20 years [11, 39, 150]. Among about 70 detected frequencies, a dozen ℓ and m values can be assigned. The star appears to be moderately rotating with an equatorial velocity of 66 ± 16 km/s. Most of the identified modes are axisymmetric but Zima et al. [150] found that one $\ell = 1$ triplet is observed with a splitting value of 6.13 µHz. Failure to find a stellar model whose frequencies match the observed frequencies of FG Vir subsists even when taking into account second-order rotational effects.

44 Tauri has been observed for a long time and is known to pulsate with at least 13 frequencies. Mode identification indicates a few possible splitting components among them which indicate that the star is a slow rotator with 26 km/s as an equatorial velocity (Poretti et al. [111], Antoci et al. [4] and references therein).

HD104237. Detecting the rotational splittings of oscillating PMS stars ought to be relatively easy as they are fast rotators and their power spectra are expected to be much simpler than those of their post-ZAMS counterparts, i.e. delta Scuti stars [136]. Determining their rotation profiles and internal structures can give us clues about temporal evolution of rotation profiles and transport of internal angular momentum. In this spirit, the binary PMS system HD104237 is being studied as one of its two components has been identified as a δ Scuti star [1, 2, 44, 64].

8.3 β Cephei Stars

More massive than δ Scuti stars, B-type main sequence stars oscillate with periods between 1.5 and 8 h corresponding to low-radial-order p- and g-modes. Their projected rotational velocities can reach up to 300 km/s. Rotational splittings for $\ell = 1$ and $\ell = 2$ modes seem to have been detected for a few stars (see, for instance, Fig. 1 of Pigulski [105]).

HD129929, a $\sim 8-9\,M_\odot$ β Cephei star. This star is known to oscillate with at least six identified frequencies. From one $\ell = 1$ multiplet, the rotational splitting yields a rotational velocity of 3.61 km/s when assuming a uniform rotation. On the other hand, two successive components of a $\ell = 2$ multiplet indicate a rotational velocity of 4.20 km/s. This leads to the likely conclusion that the rotation varies with depth with a ratio of the angular velocity at the centre to that at the surface $\Omega_c/\Omega_s = 3.6$ [43].

ν **Eridani** is another β Cepheid which shows a rich frequency spectrum but with only a few of them identified (see Fig. 1 of Ausseloos et al. [5]). Pamyatnykh et al. [104] used the identified $\ell = 1$, $g1$ triplet and two components of the $\ell = 1$, $p1$ to determine a nonuniform rotation for this star with $\Omega_c/\Omega_s \sim 3$, close to what is found for HD 129929. Using an assumed depth-dependent profile, Suárez et al. [135] computed the splittings and their associated departures from equidistant in order to reproduce the observed 3 $\ell = 1$ multiplets obtained by Jerzykiewicz et al. [74]. The observed asymmetries are in favour of a nonuniform rotation profile although they are too small to allow a definitive conclusion.

θ **Ophuichi** is a $\sim 8 - 9 M_\odot$ β Cepheid with seven identified frequencies [68]: a $\ell = 1, p1$ triplet and four components of a $\ell = 2, g1$ quintuplet (Fig. 6 of Handler et al. [68]). The splittings indicate a rotation velocity of 29 ± 7 km/s, large enough to cause detected departure from equal splittings [12].

9 Inversion for Rotation

We consider the simplified problem of determining the depth-dependent rotation law of a star from the measurements of its rotational splittings. For more general 2D inversions which yields $\Omega(\theta, r)$ as well as for more details on inversion methods and results on the solar internal rotation, we refer the reader to general such as reviews Gough [54]; Gough and Thompson [56], Thompson et al. [142]. According to Eq. (46), a 1D rotation profile is related to the splittings by the integral relation

$$\delta\omega_j + \delta(\delta\omega_j) = \int_0^R K_j(r)\ \Omega(r)dr \qquad (99)$$

where for shortness j represents the set of values (n, ℓ) and $\delta(\delta\omega_j)$ is the incertainty of the measured splitting values. In general, only a finite number, N, of discrete splitting values associated with N modes are available. The measurements are polluted with some observational errors δS_j which are assumed uncorrelated with a variance σ_j^2. In order to derive the rotation profile from the observed splittings, one must invert the relation Eq. (99). As the data are related to the rotation profile by a linear functional, one is confronted with a *linear inverse problem.*

9.1 Splittings of High-Frequency p-Modes and the Abel Equation

For modes with high radial orders or degrees, the eigenfunctions vary more rapidly than the equilibrium quantities; hence as the scale of spatial variations of the eigenfunctions is smaller than that of the basic state, a local analysis is valid. One then assumes that the solution can be cast under the form of a plane wave: $v' \propto e^{i(\omega t - k \cdot r)}$, with k the wavenumber, and $|k|^2 = k_r^2 + k_h^2$, with k_r and k_h the radial and horizontal components. One can also neglect the perturbation of the gravitational field (Cowling approximation) as well as the derivatives of the equilibrium quantities. For high-frequency p-modes whose nature is mainly acoustic, the eigenfrequency is much larger than the Brünt–Väissälä frequency $(\omega^2 \gg N^2)$ which therefore can be neglected in the problem considered here. With these approximations, one recovers the expected dispersion relation for an acoustic wave $\omega^2 = k^2 c_s^2$ where $k^2 = k_r^2 + k_h^2$ and $k_h = S_\ell^2/c_s^2 = \Lambda/r^2$ with $\Lambda = \ell(\ell+1)$ and $c_s(r)$ is the adiabatic sound speed profile; $S_\ell^2 = c_s^2 \Lambda/r^2$ is the squared Lamb frequency. One then has

$$k_r^2(r) \approx \frac{1}{c_s^2}(\omega^2 - S_\ell^2) = \frac{\omega^2}{c_s^2} - \frac{\Lambda}{r^2}$$

Let us recall first that for a 1D acoustic wave with a wavenumber k with a node at one end $(r = 0)$ and a free boundary at the other end $(r = R)$, the resonant condition for the existence of a normal mode in a homogeneous medium is $kR = n\pi - \pi/2$ for some integer n. In the stellar 3D (inhomogeneous) medium, this condition becomes

$$\int_{r_t}^{R} k_r(r)\ dr = \left(n - \frac{1}{2}\right)\pi \tag{100}$$

where r_t is the radius of the mode inner turning point.

Taking for the mode frequency, the frequency corrected for the first-order effect of rotation, $\omega \sim \omega_0 + (\delta\omega + m\Omega)$ where ω_0 is the zeroth-order frequency (i.e. the frequency of the mode in the absence of rotation), one obtains

$$k_r(r) = Q + \left(\frac{\omega^{(0)}}{c_s^2 Q}\right)(\delta\omega + m\Omega)$$

with

$$Q = \left(\frac{\omega^{(0)2}}{c_s^2} - \frac{\Lambda^2}{r^2}\right)^{1/2} = \frac{\omega^{(0)}}{c_s}\left(1 - \frac{a^2}{W^2}\right)^{1/2}$$

and $a = c_s(r)/r$ and $W = \omega_0/\Lambda$. Recalling that at zeroth order, one has

$$\int_{r_t}^{R} Q(r)dr = (n - \frac{1}{2})\pi \qquad ,$$

the resonant condition Eq. (100) becomes

$$\delta\omega = \frac{1}{S} \int_{r_t}^{R} m\ \Omega(r) \left(W^2 - a^2\right)^{-1/2} \frac{dr}{c_s} \tag{101}$$

with

$$S = \int_{r_t}^{R} \left(W^2 - a^2\right)^{-1/2} \frac{dr}{c_s}$$

Note that for very high radial order modes, $a/W \ll 1$ and

$$\delta\omega = m \int_{r_t}^{R} \Omega(r) \frac{dr}{c_s} \left(\int_{r_t}^{R} \frac{dr}{c_s}\right)^{-1} \tag{102}$$

which shows that the perturbation of the frequency by the rotation can be expressed as an average of the rotation weighted by the acoustic propagation time in the radial direction inside the cavity.

Equation (101) is of Abel type. Indeed the Abel (integral) equation is defined as

$$\int_0^x \frac{f(y)}{(x-y)^\alpha} dy = g(x) \qquad 0 < \alpha < 1 \quad g(0) = 0 \tag{103}$$

and has the formal solution [54]:

$$f(y) = \frac{1}{\pi} \sin(\pi\alpha) \frac{d}{dx} \int_0^y \frac{g(x)}{(y-x)^{1-\alpha}} dx \tag{104}$$

Equation (101) then admits an analytical inverse solution. The angular velocity is given by

$$\Omega(r) = -\frac{2a}{\pi} \frac{d}{d\ln r} \int_{a_s}^{a} (a^2 - W^2)^{-1/2} \mathcal{H}(W) dW \tag{105}$$

where $a_s = a(R)$ and $\mathcal{H}(w)$ represents the data set derived from the observed splittings.

The Abel equation, Eq. (103), is known to be an ill-posed inverse problem in the sense of Hadamard [67]. The existence of an analytical solution to the inverse problem does not suppress the ill-posed nature of the numerical inversion, particularly when the observational data are noisy.

Ill-posed nature of an inverse problem. A *well-posed problem in the sense of* Hadamard [67] is defined as follows: let us consider the equation $\mathcal{H}(f) = g$ where $\mathcal{H}$ is an operator $\in \mathcal{F} \longrightarrow \mathcal{O}$ where $\mathcal{F}$ and $\mathcal{O}$ are closed vectorial spaces and more specifically here the integral equation

$$\int_0^R K(x,y)f(y)dy = g(x) \tag{106}$$

The three Hadamard conditions are

- existence of a solution: $\forall g \in \mathcal{O}, \exists f \in \mathcal{F}$ such that $\mathcal{H}(f) = g$
- unicity of the solution: $\forall g \in \mathcal{O}, \exists$ at most one $f \in \mathcal{F}$ such that $\mathcal{H}(f) = g$
- stability of the solution with respect to the data: $\forall f_n$, any sequence $\in \mathcal{F}$

such that

$$\lim_{n \to +\infty} \mathcal{H}(f_n) = \mathcal{H}(f), \qquad \text{then} \qquad \lim_{n \to +\infty} f_n = f$$

This last condition can be understood in a more practical sense as

$$\left| \frac{\delta g}{g} \right| << 1 \to \left| \frac{\delta f}{f} \right| << 1$$

Existence and unicity are the classical conditions for solving an equation and the stability condition, on the other hand, is an additional condition which requires that, for a well-posed problem, any small perturbation of the data can only lead to a small error on the solution. This condition is barely verified in practice for inverse problems. In particular, inversions of integral equations often are ill-posed problems: noise, even small, in the data can cause the derived solution to be very far from the true one.

9.1.1 Nonuniqueness and Sensitivity to Noise

One must recall that the inverse problem Eq. (106) admits the set of solutions $f(y) + h(y)$ provided

$$\int_0^R K(x, y) h(y) dy = 0$$

Furthermore, the Riemann–Lebesgue theorem states that

$$\int_a^b K(x, y) \cos(my) dy = 0; \quad \int_a^b K(x, y) \sin(my) dy = 0$$

when $m \to \infty$, valid for any integrable $K(x, y)$. Consequences are the following:

(i) On the one hand, high-frequency components of the solution $f(y)$ are not accessible as they are smoothed by the kernel $K(x, y)$ to an arbitrary small amplitude level in the data. As an illustration, let some perturbation of the source function f in Eq. (103) be of the form $\delta f(y) = A \sin(2\sqrt{\lambda y})$ [36]. It causes a perturbation of the data of the form

$$\delta g(x) = A \sqrt{\pi} \Gamma(1 - \alpha) \lambda^\beta x^{(3/2 - \alpha)/2} J_{3/2 - \alpha}(2\sqrt{\lambda y}) \tag{107}$$

where $\beta = (\alpha - 1/2)/2$ and $J_{3/2 - \alpha}$ is a Bessel function of first kind of order $3/2 - \alpha$. For high frequencies (i.e. $\lambda \longrightarrow \infty$), one can use $|J_\nu(z)| \sim |2/\pi z|^{1/2}$ and Eq. (107) becomes $|\delta g(x)| \sim \Gamma(1 - \alpha) \lambda^{(\alpha - 1)/2} x^{(1 - \alpha)/2}$ so that for any amplitude A and $\forall \alpha < 1$, $\delta g(x) \longrightarrow 0$ when $\lambda \longrightarrow \infty$. Any discontinuity

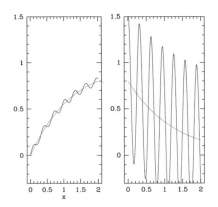

Fig. 1. The *left* panel displays the noise-free data function $g1$ and the slightly noisy data function $g2$ where the perturbed part is represented by a small (5%) amplitude oscillation. The *right* panel shows the true inverse solution, $f1$, and the inverse solution of the noisy function $g2$, $f2$, which is dominated by a spurious large (100%) amplitude oscillation (adapted from Craig and Brown [36])

can be represented by an infinite Fourier series but all the high-frequency components are smoothed out and contribute to the data at a small amplitude level; they cannot be distinguished from the high-frequency noise in the data. The smoothing properties of the operators and/or kernels in the process $f \to g$ remove any real discontinuity of the solution f and which is therefore difficult, if not impossible, to bring out from the data.

(ii) On the other hand, high-frequency small perturbations in the data (noise) appear as large-amplitude oscillating components in the calculated source. These large amplitude contributions can dominate the inverse solution even if they are totally spurious as due to small perturbations in the data due to noise. An example given by Craig and Brown [36] illustrates this issue (cf. Fig. 1). One considers the particular kernel $K(x, y) = 1$ in Eq. (106) (differentiation). Let two functions $g_1(y)$ and $g_2(y)$ differ only by the addition of a sinusoidal component, Δg, with a given frequency. We choose for instance $g_1(x) = 1 - e^{-\alpha x}$; $g_2(x) = g_1(x) + \beta \sin \omega x$. The difference appears in the source functions f_1 and f_2 as $\Delta f(y) = f_2(x) - f_1(x) - \omega \beta \cos \omega y$. Figure 1 shows the data and the source functions for $\alpha = 0.8; \beta = 0.04; and \omega = 20$. Although the difference $g_1 - g_2$ is very small, the corresponding difference in the source functions is an oscillatory behaviour with a large amplitude.

Hence since observations cannot in general be noise free, a simple 'naive' inversion is not able to give the correct inverse solution.

Discretization and a measure of the ill-posed nature of the problem. Although the first inversions for the solar rotational splittings using Eq. (101) were successful enough, the asymptotic integral relation remains an approximation. In practice, one actually considers the general relation Eq. (45) and inversions are carried out first by discretizing the integral. Let us write the resulting discretized relation as $\Delta = AX$ where Δ represents the data vector and

X is the rotation vector to be determined. The stability can be measured with the conditioning of the matrix A. If A is regular, its conditioning is defined as $cond(A) = ||A|| \, ||A^{-1}||$ where $||A||$ is a matrix norm for close A. Note that $cond(A) \geq 1$ is always verified. A well-conditioned matrix is such that $cond(A)$ remains close to 1. Using the norm $||A|| = max_j(\mu_j)$ where μ_j are the singular values of A (i.e. the square root of the eigenfrequencies of the squared symmetric matrix AA^t), one shows that $cond(A) = \mu_{max}/\mu_{min}$. A small perturbation $\delta\Delta$ of the data generates a perturbation δX of the solution, then $\delta\Delta = A(\delta X)$ and it is easy to deduce from the matrix norm properties that

$$\frac{||\delta X||}{||X||} \leq cond(A)\frac{||\delta\Delta||}{||\Delta||}$$

Accordingly, if A is well conditioned ($cond(A) \sim 1$), the stability condition is satisfied whereas for $cond(A) >> 1$, which is most often the case, $||\delta X||$ can become large compared with $||X||$ and the problem is ill posed.

Once the ill-posed nature of the inverse problem is recognized, one is then forced to carry out a regularization of the calculated solution. As seen above, helio- and astero-seismic inversions are ill-posed problems which therefore require regularization inverse techniques.

9.2 Inversion with Regularization

Several classes of inversion techniques with regularization exist. They are based on the addition of a priori information on the solution, often on the shape of the solution. One wants a solution which remains insensitive to errors in the data. Most methods involve one or two parameters which control the sensitivity of the solution to errors in the data. We briefly present two of the most currently used ones in the stellar context.

Inverse methods with a priori douceur or Tikhonov's methods. The inversion *strategy* consists in determining a solution where high-frequency variations – which cannot be identified as real or due to noise magnification – have been removed. One then seeks for a **global** solution which reproduces at best the data without reproducing the noise (seen as the high-frequency components). The procedure is to find a least square solution submitted to a regularization constraint: $min(||AX - \Delta||^2 + \gamma \, \sigma^2 \, B(X))$ where σ is the variance of the noise, γ a trade-off parameter and B a discretized regularizing function. This is equivalent to solving the normal form equation: $(A^tA + \gamma B^tB)X = A^t\Delta$ where A^t is the transposed matrix. The regularizing function $B(X)$ is often taken as the second derivative of the solution since one wishes to eliminate the high-frequency part of the solution. The trade-off parameter γ must be adjusted to realize the best compromise between a noise-insensitive but distorted solution (large γ) and a solution which is not regularized enough (small γ).

Optimally localized averages: OLA and its variants. One looks for a **local** solution, that is an averaged value of the solution over a small interval about a given radius. The SOLA method is a variant of the OLA technique [6] developed by Pijpers and Thompson [106–108]. It aims at building localized kernels $\mathcal{K}$ about a predefined value, $r = r_0$ as linear combinations of the rotational kernels $K_j(r)$: $\mathcal{K}(r, r_0) = \Sigma_j c_j(r_0) K_j(r)$. The coefficients c_j must be determined by a minimization process:

$$min_{cj} \left(\left| \int (\mathcal{K} - T)^2 \, dr \right|^2 + \gamma E^2 \right)$$

where $T(r, r_0)$ is a predefined target, E is the noise covariant matrix and γ a trade-off parameter in order to obtain a satisfying compromise between magnification of the error and the width of the target, i.e. the interval over which the solution is averaged. The SOLA method with a predefined target presents the advantage over the original OLA method that the inverse matrix is computed once and for all. Once the $c_j(r_0)$ are determined, one computes the average rotation value as $< \Omega(r) >_{r0} = \sum_j c_j(r_0) \, \delta\omega_j$. Indeed one has

$$\sum_j c_j(r_0)\delta\omega_j = \sum_j c_j(r_0) \left(\int K_j(r)\Omega(r)dr \right)$$

$$= \int \left(\sum_j c_j(r_0)K_j(r) \right) \Omega(r)dr$$

$$= \int \mathcal{K}(r, r_0)\Omega(r)dr \sim \int T(r, r_0)\Omega(r)dr$$

If the target T is a Dirac function, i.e. $T = \delta(r - r_0)$, then $\sum_j c_j\delta\omega_j = \Omega(r_0)$. However one seldom has enough information in the data to succeed in building Dirac function as localized kernels. Usually the target function takes the form of gaussian with a given width. Again a compromise must be obtained between the magnification of the error and the width of the target, i.e. the interval over which the solution is averaged.

9.3 Application to Stellar Seismology

The solar rotation profile. With the high-quality helioseismic data available for the past two decades or so, highly accurate frequency splittings were obtained and allowed to derive the 2D rotation profile in the 3/4 outer part of the Sun. See Fig. 5 of Schou et al. [126], for instance, for the result of an inversion based on SoHo data. Several 2D inversion methods have been applied with various adjustments and adaptations to the available helioseismic data sets over the years (Schou et al. [124–126], Antia et al. [3]). This has provided the internal rotation profile for the Sun, with quite a number of surprises at the time: latitudinal dependence of rotation in the convective outer region and uniform rotation in the radiative region. Only the rotation of the inner part is still not yet accessible.

The inversion results have generated many studies in order to explain these seismic results. Several studies have led to identify the shear of the Reynolds stress as responsible for differential rotation in latitude in the convective outer layers of the Sun; others have studied the importance and the role of the tachocline in the solar magnetogydrodynamical processes; see for instance Rüdiger et al. [120, 121]; Miesch et al. [99], Zahn et al. [149].

The discovery of the uniform rotation of the solar radiative zone drove the development of models able to transport angular momentum from the inner part to the outer part. It was found that type I rotationally induced mixing models are not efficient enough and predict an incorrect increase in the rotation rate with decreasing radius [138]. Internal waves, on the other hand, are very efficient in transporting the angular momentum from centre upward to the surface and enforce a uniform rotation in the radiative zone (Charbonnel and Talon [23]; Talon [140] and references therein). Figure 1 of Charbonnel and Talon [23] shows the evolution of the rotation profile for a $1.2 M_\odot$ stellar model from 0.2 up to 4.8 Gyrs when rotational-induced mixing of type I is included and the successful approach of uniform rotation when mixing of type II is included.

One current issue is the role of the magnetic field; another is to determine to what extent these results and explanations are valid for other stars than the Sun, particularly young stars which rotate faster than the Sun. In that context, information from 3D simulations will be of great help [8, 13].

Solar-like stars. The observed high-frequency solar p-modes do not give access to the core rotation. For stars slightly more massive and slightly more evolved than the Sun such as η Boo for instance, a small number of mixed modes do indeed exist in the high-frequency domain of modes excited by the turbulent convection of the outer layers. A theoretical study was carried out with SOLA

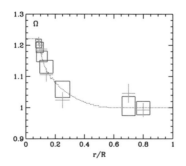

Fig. 2. Inversion for the rotation profile for a model of ϵ Cep, a δ Scuti star. Data are the splittings computed for an input stellar model (1.8 $M_\odot$ main sequence star with $T_{eff} = 7588$ K and a surface rotational velocity $v = 120$ km/s). The input-prescribed rotation profile Ω normalized to its surface value is represented by *dotted lines. Squares* show the results of a SOLA inversion using the input model; *crosses* represent a SOLA inversion using a different model than the input one with a 1.90 $M_\odot$ mass and $T_{eff} = 7906$ K with no rotation but with the same large separation 50 μHz. From Goupil et al. 2000, unpublished)

inversions for the rotation profile on simulated data for a 1.5 $M_\odot$ stellar model with a surface velocity of ~ 30 km/s and a ratio between the rotation rate at the centre to that at the surface $\Omega_c/\Omega_s \sim 3$ [86]. The incertainties were estimated in function of the expected Corot performances. Figure 8 of Lochard et al. [86] shows that the variation for the rotation profile with radius in the vicinity of the convective core – varying from $\Omega/\Omega_s = 1.5$ at $r = 0.2$ (with an incertainty on the recovered value ± 0.2) to $\Omega/\Omega_s = 2.4$ at $r \leq 0.1$ (with an incertainty on the recovered value ± 0.6) is accessible.

Inversion for rotation for δ Scuti-like oscillations. Other stars oscillate with modes with lower frequencies in the vicinity of the fundamental, driven by an opacity mechanism. These stars are slightly more massive than the Sun and develop a convective core on the main sequence which recedes with time. This generates the existence of mixed modes in the excited frequency range. Only one or two such modes are enough to give access to Ω_c, the core rotation rate [60]. One illustrative example is ϵ Cep. Assuming that one has obtained a seismic model for this star, that is a model as close as possible to reality with the same mean large separation, a SOLA inversion was performed with a data set of $\ell = 1, 6$ linearly unstable modes. The variation of the rotation profile with depth is well recovered (Fig. 2).

This theoretical study assumed a noise level according to Corot performances. The inversion was carried out with linear splittings Eq. (46) although for a real case, distortion effects on the splittings ought to be included for such a moderately rapid rotator ($v \sin i = 91$ km/s, Royer et al. [119]). Several frequencies have recently been detected for this star but its complex frequency pattern has not yet been fully elucidated [14].

Acknowledgment

I gratefully thank Rhita Maria Ouazzani for a careful reading which helped to improve grandly the manuscript.

References

1. Alecian, E., Catala, C., van't Veer-Menneret, C., Goupil, M.-J., Balona, L.: AA **442**, 993 (2005)
2. Alecian, E., Goupil, M.-J., Lebreton, Y., Dupret, M.-A., Catala, C.: AA **465**, 241 (2007)
3. Antia, H.M., Basu, S., Chitre, S.M.: MNRAS **298**, 543 (1998)
4. Antoci, V., Breger, M., Bishof, K., Garrido, R.: In: Sterken, C., Aerts, C., Appourchaux, T., Berthomieu, G., Michel, E., et al. (eds.) vol. 1306, p. 377. ESA Special Publication, PASP **349**, 181 (2006)
5. Ausseloos, M., Scuflaire, R., Thoul, A., Aerts, C.: MNRAS **355**, 352 (2004)
6. Backus, G., Gilbert, F.: Phil. Trans. R. Soc. London, A **266**, 123 (1970)

7. Baglin, A.: Auvergne, M., Barge P. Deleuil, M., Catal, C., Michel, E., Weiss, W., and the CoRoT team.: The Seismology programme of the CoRoT space mission, Proceedings of SOHO 18/GONG 2006/HELAS I, Beyond the spherical sun, ESA Special Publication, **624**, 34 (2006)
8. Ballot, J., Brun, A.S., Turck-Chièze, S.: ApJ **669**, 1190 (2007)
9. Bonanno, A., Küker, M., Paternò, L.: AA **462**, 1031 (2007)
10. Bonazzola, S., Gourgoulhon, E., Marck, J.-A., Numerical approach for high precision 3D relativistic star models. Phys. Rev D **58**, 104020 (1998)
11. Breger, M., Lenz, P., Antoci, V., et al.: A&A **435**, 955 (2005)
12. Briquet, M., Morel, T., Thoul, A., Scuflaire, R., Miglio, A., Montalbán, J., Dupret, M.-A., Aerts, C.: MNRAS **381**, 1482 (2007)
13. Brun, A.S., Miesch, M.S., Toomre, J.: ArXiv Astrophysics e-prints, astro-ph/0610073 (2006)
14. Bruntt, H., Suarez, J.C., Bedding, T.R., et al.: AA **461**, 619 (2007)
15. Burke, K.D., Thompson, M.J.: In: Proceedings of SOHO 18/GONG 2006/HELAS I, Beyond the Spherical Sun. ESA Special Publication, 624 (2006)
16. Buzasi, D. L., Bruntt, H., Bedding, T.R. et al.: AA **619**, 1072 (2005)
17. Chandrasekhar, S., Lebovitz, N.R.: ApJ **136**, 1069 (1962a)
18. Chandrasekhar, S., Lebovitz, N.R.: ApJ **136**, 1082 (1962b)
19. Chandrasekhar, S., Lebovitz, N.R.: ApJ **136**, 1105 (1962c)
20. Chandrasekhar, S., ApJ **140**, 599 (1964)
21. Chandrasekhar, S., Lebovitz, N.R.: ApJ **152**, 152 (1968)
22. Charbonneau, P.: AA **259**, 134 (1992)
23. Charbonnel, C., Talon, S.: Science **309**, 2189 (2005)
24. Chlebowski, T.: Acta Astronom. **28**, 441 (1978)
25. Christensen-Dalsgaard, J., Mullan, D.J.: MNRAS **270**, 921 (1994)
26. Christensen-Dalsgaard, J., Thompson, M.J.: AA **350**, 852 (1999)
27. Christensen-Dalsgaard, J., (CD03), Lecture note on stellar oscillation, 5th edn. http://www.phys.au.dk/ jcd/oscilnotes/) (2003)
28. Claudi, R.U., Bonanno, A., Leccia, S., et al.: AA **429**, L17 (2005)
29. Clement, M.J.: ApJ **141**, 210 (1965a)
30. Clement, M.J.: ApJ **142**, 243 (1965b)
31. Clement, M.J.: ApJ **249**, 746 (1981)
32. Clement, M.J.: ApJ **276**, 724 (1984)
33. Clement, M.J.: ApJ **339**, 1022 (1989)
34. Cowling, T.G.: Newing, R.A.: ApJ **109**, 149 (1949)
35. Cox, J.P.: Theory of stellar pulsation. Research supported by the National Science Foundation Princeton, Princeton University Press, NJ (1980)
36. Craig, I.J.D., Brown, J.C.: Inverse problems in astronomy. Adam Hilger Ltd. (1986)
37. Daszynska-Daszkiewicz, J., Dziembowski, W.A., Pamyatnykh, A.A., Goupil, M.J.: AA **392**, 151 (2002)
38. Daszynska-Daszkiewicz, J., Dziembowski, W.A., Pamyatnykh, A.A.: **407**, 999 (2003)
39. Daszynska-Daszkiewicz, J., Dziembowski, W.A., Pamyatnykh, A.A., et al.: **438**, 653 (2005)
40. Deupree, R.G.: ApJ **357**, 175 (1990)
41. Deupree, R.G.: ApJ **552**, 268 (2001)
42. Domiciano de Souza, A., Kervella, P., Jankov, S., et al.: AA **442**, 567 (2005)
43. Dupret, M.-A., Thoul, A., Scuflaire, R., et al.: AA **415**, 251 (2004)

44. Dupret, M.-A., Böhm, T., Goupil, M.-J., et al.: Commun. Asteroseismology **147**, 72 (2006)
45. Dyson, J., Schutz, B.F.: Proc. R. Soc. Lond. A Math. Phys. Sci. **368**, 389 (1979)
46. Dziembowski, W.A., Goode, P.R.: (DG92), ApJ **394**, 670 (1992)
47. Dziembowski, W.A., Goupil, M.-J.: In: Kjeldsen, H., Bedding, T.R. (eds.) The First MONS Workshop: Science with a Small Space Telescope, held in Aarhus, Denmark, June 29–30, 1998, p. 69. Aarhus Universitet (1998)
48. Eddington, A.S.: MNRAS **90**, 54 (1929)
49. Eggenberger, P., Carrier, F.: AA **449**, 293 (2006)
50. Espinosa, F., PhD thesis
51. Espinosa, F., Pérez Hernández, F., Roca Cortés, T.: In: Danesy, D., (ed.) SOHO 14 Helio- and Asteroseismology: Towards a Golden Future, vol. 559, p. 424. ESA Special Publication (2004)
52. Espinosa Lara, F., Rieutord, M.: AA **470**, 1013 (2007)
53. Fletcher, S.T., Chaplin, W.J., Elsworth, Y., Schou, J., Buzasi, D.: MNRAS **371**, 935–944 (2006)
54. Gough, D.: Sol. Phys. **100**, 65 (1985)
55. Gough, D.O., Thompson, M.J.: (GT90), MNRAS **242**, 25 (1990)
56. Gough, D.O., Thompson, M.J.: The inversion probleme, in Solar interior and atmosphere. In: Cox, A.N., Livingston, W.C., Matthews, M.S. (eds.) p. 519. University of Arizona Press (1991)
57. Gough, D.O.: Linear adiabatic stellar pulsation. In: Zahn, J.P., Zinn-Justin, J., (eds.) Astrophysical Fluid Dynamics, Les Houches Session XLVII 1987, p. 399. Elsevier Science Publishers B.V. (1993)
58. Goupil, M.J., Michel, E.: Lebreton, Y., Baglin, A.: AA **268**, 546 (1993)
59. Goupil, M.J., Michel, E., Cassisi, S., et al.: In: Feast, M.W. (eds.) IAU Colloq. 155: Astrophysical Applications of Stellar Pulsation, Astronomical Society of the Pacific Conference Series. **83**, 453 (1995)
60. Goupil, M.-J., Dziembowski, W.A., Goode, P.R., Michel, E.: AA **305**, 487 (1996)
61. Goupil, M.-J., Dziembowski, W.A., Pamyatnykh, A.A., Talon, S.: In: Breger, M., Montgomery, M. (eds.) Delta Scuti and Related Stars, Astronomical Society of the Pacific Conference Series. **210**, 267 (2000)
62. Goupil, M.J., Talon, S.: In: Aerts, C., Bedding, T.R., Christensen-Dalsgaard, J. (eds.) IAU Colloq. 185: Radial and Nonradial Pulsations as Probes of Stellar Physics, Astronomical Society of the Pacific Conference Series. **259**, 306 (2002)
63. Goupil, M.J., Samadi, R., Lochard, J., et al.: In: Favata, F., Aigrain, S., Wilson, A. (eds.) Stellar Structure and Habitable Planet Finding, vol. 538, p. 133. ESA Special Publication (2004)
64. Goupil, M.-J., Dupret, M.A., Samadi, R., et al.: JApA **26**, 249 (2005)
65. Goupil, M.J., Moya, A., Suarez, J.C., et al.: Why bothering to measure stellar rotation with CoRoT? ESA Special Publication **1306**, 51 (2006)
66. Goupil, M.J., Dupret, M.A.: EAS Publications Series. **26**, 93 (2007)
67. Hadamard, J.: Lectures on Cauchy's problem in linear partial differential equations'. Yale University Press, New Haven, CT (1923)
68. Handler, G., Shobbrook, R.R., Mokgwetsi, T.: MNRAS **362**, 612 (2005)
69. Hansen, C.J., Cox, J.P., Carroll, B.W.: ApJ **226**, 210 (1978)
70. Hansen, C.J., Cox, J.P., van Horn, H.M.: ApJ **217**, 151 (1977)
71. Hekker, S., Arentoft, T., Kjeldsen, H., et al.: ArXiv e-prints, 0710.3772 (1978)
72. Jackson, S., MacGregor, K.B., Skumanich, A.: ApJ **606**, 1196 (2004)

73. Jackson, S., MacGregor, K.B., Skumanich, A.: ApJS **156**, 245 (2005)
74. Jerzykiewicz, M., Handler, G., Shobbrook, R.R., et al.: MNRAS **360**, 619 (2005)
75. Karami, K., Christensen-Dalsgaard, J., Pijpers, F.P., et al.: ArXiv Astrophysics e-prints, astro-ph/0502194 (2005)
76. Kippenhahn, R., Meyer-Hofmeister, E., Thomas, H.: AA **5**, 155 (1970)
77. Kippenhahn, R., Weigert, A.: In: Stellar Structure and Evolution, Springer-Verlag (1994)
78. Lebovitz, N.R.: ApJ **160**, 701 (1970a)
79. Lebovitz, N.R.: Ap&SS **9**, 398 (1970b)
80. Leccia, S., Kjeldsen, H., Bonanno, A., et al.: AA **464**, 1059 (2007)
81. Ledoux, P.: ApJ **102**, 143 (1945)
82. Ledoux, P.: ApJ **114**, 373 (1951)
83. Ledoux, P., Walraven, T.: Handbuch der Physik **51**, 353 (1958)
84. Lee, U., Saio, H.: MNRAS **221**, 365 (1986)
85. Lignières, F., Rieutord, M., Reese, D.: AA **455**, 607 (2006)
86. Lochard, J., Samadi, R., Goupil, M.J.: AA **438**, 939 (2005)
87. Lovekin, C.C., Deupree, R.G.: Mem. S.A., It, **77**, 137 (2006)
88. Lynden-Bell, D., Ostriker, J.P.: MNRAS **136**, 293 (1967)
89. Maeder, A., Zahn, J.-P., AA **334**, 1000 (1998)
90. Maeder, A.: AA **399**, 263 (2003)
91. Mathis, S., Palacios, A., Zahn, J.-P.: AA **425**, 243 (2004)
92. Mathis, S., Zahn, J.-P.: AA **425**, 229 (2004)
93. Mathis, S., Decressin, T., Palacios, A., et al.: In: Proceedings of SOHO 18/GONG 2006/HELAS I, Beyond the Spherical Sun, vol. 624, ESA Special Publication (2006)
94. Mathis, S., Eggenberger, P., Decressin, T., et al.: In: EAS Publications Series, vol. 26, p. 65 (2007a)
95. Mathis, S., Palacios, A., Zahn, J.-P.: AA **462**, 1063 (2007b)
96. Mestel, L.: In: Thompson, M.J., Christensen-Dalsgaard, J. (eds.) Stellar Astrophysical Fluid Dynamics, p. 75. Cambridge University Press, ISBN 0-521-81809-5, (2003)
97. Meynet, G., Maeder, A.: AA **361**, 101 (2000)
98. Michel, E., Chevreton, M., Goupil, M.J., et al.: 4th SOHO Workshop Helioseismology, vol. 2, p. 533 (1995)
99. Miesch, M.S., Brun, A.S., DeRosa, M.L., Toomre, J.: American Astronomical Society Meeting Abstracts 210, 17 (2007)
100. Monnier, J.D., Zhao, M., Pedretti, E., et al.: Science **317**, 342 (2007)
101. Mosser, B., Bouchy, F., Catala, C., et al.: AA **431**, L13 (2005)
102. Palacios, A., Talon, S., Charbonnel, C., Forestini, M.: AA **399**, 603 (2003)
103. Pamyatnykh, A.A.: PASP **284**, 97 (2003)
104. Pamyatnykh, A.A., Handler, G., Dziembowski, W.A.: MNRAS **350**, 1022 (2004)
105. Pigulski, A.: Coast **150**, 159 (2007)
106. Pijpers, F.P., Thompson, M.J.: MNRAS **262**, L33 (1992)
107. Pijpers, F.P., Thompson, M.J.: AA **281**, 231 (1994)
108. Pijpers, F.P., Thompson, M.J.: MNRAS **279**, 498 (1996)
109. Pijpers, F.P.: AA **326**, 1235 (1997)
110. Pijpers, F.P.: MNRAS **297**, L76 (1998)
111. Poretti, E., Mantegazza, L., Riboni, E.: AA **256**, 113 (1992)
112. Reese, D.: PhD thesis, AA (Universit'e Toulouse III - Paul Sabatier) (2006)
113. Reese, D., Lignières, F., Rieutord, M.: AA **455**, 621 (2006)

114. Rieutord, M.: AA **451**, 1025 (2006a)
115. Rieutord, M.: ArXiv Astrophysics e-prints, astro-ph/0608431 (2006b)
116. Rieutord, M.: ArXiv Astrophysics e-prints, astro-ph/0702384 (2007)
117. Roxburgh, I.W.: AA **428**, 171 (2004)
118. Roxburgh, I.W.: AA **454**, 883 (2006)
119. Royer, F., Grenier, S., Baylac, M.-O., Gómez, A.E., Zorec, J.: AA **393**, 897 (2002)
120. Rüdiger, G., Küker, M., Chan, K.L.: AA **399**, 743 (2003)
121. Rüdiger, G., Kitchatinov, L.L., Arlt, R.: AA **444**, L53 (2005)
122. Saio, H.: ApJ **244**, 299 (1981)
123. Saio, H.: In: Aerts, C., Bedding, T.R., Christensen-Dalsgaard, J. (eds.) IAU Colloq. 185: Radial and Nonradial Pulsationsn as Probes of Stellar Physics, Astronomical Society of the Pacific Conference Series, vol. 259, p. 177 (2002)
124. Schou, J., Christensen-Dalsgaard, J., Thompson, M.J.: ApJ **433**, 389 (1994a)
125. Schou, J., Brown, T.M.: ApJ **434**, 378 (1994b)
126. Schou, J., Antia, H.M., Basu, S., et al.: ApJ **505**, 390 (1998)
127. Schutz, B.F.: MNRAS **190**, 7 (1980a)
128. Schutz, B.F.: MNRAS **190**, 21 (1980b)
129. Simon, R.: AA **2**, 390 (1969)
130. Soufi, F., Goupil, M.J., Dziembowski, W.A.: AA **334**, 911 (1998)
131. Suárez, J.C., Bruntt, H., Buzasi, D.: AA **438**, 633 (2005)
132. Suárez, J.C., Garrido, R., Goupil, M.J.: AA **447**, 649 (2006a)
133. Suárez, J.C., Goupil, M.J., Morel, P.: AA **449**, 673 (2006b)
134. Suárez, J.C.: EAS Publications Series, vol. 26, p. 121 (2007)
135. Suárez, J.C., Garrido, R., Moya, A.: AA **474**, 971 (2007)
136. Suran, M., Goupil, M., Baglin, A., et al.: AA **372**, 233 (2001)
137. Sweet, P.A.: MNRAS **110**, 548 (1950)
138. Talon, S., Zahn, J.-P., Maeder, A., Meynet, G.: AA **322**, 209 (1997)
139. Talon, S.: In: Proceedings of SOHO 18/GONG 2006/HELAS I, Beyond the Spherical Sun, vol. 624, p. 37. ESA Special Publication, (2006)
140. Talon, S.: In: Proceedings of the Aussois School "Stellar Nucleosynthesis: 50 years after B2FH" ArXiv e-prints, vol. 708 (2007)
141. Tassoul, J.-L.: Theory of Rotating Stars, Princeton Series in Astrophysics, University Press, Princeton (1978)
142. Thompson, M.J., Christensen-Dalsgaard, J., Miesh, M.S., Toomre, J.: The internal rotation of the Sun. Ann. Rev. Astron. Astrophys. **41**, 599 (2003)
143. Unno, W., Osaki, Y., Ando, H., et al.: Nonradial Oscillations of Stars, 2nd edn. University of Tokyo Press, Tokyo (1989)
144. Vogt, H.: Astron. Nachr. **234**, 93 (1929)
145. von Zeipel, H.: MNRAS **84**, 665 (1924a)
146. von Zeipel, H.: MNRAS **84**, 684 (1924b)
147. Zahn, J.-P.: AA **265**, 115 (1992)
148. Zahn, J.-P.: In: Thompson, M.J., Christensen-Dalsgaard, J. (eds.) Stellar Astrophysical Fluid Dynamics, p. 205. ISBN 0-521-81809-5, Cambridge University Press (2003)
149. Zahn, J.-P., Brun, A.S., Mathis, S.: AA **474**, 145 (2007)
150. Zima, W., Wright, D., Bentley, P.L., et al.: AA **455**, 235 (2006)

Approaching the Low-Frequency Spectrum of Rotating Stars

M. Rieutord

Laboratoire d'Astrophysique de Toulouse-Tarbes, CNRS et Université de Toulouse, 14 avenue E. Belin, 31400 Toulouse, France
rieutord@ast.obs-mip.fr

Abstract In this lecture I present the basic knowledge needed to understand the properties of the low-frequency spectrum of rotating stars. This spectrum is a mixture of inertial and gravity modes. These modes both have singularities in the limit of vanishing diffusion for a generic container. I explain the nature and the role of these singularities; I also discuss the way these modes can be computed and the actual difficulties that need to be circumvented to get sensible results.

1 Introduction

Rapidly rotating stars have benefitted from a renewed interest from stellar physicists as they have popped up in the observational fields of interferometry and asteroseismology.

Recent progress in interferometry allowed observers to measure the shape of a nearby star on the background sky, and for instance detect directly its centrifugal distortion. First successes were obtained by van Belle et al. [18] on Altair, but recent works give spectacular results on stars like Achernar [6], Altair [7,10,11] and Vega [1]. Observations not only give the angular diameters of the stars, but can also determine the orientation of the spin axis, thanks to the measurement of the brightness distribution on the stellar surface.

These new observations are very important for stellar theory, because beyond the determination of the rotational distortion these data give access to the mass distribution inside a star, and all the physics which controls it.

Rotation long appeared as a key parameter in asteroseismology, as it permits the identification of modes by the famous rotational splitting. However, the recent launch of the CoRoT mission, the future launch of the KEPLER one, will strongly increase the precision of the measurements of stellar eigenfrequencies. The precision will be such that a parameter like rotation must perfectly be taken into account in the models, so that other quantities like density and temperature can be precisely constrained. It turns out that many stars thought to be not rapidly rotating now fall in this category as the influence of rotation cannot be taken into account through a simple perturbative method. Sometimes, like

Rieutord, M.: *Approaching the Low-Frequency Spectrum of Rotating Stars.* Lect. Notes
Phys. **765**, 101–121 (2009)
DOI 10.1007/978-3-540-87831-5_4   © Springer-Verlag Berlin Heidelberg 2009

for the slowly oscillating γ-Doradus stars, the rotation frequency is of the same order of magnitude as the excited eigenfrequencies. There too, rotation needs to be accounted for by direct, non-perturbative methods.

In the foregoing examples, two effects of rotation mix: the centrifugal distortion of the star and the Coriolis acceleration. The first one modifies the shape of the star and mainly affects the high-frequency acoustic oscillations, while the second one changes the low-frequency part of the spectrum. In this series of lectures, the former is discussed by M.-J. Goupil, whereas we concentrate on the latter.

As a first step we shall discuss the case of plane waves propagating in a rotating fluid; we will then naturally move on to the eigenmodes of rotating fluids, the so-called inertial modes, and also present the gravity modes, which share similar fundamental properties. This will bring us to the Poincaré equation, which controls these types of modes. The Poincaré equation being of hyperbolic type, we need to discuss in detail the various consequences of this property. We can then discuss the more complex case of gravito-inertial modes. We end this lecture with an introduction to the numerical methods which can be used to compute this part of the eigenspectrum of a rotating star.

2 Waves in a Rotating Fluid

2.1 Inertial Waves

Inertial waves owe their existence to the Coriolis acceleration which serves as a restoring force and ensures the conservation of angular momentum. Let us consider the motion of a fluid particle under the action of this force. Its velocity verifies that

$$\frac{d\mathbf{v}}{dt} + 2\mathbf{\Omega} \wedge \mathbf{v} = \mathbf{0}$$

This equation is easily solved, and one finds that

$$v_x = v_0 \cos(2\Omega t) \qquad \text{and} \qquad v_y = v_0 \sin(2\Omega t)$$

if at $t = 0, v_x = v_0$ and $v_y = 0$. The trajectory of the particle is also easily derived:

$$x = x_0 + \frac{v_0}{2\Omega} \sin(2\Omega t) \qquad \text{and} \qquad y = y_0 - \frac{v_0}{2\Omega} \cos(2\Omega t)$$

These expressions show that particles have a circular motion. The Coriolis acceleration thus brings the particles back to their equilibrium position after they have followed a circular trajectory of (Rossby) radius $v_0/2\Omega$.

2.2 Dispersion Relation

Let us now consider waves propagating in an incompressible inviscid rotating fluid. The linearized equations of motion for disturbances read as follows:

$$\frac{\partial \mathbf{v}}{\partial t} + 2\boldsymbol{\Omega} \wedge \mathbf{v} = -\frac{1}{\rho}\boldsymbol{\nabla}P, \qquad \boldsymbol{\nabla}\cdot\mathbf{v} = 0$$

Choosing $(2\Omega)^{-1}$ as the time scale and L as the length scale, we may rewrite these equations with dimensionless variables as

$$\frac{\partial \mathbf{u}}{\partial \tau} + \mathbf{e}_z \wedge \mathbf{u} = -\boldsymbol{\nabla}p, \qquad \boldsymbol{\nabla}\cdot\mathbf{u} = 0 \tag{1}$$

Assuming that these waves are plane waves, i.e.

$$(p, \mathbf{u}) = (p, \mathbf{u})_0 e^{i(\omega\tau - \mathbf{k}\cdot\mathbf{x})}$$

incompressibility implies that

$$\mathbf{k}\cdot\mathbf{u} = 0 \tag{2}$$

showing that the waves are transverse. The equation of momentum gives

$$i\omega\mathbf{u} + \mathbf{e}_z \wedge \mathbf{u} = i\mathbf{k}P$$

from which we derive

$$\begin{cases} \mathbf{e}_z \cdot (\mathbf{u}\wedge\mathbf{k}) = ik^2 P \\ i\omega u_z = ik_z P \\ i\omega\mathbf{k}\wedge\mathbf{u} = k_z\mathbf{u} \end{cases} \tag{3}$$

The dispersion relation of the waves follows from the elimination of the amplitudes:

$$\omega^2 = \frac{k_z^2}{k^2} \tag{4}$$

From this dispersion relation we first see that the pulsation of the waves is bounded up by the Coriolis frequency 2Ω, showing that the associated oscillations occupy the low-frequency part of the spectrum.

These waves propagate very anisotropically. Let us first derive the phase velocity; this is

$$\mathbf{v}_\phi = \frac{\omega}{k}\mathbf{e}_k = \frac{k_z}{k^3}\mathbf{k} \tag{5}$$

which shows that the phase prefers propagating along the rotation axis. Let us now compute the group velocity:

$$\mathbf{v}_g = \boldsymbol{\nabla}_k\omega(\mathbf{k}) = \frac{\mathbf{k}\wedge(\mathbf{e}_z\wedge\mathbf{k})}{k^3} \tag{6}$$

This relation shows that the group velocity is orthogonal to the phase velocity! Energy travels perpendicularly to the phase.

3 Inertial Modes

3.1 General Properties

The plane wave solution is acceptable only if the wavelength is very small compared to the size of the container. This is not necessarily the case, especially in asteroseismology where one is interested in the global oscillations of stars. We thus need to consider the eigenmodes associated with these waves; still using the simplified set-up of the incompressible inviscid fluid, the eigenfunctions verify that

$$\begin{cases} i\omega\mathbf{u} + \mathbf{e}_z \wedge \mathbf{u} = -\boldsymbol{\nabla} P \\ \boldsymbol{\nabla} \cdot \mathbf{u} = 0 \\ \mathbf{u} \cdot \mathbf{n} = 0 \qquad \text{on the boundary} \end{cases} \tag{7}$$

From this system we first derive the orthogonality property of these modes; if ω_n and ω_m are two different pulsations then

$$\int_{(V)} \mathbf{u}_n \cdot \mathbf{u}_m^* \, dV = 0 \tag{8}$$

because

$$\begin{cases} i\omega_n\mathbf{u}_n + \mathbf{e}_z \wedge \mathbf{u}_n = -\boldsymbol{\nabla} P_n \\ -i\omega_m\mathbf{u}_m^* + \mathbf{e}_z \wedge \mathbf{u}_m^* = -\boldsymbol{\nabla} P_m^* \end{cases} \tag{9}$$

which leads to

$$i(\omega_n - \omega_m) \int_{(V)} \mathbf{u}_n \cdot \mathbf{u}_m^* dV = 0$$

Moreover, as expected from the dispersion relation the spectrum is bounded: $\omega \leq 1$. This comes from

$$\omega = \frac{\int_{(V)} \mathrm{Im}[(\mathbf{u}^* \wedge \mathbf{u}) \cdot \mathbf{e}_z] dV}{\int_{(V)} |\mathbf{u}|^2 dV}.$$

Using several times Schwarz inequality, it turns out that

$$|\omega| \leq \frac{\int_{(V)} |\mathrm{Im}[(\mathbf{u}^* \wedge \mathbf{u}) \cdot \mathbf{e}_z]| dV}{\int_{(V)} |\mathbf{u}|^2 dV} \leq 1 \tag{10}$$

since $|\mathrm{Im}[(\mathbf{u}^* \wedge \mathbf{u}) \cdot \mathbf{e}_z]| \leq \|\mathbf{u}^* \wedge \mathbf{u}\| \leq \|\mathbf{u}\|^2$. Thus again we find that inertial oscillations have a period larger than the semi-period of rotation.

Finally, let us note that when the spectrum exists[1] it is dense in the interval $[0, 1]$. A classical example of such a spectrum is

$$\frac{n}{\sqrt{n^2 + m^2}}, \qquad (n, m) \in \mathbb{N}^2$$

It is dense since any frequency in $[0, 1]$ can be approximated to any precision by a pair of integers.

---

[1] Mentioned without precision, the spectrum means the point spectrum of an operator, that is, all the elements $\lambda \in \mathbb{C}$ such that $\mathcal{L}f = \lambda f$, where f is square-integrable. For the Poincaré equation, the point spectrum is usually empty!

3.2 Rossby Waves

When discussing waves in rotating fluids one often thinks of Rossby waves. What are they? Just a sort of inertial modes actually. As they play an important part in planetary atmospheres, they are often called planetary waves.

The idea is the following: we are looking for waves propagating in a very thin pellicula like the atmosphere of the Earth. We are seeking two-dimensional solutions (vertical motions are inhibited or much smaller than horizontal ones). The dispersion relation of such waves cannot be extracted from the one of the inertial waves since we impose on these new waves an additional constraint, namely $v_z = 0$. As any dispersion relation requires a simplification by an amplitude; this amplitude cannot be zero; we thus need to derive the dispersion relation from the beginning. Equations of motion are

$$\begin{cases} i\omega\mathbf{v} + 2\boldsymbol{\Omega}(y) \wedge \mathbf{v} = -\boldsymbol{\nabla}P \\ \boldsymbol{\nabla} \cdot \mathbf{v} = 0 \end{cases} \tag{11}$$

Note that now the rotation vector depends on y which is the north–south coordinate. Here we take a local frame where the x-axis points to the east and the z-axis to the local zenith. Moreover, for two-dimensional motions, the vertical component of the rotation vector is the only useful component. We thus write

$$i\omega\mathbf{v} + 2\Omega(y)\mathbf{e}_z \wedge \mathbf{v} = -\boldsymbol{\nabla}P$$

where $\Omega(y) = \Omega \sin \lambda(y)$ and λ is the latitude; explicitly

$$\begin{cases} i\omega v_x - 2\Omega(y)v_y = -\dfrac{\partial P}{\partial x} \\[2mm] i\omega v_y + 2\Omega(y)v_x = -\dfrac{\partial P}{\partial y} \\[2mm] \dfrac{\partial v_x}{\partial x} + \dfrac{\partial v_y}{\partial y} = 0 \end{cases} \tag{12}$$

Pressure is eliminated for the vertical vorticity ζ; thus

$$i\omega\zeta = -2v_y \frac{d\Omega}{dy} \tag{13}$$

This equation shows that the latitude dependence of the rotation vector is crucial. We may now assume that $d\Omega/dy$ is constant; this is the so-called β-plane approximation, β being the gradient of planetary vorticity. With this assumption, and setting $(v_x, v_y) \propto \exp[i\omega t - ik_x x - ik_y y]$, we easily get the dispersion relation of the Rossby waves:

$$\omega = -\frac{2k_x}{k_x^2 + k_y^2}\left(\frac{d\Omega}{dy}\right) \tag{14}$$

This relation shows that $\omega/k_x < 0$ since $d\Omega/dy > 0$; thus Rossby waves are retrograde: they propagate in a counter-rotating way, to the west. The expression of their group velocity, namely,

$$\mathbf{v}_g = 2\frac{d\Omega}{dy}\left((k_y^2 - k_x^2)\mathbf{e}_x + 2k_xk_y\mathbf{e}_y\right)/k^4$$

shows that energy propagates in all the directions.

The dispersion relation of these waves shows that the latitudinal variation of the rotation rate is crucial. Moreover, we may observe from the momentum equation that even if the velocity field is that of a plane wave, this is not the case of the pressure perturbation since $\partial P/\partial x \neq ik_x P$.

As mentioned in introduction, it is clear that Rossby waves are just a specific type of inertial *mode* which meet some specific constraints like bidimensionality.

3.3 Planetary Waves

Let us consider now a global analysis of the Rossby perturbations on the whole surface of the sphere. We would call these modes *planetary modes*. Since the flow is incompressible and two dimensional, it can be described by a stream function $\chi(\theta, \varphi)$, such that

$$\mathbf{v} = \boldsymbol{\nabla} \wedge (\chi\mathbf{e}_r)$$

$\mathbf{e}_r$ being the radial unit vector of spherical coordinates. We obtain the equation controlling χ by applying the operator $\mathbf{e}_r \cdot \boldsymbol{\nabla}\wedge$ to (7). It turns out that

$$i\omega\mathbf{e}_r \cdot \boldsymbol{\nabla} \wedge^2 (\chi\mathbf{e}_r) + \mathbf{e}_r \cdot \boldsymbol{\nabla} \wedge (\mathbf{e}_z \wedge \mathbf{u}) = 0$$

which leads to

$$i\omega\Delta\chi + \frac{\partial\chi}{\partial\varphi} = 0$$

Now, if we expand the solutions onto spherical harmonics, namely

$$\chi = \sum_{\ell,m} \chi_m^\ell Y_\ell^m$$

we find that an eigenmode corresponds to each harmonic, with the eigen frequency $\omega_{\ell m}$ following the dispersion relation:

$$\omega_{\ell m} = \frac{m}{\ell(\ell+1)} \tag{15}$$

Note that we used the equation verified by spherical harmonics $\Delta Y_\ell^m = -\ell(\ell+1)Y_\ell^m$. We observe that the phase angular velocity is $-\omega/m = -1/\ell(\ell+1)$ and always negative.[2] Thus, just like Rossby waves, planetary waves propagate to the west.

---

[2] We indeed assumed that χ is proportional to $e^{i(\omega t+m\varphi)}$.

4 The Poincaré Equation

Taking the divergence of the momentum equation in (7), we find the equation of the pressure perturbation, namely

$$\Delta P - \frac{1}{\omega^2} \frac{\partial^2 P}{\partial z^2} = 0 \tag{16}$$

well known under the name of *Poincaré equation* since Cartan [2]. Since $\omega \leq 1$ this equation is spatially hyperbolic.

Before investigating the properties of this equation, let us make a stop on gravity modes, which also need the solutions of Poincaré equation.

4.1 A Brief Stop on Gravity Modes

Let us consider an incompressible stably stratified fluid (we use the Boussinesq approximation). Disturbances of the equilibrium verify the set of equations:

$$\begin{cases} \dfrac{\partial \delta \mathbf{v}}{\partial t} = -\dfrac{1}{\rho} \boldsymbol{\nabla} \delta p + \dfrac{\delta \rho}{\rho} \mathbf{g} \\[2mm] \dfrac{\partial \delta T}{\partial t} + \delta \mathbf{v} \cdot \boldsymbol{\nabla} T_0 = 0 \\[2mm] \boldsymbol{\nabla} \cdot \delta \mathbf{v} = 0 \end{cases} \tag{17}$$

where T_0 is the background temperature which we suppose to vary linearly with height z. We set

$$T_0 = T_{00} + \beta z \qquad \text{with} \quad \beta > 0$$

For small variations of temperature

$$\frac{\delta \rho}{\rho} = -\alpha \delta T$$

where α is the dilation coefficient of the fluid. The equations may be rewritten as

$$\begin{cases} i\omega \delta \mathbf{v} = -\boldsymbol{\nabla} \delta p - \dfrac{N^2}{i\omega} \delta v_z \mathbf{e}_z \\[2mm] \boldsymbol{\nabla} \cdot \delta \mathbf{v} = 0 \end{cases} \tag{18}$$

where we introduced the squared Brunt–Väisälä frequency, namely $N^2 = \alpha \beta g$. Assuming that N is constant and eliminating the velocity, we find that

$$\Delta_{x,y} \delta p - \left(\frac{\omega^2}{N^2 - \omega^2} \right) \frac{\partial^2 \delta p}{\partial z^2} = 0$$

which is the Poincaré equation, here again (note that for gravity modes $\omega \leq N$).

4.2 Properties of the Solutions of the Poincaré Equation

The first important point is that Poincaré equation is hyperbolic with respect to space coordinates.

4.2.1 A Reminder About Hyperbolic Equations

Second-order partial differential equations are divided into four categories: the elliptic, hyperbolic, parabolic and mixed types. This division is based on a property of the coefficients of the second-order partial derivatives. Let us consider the general form:

$$A(x,y)\frac{\partial^2 f}{\partial x^2} + B(x,y)\frac{\partial^2 f}{\partial x \partial y} + C(x,y)\frac{\partial^2 f}{\partial y^2} + \cdots = 0$$

The function $D(x,y) = B^2 - 4AC$ determines the type of the equation. If everywhere in the definition domain of f

- $D(x,y) > 0$, the equation is hyperbolic
- $D(x,y) = 0$, the equation is parabolic
- $D(x,y) < 0$, the equation is elliptic

whereas if $D(x,y)$ changes sign in the domain, the equation is said to be of *mixed type*. Examples:

- The wave equation is hyperbolic
- The heat equation is parabolic
- The equation of a potential is elliptic
- Tricomi's equation is of mixed type

4.2.2 The Consequences of Hyperbolicity

The Poincaré problem is ill posed in the sense of Hadamard: a hyperbolic problem is well posed when associated with Cauchy-type conditions, i.e. *initial conditions*. Boundary conditions are usually impossible to satisfy with $\mathcal{C}^\infty$ functions. We detail now some implications of ill-posedness.

(1) **Under-determination**
A first consequence of ill-posedness is that solutions are not fully determined. For instance, let us consider the wave equation

$$\frac{\partial^2 f}{\partial x^2} - \frac{1}{c^2}\frac{\partial^2 f}{\partial t^2} = 0$$

We all know that the general solution of this equation may be written as

$$f(x,t) = \Phi(x - ct) + \Psi(x + ct)$$

where Φ and Ψ are arbitrary functions. To be fully determined, they need two initial conditions, for instance,

$$f(x,0) = \cos x \qquad \text{and} \qquad \left(\frac{\partial f}{\partial t}\right)_0 = 0$$

which leads to

$$f(x,t) = \frac{1}{2}\left[\cos(x - ct) + \cos(x + ct)\right]$$

Now, just imagine that instead of asking for two initial conditions to be met, we had been asking for one initial condition and a condition at some later time, just like

$$f(x,0) = I(x) \qquad \text{and} \qquad f(x,T) = F(x)$$

where $I(x)$ and $F(x)$ are given data. This problem is mathematically ill posed and the solution cannot be fully specified. Indeed, we find that Ψ just has to satisfy

$$\Psi(x) = \Psi(x + 2cT) + F(x + 2cT) - I(x)$$

which means that this function needs to be given in the interval $[0, 2cT]$.

(2) **Infinite degenerescence**
An ill-posed problem may also have infinitely degenerate eigenvalues. For example, if one solves the (two-dimensional) Poincaré equation in a rectangle, namely

$$\frac{\partial^2 \psi}{\partial x^2} + \left(1 - \frac{1}{\omega^2}\right)\frac{\partial^2 \psi}{\partial z^2} = 0, \qquad \psi = 0 \quad \text{on} \quad \partial D$$

the classical solution is

$$\psi_{mn}(x, z) = A_{mn} \sin m\pi x \, \sin n\pi z$$

and

$$\omega_{mn}^2 = \frac{n^2}{m^2 + n^2} \tag{19}$$

The eigenvalues are infinitely degenerate because

$$\omega_{mn} = \omega_{jm,jn} \qquad \forall j \in \mathbb{N}$$

and each eigenmode is arbitrary

$$\psi_{mn}(x, z) = \sum_{j=1}^{\infty} a_j \sin jm\pi x \sin jn\pi z$$

(3) **Singularities**
If $D(x, y) = B^2 - 4AC > 0$, we can "factorize" the second-order terms; hence,

$$A\frac{\partial^2 f}{\partial x^2} + B\frac{\partial^2 f}{\partial x \partial y} + C\frac{\partial^2 f}{\partial y^2} + \cdots = 0$$

is changed into

$$\left(a\frac{\partial}{\partial x} + b\frac{\partial}{\partial y}\right)\left(a'\frac{\partial}{\partial x} + b'\frac{\partial}{\partial y}\right) + \cdots = 0$$

which means that there exists a coordinate system (u, v) such that

$$\frac{\partial^2 f}{\partial u \partial v} + \cdots = 0$$

with

$$\frac{\partial}{\partial u} = a\frac{\partial}{\partial x} + b\frac{\partial}{\partial y}$$

$$\frac{\partial}{\partial v} = a'\frac{\partial}{\partial x} + b'\frac{\partial}{\partial y}$$

(u, v) are the *characteristic coordinates* and the curves $u = \text{Cte}$ and $v = \text{Cte}$ are the characteristic curves. They are two independent families of curves determined by the equations

$$\frac{dy}{dx} = \frac{b}{a} \quad \text{and} \quad \frac{dy}{dx} = \frac{b'}{a'}$$

Let us now illustrate the foregoing discussion by an example where we impose boundary conditions to a wave-type equation. A typical situation is illustrated in Fig. 1. From this figure we see that

- If Φ is given in $[0, 2cT]$ then it is known in $[2cT, 4cT]$; for instance, if we give Φ_0 at $x = 0, t = 0$, then we determine Ψ_1 at $t = T$, from which we find Φ_2 at $t = 0$ and $x = 0 + 2cT$.
- If Φ is known in $[0, 2cT]$, then $f(x, t)$ is known at any point of the domain. As shown in Fig. 1, the values needed to make $f(M)$ are those of $\Phi_{M'}$ and $\Phi_{M''}$.
- If we isolate a rectangle, as in Fig. 2, limited on the sides by line segments at $x = \text{Cst}$ where f is given, we immediately find the possibility of a contradiction between the values of f on one side and those on the other side. Indeed, in this figure we see that the values of Φ_R and Φ_G, given on the red segment $[0, 2cT]$, control the value of f at points M_0 and M_1. However, the values of f at these points are data, which need not be compatible with both Φ_R and Φ_G. If they are not, the solution must have a discontinuity somewhere.

4.2.3 Back to Poincaré

In three dimensions characteristic lines are replaced by characteristic surfaces. For the Poincaré equation, the characteristic surfaces are cones of equation

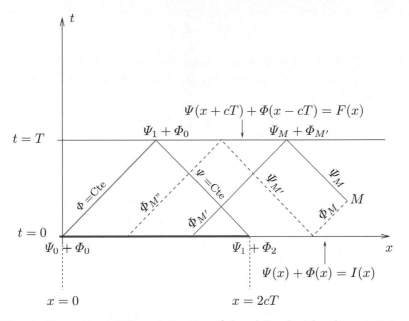

Fig. 1. Illustration of the propagation of Φ and Ψ values by characteristics

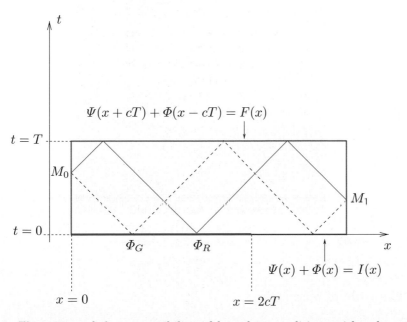

Fig. 2. Illustration of the compatibility of boundary conditions with a hyperbolic equation. If Φ_G and Φ_R are given, the values of the solution are known in M_0 and M_1. But $f(M_0)$ and $f(M_1)$ are independent data which may not result from the same values of Φ_G and Φ_R

$$z = \pm\sqrt{\frac{1 - \omega^2}{\omega^2}}r + C \qquad (20)$$

Of course, in the meridional plane, they appear as straight lines which make an angle $\vartheta = \arcsin\omega$ with the rotation axis. These lines show the direction of energy propagation.

4.2.4 Attractors

Characteristics bounce on the boundaries of the domain and we may follow their trajectories just like those of a dynamical system in a phase space. However, this dynamics is very simple since the direction of the characteristics has only two possibilities: $\pm\vartheta$. In general, the trajectories converge towards an attractor as shown in Fig. 3. Attractors are specific to a container. The case of the spherical shell, more appropriate to astrophysical bodies, has been investigated in detail in Rieutord et al. [13]. In the inertial frequency range $[0, 2\Omega]$, they may be characterized by the variations of their Lyapunov exponents.

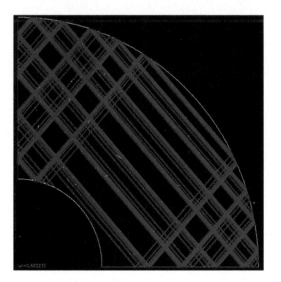

Fig. 3. Convergence of characteristics towards an attractor

4.3 The Role of Viscosity

The foregoing discussion ignored viscosity; however, as one can imagine, singularities are smoothed out by viscosity. Mathematically, when viscosity is restored, the problem is well posed and solutions are said to be regularized. Nevertheless, one may wonder whether the singularities associated with attractors leave some signature in the viscous solutions. The answer is definitely yes, provided the

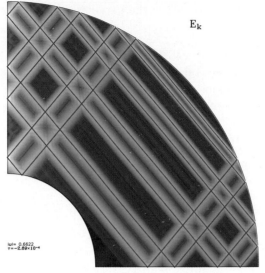

E_k

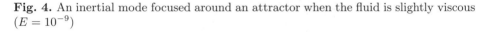

|ω|= 0.6622
τ=-2.69×10⁻⁴

Nr=450 L=1300 M=0⁻ E=1.0×10⁻⁹ η=0.350 CL=ff

Fig. 4. An inertial mode focused around an attractor when the fluid is slightly viscous $(E = 10^{-9})$

viscosity is sufficiently small. As shown in Fig. 4, the eigenmode concentrates along the path of characteristics defined by the attractor: it generates an oscillating detached shear layer. The physical interpretation is that a wave packet launched randomly in the container will rapidly be focused along the attractor. As time advances, it gets closer and closer to the attractor while its wavenumber increases until diffusion effects are strong enough to balance the contraction of the mapping made by the characteristics. This scenario has been used to derive analytical solutions of eigenmodes controlled by attractors (see Rieutord et al. [14]) in a two-dimensional case. In a thin shell, representing, for instance, the Earth's atmosphere, attractors may be confined to equatorial regions and three-dimensional equations can be simplified into two-dimensional ones.

The structure of a detached shear layer (in the two-dimensional case) can be derived from the system

$$\begin{cases} \lambda \mathbf{u} + \mathbf{e}_z \wedge \mathbf{u} = -\boldsymbol{\nabla} P + E \Delta \mathbf{u} \\ \boldsymbol{\nabla} \cdot \mathbf{u} = 0 \end{cases} \tag{21}$$

which controls the shape of inertial modes when there is viscosity (here given by the Ekman number E). It turns out that the velocity in a shear layer is determined by the following differential equation:

$$\frac{d^2 u}{dz^2} - \left[\frac{1}{4} z^2 + e^{i\pi/4} B \left(\frac{p}{2\alpha_0 A f_{20}} \right)^{1/2} (\tau_1 \pm i\omega_1) \right] u = 0$$

where z is the coordinate across the layer. This is actually the Schrödinger equation of a quantum particle trapped in a parabolic well, i.e. the famous harmonic oscillator. Its solutions are the parabolic cylinder functions:

$$u = U(a, z), \quad a = -n - \frac{1}{2} = e^{i\pi/4} B \left(\frac{p}{2\alpha_0 A f_{20}} \right)^{1/2} (\tau_1 \pm i\omega_1)$$

As shown in Rieutord et al. [14], the matching between these analytical solutions and the numerical ones is perfect. These solutions give for the first time (to our knowledge) an explicit example of the regularization of an operator.

4.4 The Critical Latitude Singularity

The singularities issued from the characteristics attractors are very strong; solutions are neither integrable nor square-integrable. However, they are not the only singularities: some are associated with the boundary conditions. Indeed, for the inviscid case the velocity verifies $\mathbf{v} \cdot \mathbf{n} = 0$ on the boundaries, which is equivalent to

$$-\omega^2 \mathbf{n} \cdot \boldsymbol{\nabla} P + (\mathbf{n} \cdot \mathbf{e}_z)(\mathbf{e}_z \cdot \boldsymbol{\nabla} P) + i\omega (\mathbf{e}_z \wedge \mathbf{n}) \cdot \boldsymbol{\nabla} P = 0 \qquad (22)$$

for the pressure. This condition is neither of Dirichlet type nor of Neumann type. It is referred to as "oblique derivatives". It generates singularities at the so-called critical latitude which is where the characteristics are tangential to the boundary. This singularity is weaker than the attractor one [13] and usually manifests itself in a thickening of the Ekman boundary layer (e.g. Roberts and Stewartson [15]).

4.5 A Remark on Gravity Modes

We have shown in Sect. 4.1 that gravity modes are also governed by a hyperbolic equation. One may thus wonder why singularities of attractors have never been mentioned in the astrophysical literature. The reason is that gravity modes have essentially been considered in non-rotating stars that are taken as perfect spheres. In such a geometry, the spherical symmetry of the problem makes the partial differential equations separable. Solutions are just the product of one-dimensional solutions which are regular. Singularities disappear. We see that this situation is very specific and that singularities are rather the rule than the exception.

Finally, let us mention that singular gravity modes associated with attractors have been observed experimentally by Maas et al. [9] and are not a pure conjecture of theoretical work!

5 The Gravito-inertial Modes

In stars and other natural systems rotation and stable stratification usually act together. Since gravity modes and inertial modes are low-frequency modes, they always combine in the spectral range $[0, \sqrt{(2\Omega)^2 + N_{\max}^2}]$. In slowly rotating stars, $N_{\max} \gg 2\Omega$ and there is a large number of gravity modes which are little affected by rotation; however, in rapidly rotating stars like γ-Dor, the

Brunt–Väisälä frequency, and the Coriolis frequency are of the same order of magnitude. The low-frequency modes need therefore a non-perturbative approach.

5.1 The Mathematical Side

The first question to be answered is how Poincaré equation is transformed when a stable stratification is combined with rotation. A simple way towards the answer is to consider a rotating radially stratified fluid in a sphere, and its small amplitude perturbations (e.g. Dintrans et al. [5]). Time-periodic disturbances are solutions of

$$\begin{cases} \boldsymbol{\nabla} \cdot \mathbf{v} = 0 \\ i\omega\mathbf{v} + 2\boldsymbol{\Omega} \wedge \mathbf{v} = -\boldsymbol{\nabla}p - \alpha T\mathbf{g} \\ i\omega T + \mathbf{v} \cdot \boldsymbol{\nabla}T_0 = 0 \end{cases} \tag{23}$$

where α is the dilation coefficient and T_0 the background temperature which we take such that $\boldsymbol{\nabla}T_0 = \beta(r)\mathbf{e}_r$. The local gravity is $\mathbf{g} = -g(r)\mathbf{e}_r$. When the temperature fluctuation is eliminated in favour of the velocity, the momentum equation reads as follows:

$$i\omega\mathbf{v} + 2\boldsymbol{\Omega} \wedge \mathbf{v} = -\boldsymbol{\nabla}p + \frac{N^2(r)}{\omega}iv_r\mathbf{e}_r \tag{24}$$

Taking the divergence of this equation, we get the generalization of the Poincaré equation. We write its second-order terms for axisymmetric perturbations:

$$(\omega^2 - N^2(r)\cos^2\theta)\frac{\partial^2 P}{\partial s^2} + 2N^2(r)\sin\theta\cos\theta\frac{\partial^2 P}{\partial s\partial z}$$

$$+(\omega^2 - N^2(r)\sin^2\theta)\frac{\partial^2 P}{\partial z^2} + \cdots = 0$$

where s is the radial cylindrical coordinate. These terms show the nature of the operator: it is of mixed type. The equation of the critical surfaces which separate the hyperbolic regions from the elliptic ones is

$$\omega^4 - (N^2(r) + 4\Omega^2)\omega^2 + 4\Omega^2 N^2(r)\cos^2\theta = 0$$

In some standard models (e.g. Rieutord [12]; Chandrasekhar [3]), $N \propto r$. In this case, critical surfaces are ellipsoid or hyperboloids.

In the hyperbolic regions, characteristics are no longer straight lines (in the meridional plane), but curves given by the following differential equation (e.g. Friedlander and Siegmann [8] Dintrans et al. [5]):

$$\frac{dz}{ds} = \frac{zsN^2 \pm \xi^{1/2}}{\omega^2 - N^2z^2}, \qquad \xi = \omega^2 N^2 s^2 + (1 - \omega^2)(\omega^2 - N^2z^2) \tag{25}$$

Nevertheless, as before, the general rule is that they focus onto attractors. The novelty is that they are of two kinds: either limit cycles of characteristics as before or a wedge made by a critical surface and a boundary. We give two examples of these modes in Fig. 5, for a Boussinesq model; others may be found in Dintrans et al. [5].

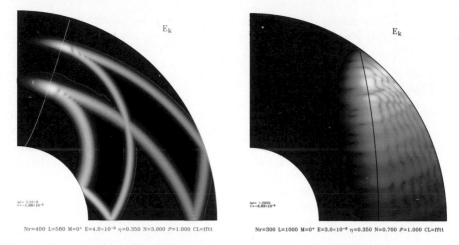

Fig. 5. Two examples of gravito-inertial modes confined in the hyperbolic region of the domain, shown by their kinetic energy amplitude in a meridional plane

5.2 Gravito-inertial Modes in Stellar Models

The foregoing discussion may be extended to more realistic models of stars. This exercise was done in Dintrans and Rieutord [4] where we computed the gravito-inertial modes in a model of a 1.8 $M_\odot$-ZAMS star. This mass is typical of the γ-Doradus stars.

By scanning the gravito-inertial frequency band, attractors have also been detected. They are limit-cycle attractors. However, we noticed some differences with the Boussinesq case, namely, the frequency bands where attractors exist are noticeably narrower. The origin of this property is not clear yet, and may come from the "distance" between the Brunt–Väisälä frequency, and the Coriolis frequency, i.e. for large-scale gravito-inertial modes, gravity dominates over rotation because $N_{\mathrm{max}} \gg 2\Omega$.

6 How Can We Compute These Modes?

To end this lecture, I would like to briefly address the numerical side of the subject. This is indeed a delicate question as the problem is two dimensional and therefore involves large matrices.

The general form of the problem may be appreciated with the example of inertial modes. If we take the curl of the momentum equation in (21), we find that the velocity field verifies

$$\lambda \boldsymbol{\nabla} \wedge \mathbf{v} = E \Delta \boldsymbol{\nabla} \wedge \mathbf{v} + \boldsymbol{\nabla} \wedge (\mathbf{e}_z \wedge \mathbf{v}), \qquad \boldsymbol{\nabla} \cdot \mathbf{v} = 0$$

completed by boundary conditions. In a more symbolic form, this problem is a generalized eigenvalue problem, like

$$\mathcal{L}(f) = \lambda\mathcal{M}(f)$$

where $\mathcal{L}$ and $\mathcal{M}$ are partial differential operators and λ is the eigenvalue.

6.1 The Grid

The first step in the numerical resolution is to decide about the discretization. In all the examples shown, we used spectral methods. These methods are indeed very appropriate since a discrete approximation of the functions is made in the most compact way. Thus, matrices have the smallest size for the required precision. This will appear as a key parameter.

Hence, for the horizontal part, we use a spherical harmonic expansion, while for the radial dependence, we use the Gauss–Lobatto grid, which is associated with Chebyshev polynomials. Typically, a function $f(r, \theta, \varphi)$ is discretized in the following way:

$$f(r, \theta, \varphi) \equiv \sum_{\ell, m} f_m^\ell(r_i) Y_\ell^m(\theta, \varphi)$$

The set of $f_m^\ell(r_i)$ constitutes the discrete representation.

6.2 The Generalized Eigenvalue Problem

Once the discretization is fixed, the eigenvalue problem takes the form of the algebraic generalized eigenvalue problem $[A]\mathbf{x} = \lambda[B]\mathbf{x}$. This new problem can be solved numerically in three ways, typically. The first one is the brute force of the QR (real) or QZ (complex) algorithm where all the eigenvalues of the system are computed. Obviously, this can be done for small-size matrices only; the reasons are that the QR/QZ algorithm uses full matrices; thus the memory requirement is rapidly prohibitive as well as the computing time which grows like N^4, N being the rank of the matrices.

When large sizes are necessary (for small diffusivities for instance), methods which determine a few eigenvalues of the spectrum are to be preferred. Indeed, the determination of the full spectrum does not make sense physically since we are usually interested in the least-damped modes; these modes are the ones which may be observed.

The method that we advise belongs to the Krylov-type methods, which iteratively determine the eigenvalues of some low-dimensional sub-space. We use the Arnoldi–Chebyshev algorithm (e.g. Valdettaro et al. [17]). In the same vein let us also mention the Jacobi–Davidson method which has been investigated more recently and may be more appropriate to parallel computers (e.g. Sleijpen and Van der Vorst [16]).

6.3 Errors

However, once a solution is obtained, one needs to be sure that it is valid and not a spurious one. In other words, we must ascertain that the numerical error is negligible.

This error contains two independent sources of errors: the truncation error and the round-off error. The truncation error is the most obvious: spectral solutions are expansions in polynomials of higher and higher orders. Numerically, we use a finite number of such polynomials (equivalently, we use a finite number of grid points) and some difference remains with the exact solution. This *truncation* or *spectral* error is easily appreciated with a spectrum of the numerical solution as the ones shown in Fig. 6. These two spectra represent the numerical solution of Fig. 4. We see the convergence of the solution on the Chebyshev polynomials and spherical harmonics basis. The truncation error is therefore $\sim 10^{-5}$ (and less for other parts of the solution).

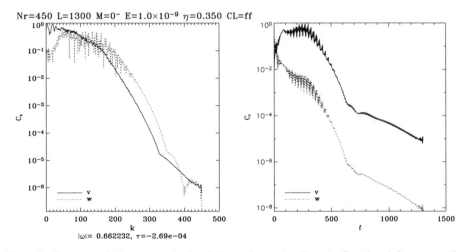

Fig. 6. Spectra of the numerical solution shown in Fig. 4. On the *left* we see the amplitude of the Chebyshev coefficients (for the radial and azimuthal velocities). On the *right*, we show the amplitudes of the spherical harmonics components for the same quantities

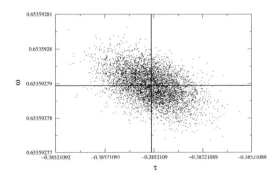

Fig. 7. Effects of the round-off error on the computation of an eigenvalue. The correct value, given by the cross, has been computed using extended precision computations (28 digits)

In the foregoing example, we may say that the numerical solution is spectrally converged. However, this solution may be pure junk if round-off errors dominate. Indeed, all calculations are done with a finite number of digits (typically 16 in double precision); unfortunately, it is quite common in numerical linear algebra that the 10^{-16} errors on the data are amplified by a factor 10^{16} on the result. This comes from the conditioning of the matrices, which may be very bad. Such huge amplifications occur especially with ill-conditioned operators of large size. One way of estimating the round-off error on a numerical result consists in modifying the input data with a 10^{-16} noise. An example is given in Fig. 7 for the computation of an eigenvalue. A more detailed discussion, with examples, may be found in Valdettaro et al. [17].

7 Conclusions

To conclude this lecture, I would like to stress a few properties of the low-frequency spectrum of rotating stars, and some points in computing the associated eigenvalues and eigenmodes.

- In a general set-up, gravito-inertial modes, as understood by physicists, do not exist in an adiabatic approach. This is because the operator governing the eigenvalue problem is either spatially hyperbolic or of mixed type and is rarely compatible with boundary conditions.
- Diffusivities (viscosity or heat diffusion) therefore play an important part in the dynamics of the system. They regularize the singularities and control the size of the associated shear layers.
- When computing such modes, the singularity of the adiabatic limit shows up in the bad conditioning of the matrices of the associated linear systems. A careful control of the round-off error is therefore needed to obtain sensible results. Our experience is that the diffusion coefficients like viscosity are quite appropriate to improve the conditioning of the linear operators. Relying on numerical diffusion would be very hazardous.

Realistic numbers are often beyond reach of numerical solutions in astrophysical problems (just think of the Reynolds number in the convective zones of stars). In the case of the eigenspectrum of a star, it turns out that the recent progress of computers' power together with that of numerical methods makes the astrophysical regime within (indirect) reach of the calculations. Indeed, although the brute force computation is often not possible, the determination of the asymptotic laws governing a given eigenmode when diffusion numbers become small is possible in most cases. Thus, difficulties are to be expected from the models rather than from the eigenmode computation, although, as we have shown, this calculation is not an easy game.

Acknowledgment

I am very grateful to Coralie Neiner and Jean-Pierre Rozelot for the smooth organization of this school and the fruitful time spent there. The high-resolution computations, illustrating this lecture, have been realized on the NEC-SX8 of the "Institut du Développement et des Ressources en Informatique Scientifique" (IDRIS) which is gratefully acknowledged.

References

1. Aufdenberg, J.P., Mérand, A., Coudé du Foresto, V., Absil, O., Di Folco, E., Kervella, P., Ridgway, S.T., Berger, D.H., Brummelaar,T.A.t., McAlister, H.A., Sturmann, J., Sturmann, L., Turner, N.H. First Results from the CHARA array. VII. Long-baseline interferometric measurements of Vega consistent with a pole-on, rapidly rotating star. ApJ **645**, 664–675 (2006)
2. Cartan, E.: Sur les petites oscillations d'une masse fluide. Bull. Sci. Math. **46**, 317–352, 356–369 (1922)
3. Chandrasekhar, S.: Hydrodynamic and Hydromagnetic Stability. Clarendon Press, Oxford (1961)
4. Dintrans, B., Rieutord, M.: Oscillations of a rotating star: a non-perturbative theory. A&A **354**, 86–98 (2000)
5. Dintrans, B., Rieutord, M., Valdettaro, L.: Gravito-inertial waves in a rotating stratified sphere or spherical shell. J. Fluid Mech. **398**, 271–297 (1999)
6. Domiciano de Souza, A., Kervella, P., Jankov, S., Abe, L., Vakili, F., di Folco, E., Paresce, F. The spinning-top Be star Achernar from VLTI-VINCI. A&A **407**, L47–L50 (2003)
7. Domiciano de Souza, A., Kervella, P., Jankov, S., Vakili, F., Ohishi, N., Nordgren, T.E., Abe, L.: Gravitational-darkening of Altair from interferometry. A&A **442**, 567–578 (2005)
8. Friedlander, S., Siegmann, W.: Internal waves in a rotating stratified fluid in an arbitrary gravitational field. Geophys. Astrophys. Fluid Dyn. **19**, 267–291 (1982)
9. Maas, L., Benielli, D., Sommeria, J., Lam, F.P.: Observation of an internal wave attractor in a confined, stably stratified fluid. Nature **388**, 557–561 (1997)
10. Monnier, J.D., Zhao, M., Pedretti, E., Thureau, N., Ireland, M., Muirhead, P., Berger, J.P., Millan-Gabet, R., Van Belle, G., ten Brummelaar, T., McAlister, H., Ridgway, S., Turner, N., Sturmann, L., Sturmann, J., Berger, D.: Imaging the Surface of Altair. Science **317**, 342–, arXiv:0706.0867 (2007)
11. Peterson, D., Hummel, C., Pauls, T., Armstrong, J., Benson, J., Gilbreath, G., Hindsley, R., Hutter, D., Johnston, K., Mozurkewich, D., Schmitt, H.: Resolving the effects of rotation in Altair with long-baseline interferometry. ApJ **636**, 1087–1097 (2006)
12. Rieutord, M.: The dynamics of the radiative envelope of rapidly rotating stars. i. a spherical boussinesq model. A&A **451**, 1025–1036 (2006)
13. Rieutord, M., Georgeot, B., Valdettaro, L.: Inertial waves in a rotating spherical shell: attractors and asymptotic spectrum. J. Fluid Mech. **435**, 103–144 (2001)
14. Rieutord, M., Valdettaro, L., Georgeot, B.: Analysis of singular inertial modes in a spherical shell: the slender toroidal shell model. J. Fluid Mech. **463**, 345–360 (2002)
15. Roberts, P., Stewartson, K.: On the stability of a maclaurin spheroid of small viscosity. ApJ **137**, 777–790 (1963)
16. Sleijpen, G., Van der Vorst, H.: A Jacobi-Davidson iteration method for linear eigenvalue problems. SIAM Rev. **42**, 267–293 (2000)

17. Valdettaro, L., Rieutord, M., Braconnier, T., Fraysse, V.: Convergence and round-off errors in a two-dimensional eigenvalue problem using spectral methods and arnoldi-chebyshev algorithm. J. Comput. Appl. Math. **205**, 382–393, physics/0604219 (2007)
18. van Belle, G.T., Ciardi, D.R., Thompson, R.R., Akeson, R.L., Lada, E.A.: Altair's Oblateness and Rotation Velocity from Long-Baseline Interferometry. ApJ **559**, 1155–1164 (2001)

The Rotation of the Solar Core

S. Turck-Chièze

Service d'Astrophysique/IRFU/CEA and UMR AIM
CE Saclay, 91191 Gif sur Yvette, France
sylvaine.turck-chieze@cea.fr

Abstract In order to get a unified representation of stars, one needs to introduce the internal dynamical processes in the stellar structure equations. The validation of these complex equations supposes a proper reproduction of the helioseismic observations. Indeed the helioseismic discipline can provide today a crucial insight into the solar internal rotation profile, thanks to acoustic modes and to the first gravity mode detections. This information largely improves the previous situation where only external stellar rotation rates or abundance anomalies were accessible. In this review, I summarize first the theoretical studies and our recent results; then I present the respective role of acoustic and gravity modes and I show the solar rotation profile deduced from the instruments onboard SoHO with its uncertainties. The confrontation of these results with the recent theoretical developments exhibits important differences which demonstrate that some complementary work is necessary on both observational and theoretical sides.

1 Introduction

Dynamical processes are present in stars all across the Hertzsprung–Russell diagram. They are clearly visible in young stars; these stars are generally rapid rotators which then decelerate after their dissociation from their initial disk. Rotation plays also a major role in massive stars and more specifically in the final stages of evolution. It is also possible to measure now strong deformation of stars which can no more be treated like spherical objects; they are generally strong rotators with often strong stellar winds. Moreover as most of the celestial bodies are rotating, dynamo action produces or amplifies magnetic field in stars, planets and galaxies. But different kinds of dynamo must be studied, and it is now believed that the initial conditions are important to understand stellar and galactic magnetic fields [1]. We need to determine magnetic field saturation growth and the stability of the magnetic configurations [2]. We have today a very poor information on this difficult ingredient of astrophysics and it is partly why it is absent from most of the present simulations. Nevertheless, its role is extremely crucial in star formation and we need to consider it to reproduce the extension of the convective zone in young stars [3, 4], and probably to model massive presupernovae. Moreover, it begins to be an important ingredient to precisely estimate the Sun–Earth relationship and to predict coming and possibly different types of solar cycle(s) [5].

So there is a strong motivation to introduce these dynamical processes in stellar evolution in order to use the same equations to describe the different stages of stellar evolution. Such activity has been mainly developed for specific stars, mainly massive stars [6, 7]. The formalism has been also developed for solar-like stars [8], but there must be some interplay between rotation, magnetic field and internal gravity waves [9, 10] which justifies to use some observational constraints to validate theoretical secular dynamical models. The first step is evidently the introduction of the effects resulting from the rotation of the Sun. So, in the next section I show how the solar model evolves from the standard to the dynamical one and what kind of theoretical internal rotation and sound speed we get. In Sect. 3, the potentiality of the helioseismic disciple is shown and what kind of profile we have already got from the acoustic and now gravity mode detections in the solar core. In Sect. 4, I discuss the comparison between the two approaches and the direction of improvements is discussed in Sect. 5.

2 Evolution of the Solar Modeling

Present solar luminosity, radius, mass and surface abundances have guided the determination of the ingredients of the calibrated solar structure model. The temporal evolution of the Sun depends on the nuclear reaction rates which produce the nuclear energy and on the way this energy is transported to the solar surface by radiation or convection. It is not so easy to leave this rather "simple" way to determine the solar structure model because the step beyond is difficult. Indeed we know that the four structural equations are not sufficient to describe the sudden phenomena that we observe in UV. High-performance computers are not yet able to solve the complete MHD equations in 3D for an evolving Sun, assuming simultaneously the internal dynamical motions and the whole detailed microphysics. So we need to progress step by step. At the dawn of the strong asteroseismology development, it is interesting to notice that the Sun is a unique case extremely useful to validate the introduction of the dynamical processes.

2.1 The Standard Solar Model: SSM

The ingredients of the classical structural equations are the nuclear reaction rates, the opacity coefficients, the equation of state and the initial composition. The radiative zone represents 98% of the total mass of the Sun and uses most of these ingredients. This shows its importance; the equilibrium between gravitational energy, nuclear energy production and the energy escaping by photon interaction is governed mainly by this region on long timescales. Nuclear cross-sections have been measured in laboratory during at least three decades and the extrapolation toward the stellar plasma conditions has been largely studied and measured in one specific case. The sound speed in the core extracted from SoHO has put a real constraint on the fundamental reaction proton + proton and on the Maxwellian distribution [11]. So one may consider that the reaction rates are reasonably under control now. For the two other ingredients (opacity coefficients and equation of state) the knowledge is purely theoretical and must be checked. See the reviews [12, 13] which describe the different inputs and their respective knowledge.

The conditions of temperature and density in the radiative zone ensure that the plasma is totally ionized for its main constituents: hydrogen and helium but heavier species such as iron and then silicon down to oxygen are considered as partially ionized. The bound–bound interaction of photons with matter is very efficient to evacuate the energy produced in the first radial quarter (practically half the solar mass). This kind of contribution is highly sensitive to the metal content ($\propto Z^4$), so it is necessary to calculate this interaction for *all* the elements present in the Sun (from hydrogen to iron). The small amount of iron (some 10^{-4} of hydrogen in fraction number) contributes to about one-fifth of the opacity cross-section in central conditions. This point shows the important role of the detailed knowledge of the internal composition but also how they interact on opacity coefficients and on the gravitational settling [14]. Up to now the details of the ion interactions have never been verified except indirectly through acoustic pulsation eigenmodes which probe plasma properties throughout the Sun, but the predictions (neutrinos and acoustic mode frequencies) of the standard model have been regularly compared to the observations with already great successes. The differences have been always important to progress [5, 15].

2.2 The Seismic Solar Model: SeSM

The quality of the seismic observations has allowed to build a *seismic model* which reproduces the measured solar sound speed [16, 17] in the context of classical stellar evolution. The interest of such a model comes from the idea that the framework of the standard model is certainly too crude for reproducing all the existing observables. From the seismic model, we predict observables like neutrinos or gravity modes deduced not uniquely from the classical assumptions but which reproduce also the observed sound speed [18, 19]. This model allows us to avoid any conflict with new updates such as the recent reestimate of the heavy element mass fraction which deteriorates the agreement with the observed sound speed. This update is often considered at the origin of a crisis in helioseismology these last years; in fact it encourages a lot of studies to properly estimate the internal composition and its impact on the solar model. So, there is no reason to consider the present situation better or worse than before because as far as we know, the standard model is not the final representation of the real Sun.

2.3 The Dynamical Solar Model: DSM

Dynamical processes have been first introduced to describe massive stars and then have been applied to the case of the Sun. Pinsonneault and collaborators

[20, 21] have first treated the rotation effect via a diffusion equation and predicted a large amount of differential rotation in the solar interior. Later studies, using a refined version of rotational mixing in which the advective nature of the Eddington-Sweet meridional circulation is taken into account, reached the same conclusion [22]. In these calculations, the evolution of a 1 $M_\odot$ star is followed in introducing a high rotation rate at the beginning of the calculation (typically 50 km/s) rapidly slowed down at the surface of the star to mimic what one observes in the young clusters.

Other authors computed the effect of a static fossil dipolar magnetic field on the solar rotation. They showed that this magnetic field indeed spins down the radiative zone if it is disconnected from the convection zone [23, 24]. Eggenberger et al. [25] showed that they reproduce better the flat solar radiative rotation rate by introducing the magnetic instability of the Tayler-Spruit dynamo but they do not produce a core increase and this kind of instability they use is still in debate [26]. In parallel, 3–D MHD calculations of portions of the Sun have been undertaken in order to reproduce the seismic observations. The first 3–D MHD simulations of the radiative zone have been performed, including the differential rotation of the tachocline but not the rotation of the core. They show that the fossil dipolar field diffuses outward during the main sequence and connects to the surface convection zone, imprinting its differential rotation to the radiative core [27].

2.4 Building the DSM Step by Step

Up to recently, very few works have been dedicated to detailed solar calculations including the dynamical processes except the work of Yang and Bi which attacks the problem in its different aspects but does not take into account yet the advective term in the transport of the chemical species [28]. Nevertheless, in a rotating star, some large-scale meridional circulations in both the convective and radiative zones [29] result from the fact that it is not possible to keep simultaneously the hydrostatic equilibrium and the thermal balance. These motions have an impact on the transport of momentum and chemical through the advective-diffuse equations established by [30].

Today, it becomes interesting to show the role of the different terms because more and more constraints come from helioseismology. We are presently studying the impact of the rotation history on the different observables: neutrinos, rotation profile and sound speed in resolving the different equations in two codes: the CESAM and the STAREVOL codes [31]. We notice that the evolution of the rotation along time is important for determining the present rotation profile. As it is well known that the solar magnetic torquing arrives during the premain sequence, we have computed two kinds of models illustrated in Fig. 1, first a pure theoretical model beginning during this stage where the star is not totally contracted; our objective was to reach the present superficial rotation without imposing any high initial high rotation and consequently any braking when the star and the disk are decoupled. This objective is satisfied in beginning the evolution with an early low and nearly flat rotation profile. One notices in Fig. 1

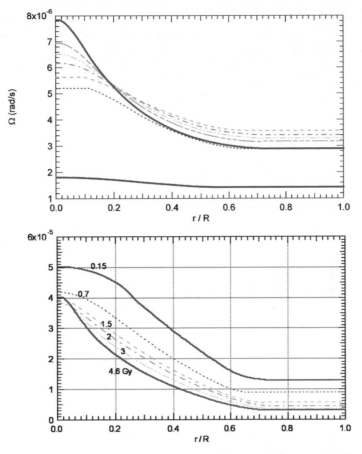

Fig. 1. Evolution of the internal solar rotation profile with different initial hypotheses. In the first figure, the initial external rotation (*red full line* between 1.8 and 1.2×10^{-6} rad/s at the *bottom*) is chosen to get the superficial rotation of 2 km/s at the present time. The solar internal rotation is shown for ages 0.010, 0.025, 0.050, 1, 2, 3, 4.6 Gyrs (from *bottom* to the *top*, initial and final profiles in *red full lines*) beginning the computation from the PMS, but ignoring any braking. In the second case, we impose a uniform 20 km/s rotation rate at the beginning of the evolution, quickly slowed down at 30 million years to reach 2 km/s at the present time. The lines correspond to age from 0.15 to 4.6 Gyrs. From [31]

(top) that such a model leads to a radial differential profile in the radiative zone which increases with time due to the role of the advective term. In the second case, Fig. 1 (bottom), we impose a rapid rotation like that observed in young stars and we decelerate the surface star at 30 million years when the star is decoupled from its disk [4, 32, 33]. In the first case, the resulting present central rotation is 5 times smaller than in the second case. This work demonstrates that the present central rotation does not reflect evidently a reduction of its birth value. The slope of the rotation profile of a 1 $M_\odot$ at the present age is rather similar for the different cases, but the central value largely depends on the history of the rotation rate during the first billion year.

The resulting solar sound speed and density profiles are examined in parallel [31]. The effect is small and slightly increases the difference with the observed profile. A quantitative estimate of this change depends largely on the initial conditions of the rotation profile and its story.

3 The Constraints Coming from Helioseismology

Helioseismology has been largely developed during the last 30 years. Two kinds of modes are important to detect: the acoustic modes and the gravity modes. The first one, excited by the superficial granulation, is very sensitive to the superficial layers, but thanks to the detection of million modes including quadrupolar, dipolar and radial modes, it is now possible to extract a static information from the surface to the core through the sound speed and density profiles. Moreover, because of the rotation of the Sun, the determination of the splitting of the modes gives access to the rotation profile down to the limit of the nuclear core. On the other side, the gravity modes, for which the restoring force is the gravity, are mainly sensitive to the real center or the whole nuclear core. To illustrate the respective role of the two kinds of modes, Fig. 2 shows [34] two rotation kernels that determine the sensitive solar regions corresponding to a quadrupolar acoustic mode of a specific order n=6 for the lowest detected acoustic mode splitting and the quadrupolar gravity mode n= −3 for the potentially detected gravity mode candidate (see below).

Helioseismology was already a mature discipline when SoHO had been launched. The theoretical framework was developed and ground networks (GONG,

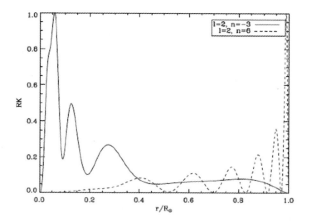

Fig. 2. Rotation kernels estimated in using the outputs of the Seismic Solar Model and the oscillation code of J. Christensen-Dalsgaard [35] for the acoustic mode $\ell = 2$, n = 6 (lowest splitting detected) and for the gravity mode ℓ =2, n = −3 (gravity mode candidate). From [34]

IRIS, BiSON) were operational. Two very important results have appeared just before the launch: the determination of the photospheric helium content [36, 37] and the determination of the depth of the convective zone [38]. In fact, before helioseismology, the solar *helium content* (the second element in mass fraction) was only deduced from theoretical solar models. Its practically cosmological estimate (0.25 in mass fraction) showed the limit of one basic hypothesis of stellar models supposing the initial composition is equal to the present photospheric composition. This determination has confirmed the need to introduce extra phenomena such as the slow atomic diffusion introduced first by [39] in 1989. This process leads today to a reduction of practically 10–15% of the He mass fraction at the solar surface in comparison with its initial value [40–42].

3.1 The Observed Solar Internal Sound Speed and Density Profiles

Then SoHO has played a dominant role in the investigation of the radiative zone. This very long and stable mission has been crucial to establish the properties of these layers down to the core because one needs very precise frequencies to scrutinize the whole radiative zone due to the intrinsic properties of the acoustic modes. Global acoustic modes of high frequencies are the easiest to observe but they have a resonant cavity that includes the outer layers, largely perturbed by the turbulence and the varying magnetic field component along the 11-year cycle. Furthermore, these modes have a reduced lifetime leading to broad peaks dominated by the stochastic excitation of the modes. One success of SoHO, obtained by measuring the variability of the Doppler velocity shifts through two instruments (GOLF: Global Oscillations at Low Frequency especially designed for this purpose and MDI: Michelson Doppler Imager), has been the capability to reach the low-frequency range of the acoustic spectrum. The corresponding modes have a higher lifetime but a smaller intensity so they benefit from exceptional conditions of GOLF/SoHO observations [43–45]. From these modes, we have extracted a very clean sound speed profile down to 0.06 $R_\odot$ and a reasonable density profile [16, 17].

Helioseismology was the key for validating the various ingredients used in the construction of the standard solar model. It is in fact interesting to notice that each phenomenon (specific nuclear rate, specific opacity coefficient, screening or Maxwellian tail distribution) has a specific influence on the sound speed profile [46]. We have indeed shown that the present sound speed does not favor any tiny variation of the Maxwellian distribution nor strong screening or large mixing in the core [11]. This is of particular importance to check the validity of the involved nuclear processes. It has also been possible from the sound speed profile in the core and due to the signature of each specific reaction rate to put an observational constraint on the value of the p–p reaction rate which was known only theoretically because of the weak character of the interaction. Its influence on the sound speed profile in the core is strong and the cross-section is well constrained nowadays within 1% with good neutrino predictions [46–48].

3.2 The Observed Solar Internal Rotation Profile

The SoHO observations allow also to test physics beyond the standard model. Helioseismic inversions of the acoustic mode rotation splittings lead to a rather flat rotation profile in the radiative zone between 0.2 and 0.65 $R_\odot$ [49, 50] shown in Fig. 3 (top). It has been very important to establish that the large differential rotation, already observed at the surface for a long time, is maintained inside the convective envelope and suddenly disappears in the transition region between

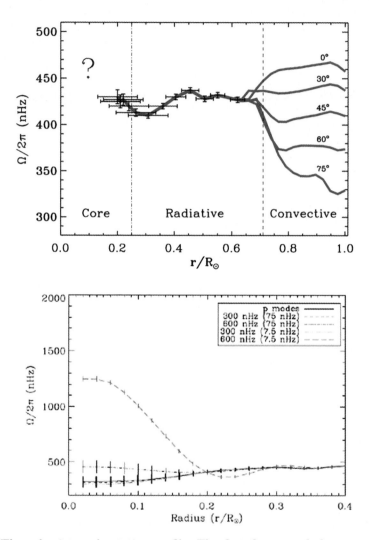

Fig. 3. The solar internal rotation profile. The first figure used the acoustic mode splittings obtained with 2088 days of observations by the instruments GOLF and MDI located aboard SoHO. From [56]. The second figure used in addition the gravity mode candidate for different values of possible splittings. From [34]

convective and radiation transport. The consequences in the dynamo process are explained below.

3.2.1 The Rotation in the Radiative Zone and the Role of the Gravity Modes

The rather flat rotation in the radiative zone down to the limit of the core is now well established. But it is impossible to determine the nuclear core rotation rate from the acoustic mode splittings (recall Fig. 2). Effectively the dipolar modes ($\ell = 1$), which partly reach this domain, are perturbed by the modification of the subsurface layers during the solar cycle, so one needs to be extremely careful on the limits of the inversion of the acoustic modes [45]. It is why the rotation of the core was unknown up to recently.

This knowledge needs the detection of gravity modes. The GOLF instrument aboard SoHO has been specifically designed to detect some of them [51, 52]. But their very low surface amplitudes have put some doubts about the capability of detecting any signal. Intensive studies have been dedicated to this field since the first years and a positive search strategy (among others) has been defined to analyze the GOLF/SoHO observations:

(1) to look for multiplets at high frequency as the main interest of these modes is to know their splittings to determine the central rotation. Such a search improves also statistically the capability of detection of very low signals and the range above 150 µHz corresponds to the region where the velocities are the greatest [53–55]. A candidate with at least three components (but possibly five or six) has been detected with more than 98% CL.
(2) to examine the properties of the asymptotic behavior of the gravity modes at low frequency and to sum the signal coming from about 20 modes to detect a clear signal [56, 57]. A signal of gravity dipole modes has been identified with more than 99.7% CL. The analysis of this signal favors a rotation of the core 3 or 5 times greater than the rest of the radiative zone.

So, GOLF has allowed the detection of some solar gravity mode signatures. Nevertheless the information we can deduce on the central rotation stays poor due to the limited number of modes detected and the ambiguity on the component identification. Figure 3 (bottom) shows an inversion done in using the acoustic mode splittings plus the splitting extracted from the pattern of the gravity mode candidate supposing that it is an $\ell = 2, n = -3$ signal [34]. Due to the ambiguity on the component labeling, two possible values of splitting are considered: 300 nHz which corresponds to the detection of only three components and leads to a flat rotation in the core and 600 nHz which corresponds to the detection of at least five components (if one supposes different axes for the rotation and the magnetic field, a quadrupole mode may have up to 10 components). It is interesting to note that the analysis of the two regions are only compatible if the core of the Sun turns quicker than the rest of the radiative zone by a factor 3–5. This solution is favored but not demonstrated. If it were true, the first analysis supports the possibility that the rotation axes of the core and of the rest of the radiative zone may be different. Of course all these assumptions need to be confirmed by improved observations. Presently, the analysis of the other instruments shows the superiority of the GOLF instrument in the gravity mode range and that one component of the GOLF candidate at 220.7 µHz appears permanently in the VIRGO data and probably also in the GONG data.

3.2.2 The Rotation in the Convective Zone Thanks to the High-Degree Acoustic Modes

Figure 4 shows a zoom of the rotation profile in the transition region and in the convective zone which represent, respectively, 4% and 2% in fraction mass. This region is very important to study and the 2D rotation profile is the first element we need for a good understanding of the solar activity leading to the 22-year solar cycle. It is now established that this cycle necessitates to consider three important regions in a mean field $\alpha - \omega$ or flux transport dynamo model:

– the leptocline (subsurface layers) where one sees a rapid change of the rotation with some inversion of variation at high latitude, and double layers which vary in opposite phases along the solar cycle [56–60].
– the whole convective zone where the differential rotation is maintained and the meridional circulation evolves [60]. Dikpati and collaborators [61] consider that a specific solar cycle depends on the velocity of this circulation and the characteristics of the cycle (amplitude and duration) on three or four previous cycles.
– the tachocline, name attributed to the transition layers between rigid rotation and differential rotation. This region sustains a strong horizontal turbulent mixing [62]. This region is a crucial region for the development and the amplification of the toroidal field which allows not only the establishment of the dynamo process but also its existence today [63].

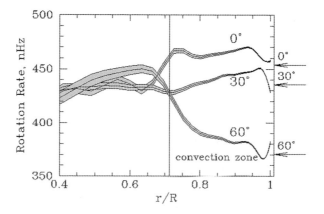

Fig. 4. A zoom on the solar rotation profile in the convective zone extracted from the acoustic mode of MDI. From [64]

4 Confrontation of the Seismic Internal Rotation Profile to the Theoretical Profiles

Despite all the previous described efforts, the observed solar rotation profile from the core to the surface remains to be guaranteed and explained in details for the region below 0.2 $R_\odot$. We have shown with the satellite SOHO that it is certainly possible to establish this profile down to the central region and the present situation puts already interesting constraints to modelers. If one compares Fig. 1 to Figs. 3 and 4, one notices that in 1D secular model one cannot reproduce the latitudinal differential rotation but one begins to introduce the transport of momentum and chemicals associated to the meridional circulations [29]. It is clear that the rotation in the radiative zone appears flatter between 0.4 and 0.6 $R_\odot$ than in the models previously described. This important fact calls for adding other processes than the effects of rotation. The absolute central rotation value and its slope in the nuclear core is very constraining; one notices already that the results we got are not reproduced by the second model we describe nor by all the already published models including only the rotation effects for which the contrast between central and surface values is greater than a factor 10 (which is rejected by our asymptotic detected dipolar gravity modes). In the first case, the contrast is smaller and the profile could appear more similar so it is interesting to pursue such study and explore in more detail how the premain sequence could influence this contrast.

The resulting solar sound speed and density profiles are examined in parallel. A quantitative estimate of their change depends largely on the initial conditions of the rotation profile and its story but these changes are not very large [31]. These studies will be pursued during the next years in introducing all the processes and their interconnection and in using 3D simulations to guide us in the understanding of some instabilities or some estimate of the real energy budget. The corresponding Dynamical Solar Model will be confronted with all the seismic and neutrino observations.

5 Perspectives

All the dynamical processes, convection, rotation and magnetic fields, and their interconnection need to be included simultaneously in solar (stellar) models as shown in Fig. 5 together with the influence of the low-frequency gravity waves. It will result in a dynamical solar structure model which must be confronted with all the present seismic observations. This new objective involves the introduction of the various terms which contribute to the angular momentum transport along the evolution. A complete formalism has been established recently which takes into account the different aspects of the transport of momentum in stellar equations [65]. This complex system of 16 equations is under implementation in different stellar evolution codes (STAREVOL, CESAM) to build step by step the Solar Dynamical Model using also the observational constraints coming from the young stars and those coming from asteroseismology.

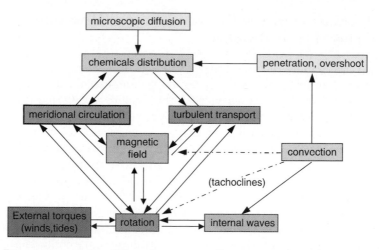

Fig. 5. Description of the different processes that one needs to incorporate in the stellar equations in order to take into account the internal dynamics of stars. From [29]

A promising process to solve the flattening of solar rotation profile in a large part of the radiative zone is the internal gravity waves (IGWs). These low-frequency traveling waves are excited at the base of the convection zone and may be a source of angular momentum redistribution, since they take momentum from the region where they are excited and deposit it where they are damped. When both prograde and retrograde waves are excited, in the presence of shear turbulence, this produces a rapidly oscillating shear layer similar to the quasi-biennial oscillation of the Earth's stratosphere. If the surface convection zone is rotating more slowly than the core as expected from surface magnetic braking, the shear layer oscillation (SLO) becomes asymmetrical and produces differential filtering that favors the penetration of low-degree, low-frequency retrograde wave at least during the first phases of the evolution [66]. These waves may then deposit their negative angular momentum in the deep interior, causing the spin-down of the solar core on evolutionary timescales. A complete formalism has been developed [67, 68] to include such an effect in stellar evolution codes. In the absence of differential rotation, wave transport becomes negligible away from the thin SLO. Such formalism has been applied to an evolving solar mass model in which the surface convection zone is slowly spun-down with time [69].

6 Conclusions

I have illustrated what role the past rotation can play in the understanding of the present one and how it is important to determine such a profile from the core to the surface in order to validate the complex interplay between the different dynamical processes. The SoHO observations have been extremely useful to progress on this profile in the different regions of the Sun. We are now confident that such a rotation profile is accessible to the observations; one needs to detect simultaneously the splitting of several individual gravity modes and of a great number of acoustic modes. In the future the mission SDO will be extremely useful to demonstrate how this profile could evolve with time and latitude along the solar cycle near the surface or at the base of the convective zone. For a definitive insight into the rotation of the solar core, one needs probably new measurements with an improved instrument using the GOLF-NG design [70]. It is certainly important after SOHO to prepare a new large space mission like DynaMICCS [71] or HIRISE proposed to ESA in the framework of the Cosmic Vision 2015–2025. These missions, thanks to a formation flying concept, can carry all the instruments which explore all the dynamical effects from the core to the corona useful for a complete 3D vision of the Sun. Such a mission will definitively establish the origins of the variability of the Sun, explain the great minima like the Maunder minimum [72] and the great maxima of solar activity and help to predict the next coming cycles.

References

1. Rüdiger, G., Hollerbach, R.: The Magnetic Universe. Wiley-VCH Verlag, GmbH & Co, KGaA (2004).
2. Spruit, H.C.: A&A **31**, 923 (2002).
3. D'Antona, F., Ventura, P., Mazzitelli, I.: ApJ **543**, L77 (2000).
4. Piau, L., Turck-Chièze, S.: ApJ **566**, 419 (2002).
5. Turck-Chièze, S., Talon, S.: Adv. Space. Res. **41**, 855 (2008).
6. Maeder, A., Meynet, G.: ARA&A **38**, 143 (2000).
7. Maeder, A., Meynet, G.: A&A **422**, 225 (2004).
8. Talon, S., Zahn, J.P.: A&A **329**, 315 (1998).
9. Mathis, S., Zahn, J.P.: A&A **425**, 229 (2004).
10. Mathis, S., Zahn, J.P.: A&A **440**, 653 (2005).
11. Turck-Chièze, S., P. Nghiem, S. Couvidat & S. Turcotte: Sol. Phys. **200**, 323 (2001).
12. Turck-Chièze, S., Däppen, W., Provost, J., Schatzman, E., Vignaud, D.: Phys. Rep. **230**, 57 (1993).
13. Turck-Chièze, S., Rep. Progr. Phys., submitted.
14. Turck-Chièze, S., Couvidat, S., Piau, L.: In: Element Stratification in Stars: 40 Years of Atomic Diffusion, EAS Publ. Series. Vol. 17, p. 149 (2005).
15. Bahcall, J.N., Serenelli, A.M., Basu, S.: ApJS **165**, 400 (2006).
16. Turck-Chièze, S., et al.: ApJ **555**, L69 (2001).
17. Couvidat, S., Turck-Chièze, S., Kosovichev, A.: ApJ **599**, 1434 (2003).
18. Turck-Chièze, S., et al.: Phys. Rev. Lett. **93**, 211102 (2004).
19. Mathur, S., Turck-Chièze, S., Couvidat, S., García, R.A.: ApJ **658**, 594 (2007).
20. Pinsonneault, M.H., Kawaler, S.D., Sofia, S., Demarque, P.: ApJ **338**, 424 (1989).
21. Chaboyer, B., Demarque, P., Pinsonneault, M.H.: ApJ **441**, 865 (1995).
22. Talon, S.:/Hydrodynamique des étoiles en rotation, PhD thesis, University Paris VII France (1997).
23. Charbonneau, P., MacGregor, K.B.: ApJ **417**, 762 (1993).

24. Barnes, G., Charbonneau, P., MacGregor, K.B.: ApJ **511**, 466 (1999).
25. Eggenberger, P., Maeder, A., Meynet, G.: A&A **440**, L9 (2005).
26. Zahn, J.P., Brun, A.S., Mathis, S.: A&A **474**, 145 (2007).
27. Brun, A.S., Zahn, J.P.: A&A **457**, 665 (2006).
28. Yang, W.M., Bi, S.L.: A&A **449**, 1161 (2006).
29. Mathis, S., Decressin, T., Palacios, A., et al.: Astron. Notes **328**(10), 1062 (2007).
30. Maeder, A., Zahn, J.P.: A&A **334**, 1000 (1998).
31. Palacios, A., Nghiem, P., Turck-Chièze, S.: ApJ submitted (2008).
32. Bouvier, J., Forestini, M., Allain, S.: A&A **326**, 1023 (1997).
33. Bouvier, J.: IAU Symposium 243, astro-ph 0712.2988 (2007).
34. Mathur, S., Eff-Darwich, A., Garcia, R.A., Turck-Chièze, S.: A&A **484**, 517 (2008).
35. Christensen-Dalsgaard, J., Gough, D.O., Thompson, M.J.: ApJ **378**, 413 (1991).
36. Vorontsov, S., et al.: Nature **349**, 49 (1991).
37. Basu, S., Antia, H.M.: JApAS **16**, 392 (1995).
38. Christensen-Dalsgaard, J.: ApJ **378**, 413.
39. Cox, A., Guzik, J., Kidman, R.B.: ApJ **342**, 1187 (1989).
40. Christensen-Dalsgaard, J.: ApJ **403** (1993).
41. Thoul, A., Bahcall, J., Loeb, A.: ApJ **421**, 828 (1994).
42. Brun, A.S., Turck-Chièze, S., Morel, P.: ApJ **506**, 913 (1998).
43. Bertello, L., et al.: ApJ **537**, L143 (2000).
44. García, R.A., et al.: Sol. Phys. **200**, 361 (2001).
45. García, R.A., et al.: Astron. Nach. **329**, 476 (2008).
46. Turck-Chièze, S., et al.: Sol. Phys. **175**, 247 (1997).
47. Turck-Chièze, S.: NuPhS **143**, 35 (2005).
48. Turck-Chièze, S.: NuPhS **145**, 17 (2005).
49. Thompson, M.J., Christensen-Dalsgaard, J., Miesch, M.S., Toomre, J.: ARA&A **41**, 599 (2003).
50. Couvidat, S., Turck-Chièze, S., et al.: ApJ **597**, L77 (2003).
51. Gabriel, A.H., Grec, G., Robillot, J.M., Roca-Cortes, T., Turck-Chièze, S., et al.: Sol. Phys. **162**, 61 (1995).
52. Turck-Chièze, S.: SF2A PNPS et PNST (2008).
53. Turck-Chièze, S., García, R.A., Couvidat, S., et al.: ApJ **604**, 455 (2004).
54. Turck-Chièze, S.: Adv. Space Res. **37**, 1569 (2006).
55. Mathur, S., Turck-Chièze, S., Couvidat, S., Garcia, R.A.: ApJ **668**, 594 (2007).
56. García, R.A., Turck-Chièze, S., Jimenez, S., et al.: Science **316**, 1591 (2007).
57. Garcia, R.A., et al.: Sol. Phys., in press.
58. Lefebvre, S., Kosovichev, A.G.: ApJ **633**, L149 (2005).
59. Lefebvre, S., Rozelot, J.P., Kosovichev, A.G.: Adv. Space Res. **40**, 1000 (2007).
60. Jouve, L., Brun, A.S.: A&A **474**, 239 (2007).
61. Dikpati, M., Gilman, P.A., de Toma, G., Ghosh, S.S.: Sol. Phys. **245**, 1 (2007).
62. Spiegel, E.A., Zahn, J.P.: A&A **265**, 106 (1992).
63. Rüdiger, G., Brandenburg, A.: A&A **296**, 557 (1995).
64. Kosovichev, A.S., Schou, J., Scherrer, P.H., et al.: Sol. Phys. **170**, 43 (1997).
65. Mathis, S., et al.: In: Straka, C.W., Lebreton, Y., Monteiro, M.J.P.F.G. (eds.) Stellar Evolution and Seismic Tools for Ateroseismology, Vol. 26, p. 65 (2007).
66. Talon, S., Kumar, P., Zahn, J.-P.: ApJL. **574**, 175 (2002).
67. Talon, S., Charbonnel, C.: A&A **440**, 981 (2005).
68. Talon, S.: In: Rieutord, M., Dubrulle, B. (eds.) Stellar Fluid Dynamics and Numerical Simulations: From the Sun to Neutron Stars, EAS Series 21, p. 105 (2006).

69. Charbonnel, C., Talon, S.: Science **309**, 2189 (2005).
70. Turck-Chièze, S.: Adv. Space Res. **38**, 1812 (2006).
71. Turck-Chièze, S., et al.: The DynaMICCS perspective, Exp. Astron., (Special Issue on ESA's Cosmic Vision) DOI10.1007/s10686-008-9111-z (2008).
72. Brandenburg, A., Spiegel, E.A.: Astron. Notes in press, astro-ph 0801.2156.

Physics of Rotation in Stellar Models

G. Meynet

Astronomical Observatory of Geneva University, CH-1290, Sauverny, Switzerland
georges.meynet@obs.unige.ch

Abstract In these lecture notes, we present the equations presently used in stellar interior models in order to compute the effects of axial rotation. We discuss the hypotheses made. We suggest that the effects of rotation might play a key role at low metallicity.

1 Physics of Rotation

Axial rotation modifies the hydrostatic equilibrium configuration by adding a centrifugal acceleration term in the hydrostatic equation, induces many instabilities driving the transport of angular momentum and of chemical species in radiative zones and changes the mass loss rates. In the present lecture notes we shall consider the case of models without magnetic fields.

1.1 Hydrostatic Effects of Rotation

The equations of stellar structure. In a rotating star, the equations of stellar structure need to be modified [23]. The usual spherical coordinates must be replaced by new coordinates characterizing the equipotentials. The classical method applies when the effective gravity can be derived from a potential $\Psi = \Phi - 1/2\Omega^2 r^2 \sin^2\theta$, i.e., when the problem is conservative. There, Φ is the gravitational potential, Ω the angular velocity and r the radius at the colatitude θ. If the rotation law is shellular (i.e., such that Ω is constant on isobaric surfaces see below), the problem is non-conservative. Most existing models of rotating stars apply, rather inconsistently, the classical scheme by [23]. However, as shown by [44], the equations of stellar structure can still be written consistently, in terms of a coordinate referring to the mass inside the isobaric surfaces.[1] Thus, the problem of the stellar structure of a differentially rotating star in a shellular rotation state can be kept one dimensional.

---

[1] For shellular rotation, the shape of the isobaric surfaces is given by the same expression as the one giving the shape of the equipotentials in conservative cases, provided some changes of variables are performed.

Meynet, G.: *Physics of Rotation in Stellar Models*. Lect. Notes Phys. **765**, 139–169 (2009)
DOI 10.1007/978-3-540-87831-5_6 © Springer-Verlag Berlin Heidelberg 2009

The Roche model. In all the derivations, we shall use the Roche model, i.e., we approximate the gravitational potential by $GM_{\bar{r}}/\bar{r}$, where $M_{\bar{r}}$ is the mass inside the isobaric surface with a mean radius $\bar{r}$. The radius $\bar{r}$ which labels each isobaric surface is defined by $\bar{r} = (V_{\bar{r}}/(4/3\pi))^{1/3}$ where $V_{\bar{r}}$ is the volume (deformed by rotation) inside the isobaric surface considered. Apart from the case of extreme rotational velocities, the parameter $\bar{r}$ is close to the average radius of an isobar, which is the radius at $P_2(\cos\vartheta) = 0$, namely for $\vartheta = 54.7°$

In the frame of the Roche model, the shape of a meridian at the surface of the star (which is an isobaric surface) is given by couples of R and θ values satisfying the following equation:

$$\frac{GM}{R} + \frac{1}{2}\Omega^2 R^2 \sin^2\theta = \frac{GM}{R_{\rm p}}, \tag{1}$$

where R is the radius at colatitude θ, Ω the angular velocity, M the mass inside the surface and $R_{\rm p}$ the polar radius. Thus the shape of the surface (as well as of any isobaric surface inside the star) is determined by three parameters M, Ω and $R_{\rm p}$. The first two M and Ω are independent variables. The third one is derived from the first two and the equations of stellar structure. Setting $x = (GM/\Omega^2)^{-1/3} R$, one can write Eq. (1) as (see [23])

$$\frac{1}{x} + \frac{1}{2}x^2 \sin^2\theta = \frac{1}{x_{\rm p}}. \tag{2}$$

With this change of variable, the shape of an equipotential is uniquely determined by only one parameter $x_{\rm p}$.

Setting

$$f = \frac{R_{\rm e}}{R_{\rm p}}, \tag{3}$$

where $R_{\rm e}$ is the equatorial radius, one easily obtains from Eq. (1) that

$$R_{\rm p} = \left(\frac{GM}{\Omega^2}\right)^{1/3}\left(\frac{2(f-1)}{f^3}\right)^{1/3} = \left(\frac{GM}{\Omega^2}\right)^{1/3} x_{\rm p}. \tag{4}$$

The above equation relates the inverse of the oblateness f to $R_{\rm p}$.

The von Zeipel theorem and its consequences. The von Zeipel theorem [60] expresses that the radiative flux $\boldsymbol{F}$ at some colatitude ϑ in a rotating star is proportional to the local effective gravity $\boldsymbol{g}_{\rm eff}$. Reference [29] has generalized this theorem to the case of shellular rotation and the expression of the flux $\boldsymbol{F}$ for a star with angular velocity Ω on the isobaric stellar surface is

$$\boldsymbol{F} = -\frac{L(P)}{4\pi GM_\star}\boldsymbol{g}_{\rm eff}[1 + \zeta(\vartheta)] \quad \text{with} \tag{5}$$

$$M_\star = M\left(1 - \frac{\Omega^2}{2\pi G\rho_{\rm m}}\right) \quad \text{and} \tag{6}$$

$$\zeta(\vartheta) = \left[\left(1 - \frac{\chi_T}{\delta}\right)\Theta + \frac{H_T}{\delta}\frac{d\Theta}{dr}\right]P_2(\cos\vartheta). \tag{7}$$

There, ρ_{m} is the internal average density, $\chi = 4acT^3/(3\kappa\rho)$ and χ_T is the partial derivative with respect to T. The quantity Θ is defined by $\Theta = \tilde{\rho}/\bar{\rho}$, i.e., the ratio of the horizontal density fluctuation to the average density on the isobar [64]. One has the thermodynamic coefficients $\delta = -(\partial \ln \rho/\partial \ln T)_{P,\mu}$; H_T is the temperature scale height. The term $\zeta(\vartheta)$, which expresses the deviations of the von Zeipel theorem due to the baroclinicity of the star, is generally very small (cf. [29]).

Let us emphasize that the flux is proportional to $\boldsymbol{g}_{\mathrm{eff}}$ and not to $\boldsymbol{g}_{\mathrm{tot}}$. This results from the fact that the equation of hydrostatic equilibrium is $\nabla P/\rho = -\boldsymbol{g}_{\mathrm{eff}}$. The effect of radiation pressure is already counted in the expression of P, which is the total pressure. We may call $M_\star$ the effective mass, i.e., the mass reduced by the centrifugal force. This is the complete form of the von Zeipel theorem in a differentially rotating star with shellular rotation, whether or not one is close to the Eddington limit.

This theorem has numerous consequences. Some of them are discussed below.

1.1.1 The Position of the Star in the HR Diagram

A fast-rotating star has stronger radiative fluxes at the pole than at the equator. Therefore the position of such a star in the HR diagram will depend on the angle between the line of sight and the rotational axis (inclination angle). If for instance that angle is $90°$, a great part of the light will come from the equatorial belt characterized by lower radiative fluxes and cooler effective temperatures, while when the star is observed pole-on most of the light will come from the hot polar region characterized by stronger fluxes and higher effective temperatures. Thus the perceived luminosity and effective temperature (and also effective gravity) of a star depend on the inclination angle. This has to be kept in mind when comparisons are made with observed quantities. Computations of the effect of the inclination angle on the emergent luminosity, colors and spectrum have been performed by [35]. The effect of the inclination angle on the determination of the effective gravity is discussed in [21]. In general, a theoretical evolutionary track is given in term of total luminosity and of an average effective temperature defined by $T_{\mathrm{eff}}^4 = L/(\sigma S(\Omega))$, where σ is Stefan's constant and $S(\Omega)$ the total actual stellar surface. The total luminosity (corresponding to the integrated flux over the surface) does not depend on the angle of view, but cannot be directly compared to the "observed" luminosity deduced from the apparent luminosity coming from the hemisphere directed toward us. Let us note however that for surface velocities inferior to about 70% of the critical velocity these effects remain quite modest. As a numerical example, the ratio $(T_{\mathrm{eff}}(\mathrm{pole}) - T_{\mathrm{eff}}(\mathrm{equator}))/T_{\mathrm{eff}}(\mathrm{equator})$ becomes superior to 10% only for $\omega > 0.7$. At break-up, the effective temperature of the polar region is about a factor two higher than that of the equatorial one.

1.1.2 The Eddington Luminosity

Let us express the total gravity at some colatitude ϑ, taking into account the radiative acceleration (cf. [29])

$$g_{rad} = \frac{1}{\rho}\nabla P_{rad} = \frac{\kappa(\vartheta)F}{c}. \tag{8}$$

Thus one has

$$g_{tot} = g_{eff} + g_{rad} = g_{eff} + \frac{\kappa(\vartheta)F}{c}. \tag{9}$$

The rotation effects appear both in g_{eff} and in F. We may also consider the local limiting flux. The condition $g_{tot} = 0$ allows us to define a limiting flux:

$$F_{lim}(\vartheta) = -\frac{c}{\kappa(\vartheta)}g_{eff}(\vartheta). \tag{10}$$

From that we may define the ratio $\Gamma_\Omega(\vartheta)$ of the actual flux (see Eq. (5)) $F(\vartheta)$ to the limiting local flux in a rotating star

$$\Gamma_\Omega(\vartheta) = \frac{F(\vartheta)}{F_{lim}(\vartheta)} = \frac{\kappa(\vartheta)\, L(P)[1 + \zeta(\vartheta)]}{4\pi cGM\left(1 - \frac{\Omega^2}{2\pi G\rho_m}\right)}. \tag{11}$$

As a matter of fact, $\Gamma_\Omega(\vartheta)$ is the local Eddington ratio and

$$L_{Edd} = \frac{4\pi cGM\left(1 - \frac{\Omega^2}{2\pi G\rho_m}\right)}{\kappa(\vartheta)\,[1 + \zeta(\vartheta)]} \tag{12}$$

is the local Eddington luminosity. For a certain angular velocity Ω on the isobaric surface, the maximum permitted luminosity of a star is reduced by rotation, with respect to the usual Eddington limit. In the above relation, $\kappa(\vartheta)$ is the largest value of the opacity on the surface of the rotating star. For O-type stars with photospheric opacities dominated by electron scattering, the opacity κ is the same everywhere on the star.

For zero rotation, the usual expressions are found: $\Gamma_\Omega(\vartheta) = \Gamma = \kappa L/4\pi cGM$ and $L_{Edd} = 4\pi cGM/\kappa$.

1.1.3 Critical Limits

Critical limits correspond to values of respectively the luminosity and/or the velocity which impose that the total gravity becomes equal to zero at least at some peculiar places at the surface. We may identify different limits [39]:

– We speak of the Eddington or Γ-limit, when rotation effects can be neglected and $g_{rad} + g_{grav} = 0$,[2] which implies that

$$\Gamma = \frac{\kappa L}{4\pi cGM} \rightarrow 1. \tag{13}$$

In that case $L = L_{Edd} = 4\pi cGM/\kappa$. The opacity κ considered here is the total opacity.

---

[2] g_{grav} is the gravitational acceleration.

- The critical velocity or Ω-limit is reached for a star with an angular velocity Ω at the surface, when the effective gravity $\boldsymbol{g}_{\text{eff}} = \boldsymbol{g}_{\text{grav}} + \boldsymbol{g}_{\text{rot}} = \boldsymbol{0}$ and in addition when radiation pressure effects can be neglected.
- *The $\Omega\Gamma$-limit is reached when the total gravity $\boldsymbol{g}_{\text{tot}} = \boldsymbol{0}$, with significant effects of both rotation and radiation.* This is the general case. It should lead to the two above cases in their respective limits.

Using relation (11), we may write the expression for the total gravity as

$$\boldsymbol{g}_{\text{tot}} = \boldsymbol{g}_{\text{eff}} \left[1 - \Gamma_\Omega(\vartheta)\right] . \tag{14}$$

This shows that the expression for the total acceleration in a rotating star is similar to the usual one, except that Γ is replaced by the local value $\Gamma_\Omega(\vartheta)$. Indeed, contrary to expressions such as $\boldsymbol{g}_{\text{tot}} = \boldsymbol{g}_{\text{eff}} (1 - \Gamma)$ often found in literature, we see that the appropriate Eddington factor given by Eq. (11) also depends on the angular velocity Ω on the isobaric surface.

From Eq. (11), we note that over the surface of a rotating star, which has a varying gravity and T_{eff}, $\Gamma_\Omega(\vartheta)$ is the highest at the latitude where $\kappa(\vartheta)$ is the largest (if we neglect the effects of $\zeta(\vartheta)$, which is justified in general). If the opacity increases with decreasing T as in hot stars, the opacity is the highest at the equator and there the limit $\Gamma_\Omega(\vartheta) = 1$ may by reached first. Thus, it is to be stressed that if the limit $\Gamma_\Omega(\vartheta) = 1$ happens to be met at the equator, it is not because $\boldsymbol{g}_{\text{eff}}$ is the lowest there, but because the opacity is the highest! The reason for no direct dependence on $\boldsymbol{g}_{\text{eff}}$ is because both terms $\boldsymbol{g}_{\text{eff}}$ cancel each other in the expression (11) of the flux ratio.

The critical velocity is reached when somewhere on the star one has $\boldsymbol{g}_{\text{tot}} = \boldsymbol{0}$, i.e.,

$$\boldsymbol{g}_{\text{eff}} \left[1 - \Gamma_\Omega(\vartheta)\right] = \boldsymbol{0}. \tag{15}$$

This equation has two roots. The first one $v_{\text{crit},1}$ is given by the usual condition $\boldsymbol{g}_{\text{eff}} = \boldsymbol{0}$, which implies the equality $\Omega^2 R_{\text{eb}}^3/(GM) = 1$ at the equator. This corresponds to an equatorial critical velocity

$$v_{\text{crit},1} = \Omega\, R_{\text{eb}} = \left(\frac{2}{3}\frac{GM}{R_{\text{pb}}}\right)^{\frac{1}{2}} . \tag{16}$$

R_{eb} and R_{pb} are, respectively, the equatorial and polar radii at the critical velocity. We notice that the critical velocity $v_{\text{crit},1}$ is independent of the Eddington factor. To this extent, this is in agreement with [17]. The basic physical reason for this independence is quite clear: the radiative flux tends toward zero when the effective gravity is zero, thus there is no effect of the radiative acceleration when this occurs.

Equation (15) has a second root, which is given by the condition $\Gamma_\Omega(\vartheta) = 1$. If we call the Eddington ratio Γ_{max} the maximum value of $\kappa(\vartheta)L(P)/(4\pi cGM)$ over the surface (in general at equator), [39] has shown that for $\Gamma_{\text{max}} < 0.639$; no value of Ω can lead to $\Gamma_\Omega(\vartheta) = 1$. Physically this means that when the star is sufficiently far from the Eddington limit, the effects of rotation on the radiative equilibrium are not sufficient for lowering the Eddington luminosity such that it may have an impact on the value of the critical velocity. In that case, Eq. (15) has only one root given by the classical expression.

For a given large enough Γ_{max} (i.e., larger than 0.639), a second root is obtained given by

$$v_{crit,2}^2 = \frac{9}{4} v_{crit,1}^2 \frac{1-\Gamma}{V'(\omega)} \frac{R_e^2(\omega)}{R_{pb}^2}. \tag{17}$$

The quantity ω is the fraction Ω/Ω_c of the angular velocity at break-up. The quantity $V'(\omega)$ is the ratio of the actual volume of a star with rotation ω to the volume of a sphere of radius R_{pb}. $V'(\omega)$ is obtained by the integration of the solutions of the surface equation for a given value of the parameter ω.

For $\Gamma_{max} > 0.639$, this second root is inferior to the first one. Thus it is encountered first and is therefore the expression of the critical that has to be used.

1.1.4 The Mass Loss Rates

Due to the von-Zeipel theorem, the radiative flux, which is the driving force for the stellar winds of massive hot stars, varies as a function of the colatitude. This effect, when accounted for in the theory of the line-driven wind theory, leads to an enhancement of the quantities of mass lost and to wind anisotropies. We shall describe in more detail these effects in Sect. 1.3.

1.1.5 Three Remarks

(1) The classical critical angular velocity or the Ω-limit (to distinguish it from the $\Omega\Gamma$-limit as defined by [39]) in the frame of the Roche model is given by

$$\Omega_{crit} = \left(\frac{2}{3}\right)^{\frac{3}{2}} \left(\frac{GM}{R_{pb}^3}\right)^{\frac{1}{2}}, \tag{18}$$

where R_{pb} is the polar radius when the surface rotates with the critical velocity. The critical velocity is given by

$$v_{crit} = \left(\frac{2}{3}\frac{GM}{R_{pb}}\right)^{\frac{1}{2}}. \tag{19}$$

Replacing Ω in Eq. (4) by $(\Omega/\Omega_{crit}) \cdot \Omega_{crit}$ and using Eq. (18), one obtains a relation between R_p, f and R_{pb}:

$$R_p = \frac{3}{2} R_{pb} \left(\frac{\Omega_{crit}}{\Omega}\right)^{2/3} \left(\frac{2(f-1)}{f^3}\right)^{1/3}. \tag{20}$$

Thus one has

$$\frac{\Omega}{\Omega_{\text{crit}}} = \left(\frac{3}{2}\right)^{3/2} \left(\frac{R_{\text{pb}}}{R_{\text{p}}}\right)^{3/2} \left(\frac{2(f-1)}{f^3}\right)^{1/2}. \tag{21}$$

With a good approximation (see below) one has $R_{\text{pb}}/R_{\text{p}} \simeq 1$ and therefore

$$\frac{\Omega}{\Omega_{\text{crit}}} \simeq \left(\frac{3}{2}\right)^{3/2} \left(\frac{2(f-1)}{f^3}\right)^{1/2}. \tag{22}$$

This equation is quite useful since it allows the determination of Ω_{crit} from quantities obtained with a model computed for Ω. This is not the case of Eq. (20) which involves R_{pb} whose knowledge can only be obtained by computing models at the critical limit. In general Eq. (22) gives a very good approximation of Ω_{crit} (see [16] for a discussion of this point).

Setting v the velocity at the equator, one has

$$\frac{v}{v_{\text{crit}}} = \frac{\Omega R_{\text{e}}}{\Omega_{\text{c}} R_{\text{eb}}} = \frac{\Omega}{\Omega_{\text{crit}}} \frac{R_{\text{e}}}{R_{\text{p}}} \frac{R_{\text{p}}}{R_{\text{pb}}} \frac{R_{\text{pb}}}{R_{\text{eb}}}, \tag{23}$$

where R_{eb} is the equatorial radius when the surface rotates with the critical velocity. Using Eq. (20) above, and the fact that in the Roche model $R_{\text{pb}}/R_{\text{eb}} = 2/3$, one obtains

$$\frac{v}{v_{\text{crit}}} = \left(\frac{\Omega}{\Omega_{\text{crit}}} 2(f-1)\right)^{1/3}. \tag{24}$$

The relations between v/v_{crit} and $\Omega/\Omega_{\text{crit}}$ obtained in the frame of the Roche model (see Eq. (24)) for the 1 and 60 $M_\odot$ stellar models at $Z = 0.02$ are shown in Fig. 1. In case we suppose that the polar radius remains constant $R_{\text{pb}}/R_{\text{p}} = 1$, then Eq. (22) can be used and one obtains a unique relation between $\Omega/\Omega_{\text{crit}}$ and v/v_{crit}, independent of the mass, metallicity and evolutionary stage considered. One sees that the values of v/v_{crit} are smaller than that of $\Omega/\Omega_{\text{crit}}$ by at most $\sim 25\%$. At the two extremes the ratios are of course equal.

An interesting quantity is the ratio of the centrifugal acceleration, a_{cen}, to the gravity, g_{e}, at the equator

$$\frac{a_{\text{cen}}}{g_{\text{e}}} = \frac{\Omega^2 R_{\text{e}}^3}{GM} = \left(\frac{\Omega}{\Omega_{\text{crit}}}\right)^2 \left(\frac{2}{3}\right)^3 f^3 \left(\frac{R_{\text{p}}}{R_{\text{pb}}}\right)^3, \tag{25}$$

where we have used Eq. (18) and divided/multiplied by R_{pb}^3. Replacing $R_{\text{pb}}/R_{\text{p}}$ by its expression deduced from Eq. (20), we obtain

$$\frac{a_{\text{cen}}}{g_{\text{e}}} = 2(f-1). \tag{26}$$

We can check that at the critical limit when $f = 3/2$, $a_{\text{cen}} = g_{\text{e}}$.

(2) It is interesting to note that the fact that the effective temperature varies as a function of the colatitude on an isobaric surface does not necessarily imply that the temperature varies as a function of the colatitude on an isobaric surface.

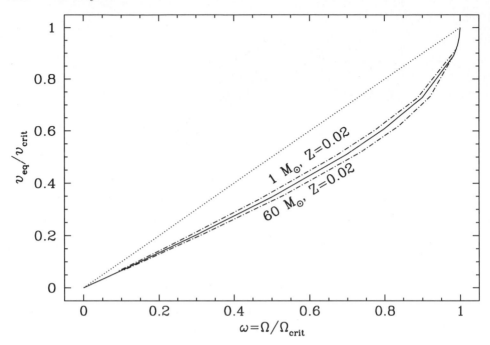

Fig. 1. Relation between v/v_{crit} and $\Omega/\Omega_{\mathrm{crit}}$ obtained in the frame of the Roche model. The *continuous line* is obtained assuming $R_{\mathrm{pb}}/R_{\mathrm{p}} = 1$ (see text and Eqs. (22) and (24)). The *dot-dashed lines* show the relations for the Z=0.02 models with 1 and 60 $M_{\odot}$ using Eq. (21). The dotted line is the line of slope 1

For instance in the conservative case, the temperature is constant on equipotentials which are also isobaric surfaces. This simply illustrates the difference between the effective temperature whose definition is related to the radiative flux ($F = \sigma T_{\mathrm{eff}}^4$) and hence to the *temperature gradient* and the temperature itself. Interestingly, one has that in a conservative case, $\Gamma_{\Omega}(\theta)$ is constant on isobaric surfaces (neglecting the term $\zeta(\vartheta)$). This means that when the $\Omega\Gamma$-limit is reached, it is reached over the whole stellar surface at the same time. This is in contrast with the Ω-limit which is reached first at the equator.

(3) Recently we have examined the effects of rotation on the thermal gradient and on the Solberg–Hoiland term by analytical developments and by numerical models [38]. Writing the criterion for convection in rotating envelopes, we show that the effects of rotation on the thermal gradient are much larger and of opposite sign to the effect of the Solberg–Hoiland criterion. On the whole, rotation favors convection in stellar envelopes at the equator and to a smaller extent at the poles. In a rotating 20 $M_{\odot}$ star at 94% of the critical angular velocity, there are two convective envelopes, the biggest one has a thickness of 13.2% of the equatorial radius. The convective layers are shown in Fig. 2. They are more extended than without rotation. In the non-rotating model, the corresponding convective zone has a thickness of only 4.6% of the radius. The occurrence of outer convection in massive stars has many consequences (see [38]).

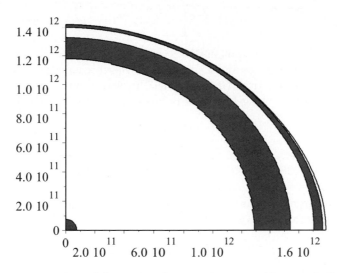

Fig. 2. 2D representation of the external convective zones (the two dark zones in the outer region) and of the convective core (the central dark region) in a model of 20 $M_\odot$ with $X = 0.70$ and $Z = 0.020$ at the end of MS evolution with fast rotation ($\Omega/\Omega_{\mathrm{crit}} = 0.94$). The axis are in units of cm. Figure taken from [38]

1.2 Transport Mechanisms of Angular Momentum and of Chemical Species

In a solid body rotation state, the first instability to set up is a thermal instability called the meridional circulation (see below). It consists of large meridional currents which transport angular momentum either inside-out or conversely from the outer regions toward the inner ones. Such meridional currents rapidly build ut gradients of the angular momentum both in the "horizontal direction" (i.e., along isobaric surface) and in the vertical one. Along isobaric surface, any gradient of Ω triggers a strong horizontal turbulence. Indeed in that direction the instability can develop without having to overcome any stable density gradient. As a consequence any gradient of Ω along isobaric surfaces is rapidly erased and the star settles into a "shellular" rotation state [64]. This means that Ω can be considered as nearly constant along isobars. In the vertical direction, where in a radiative zone, a stable density gradient counteracts any instability, the gradients of Ω are eroded on much longer timescales (see below). The equations below describe the interactions of meridional currents and of shear instabilities in a state of shellular rotation [64].

Meridional circulation. Meridional circulation is an essential mixing mechanism in rotating stars and there is a considerable literature on the subject (see Ref. in [57]). The velocity of the meridional circulation in the case of shellular rotation was derived by [64]. The velocity of meridional circulation is derived from the equation of energy conservation [43]

$$\rho T \left[\frac{\partial S}{\partial t} + (\mathbf{e}_r \dot{r} + \mathbf{U}) \cdot \nabla S \right] = \mathrm{div}(\chi \nabla T) + \rho \epsilon - \mathrm{div}\mathbf{F}_h, \qquad (27)$$

where S is the entropy per unit mass, χ the thermal conductivity, ϵ the rate of nuclear energy per unit mass and $\mathbf{F}_h$ the flux of thermal energy due to horizontal turbulence. All the quantities are expanded linearly around their average on a level surface or isobar, using Legendre polynomials $P_2(\cos\theta)$. For instance

$$T(P,\theta) = \bar{T}(P) + \tilde{T} P_2(\cos\theta).$$

Then Eq. (27) is linearized and an expression for U_2 can be deduced [64]. Using the same method [36] revised the expression for U_2 to account for expansion and contraction in non-stationary models. They also studied the effects of the μ-gradients (mean molecular weight gradients), of the horizontal turbulence and considered a general equation of state. They obtained

$$U_2(r) = \frac{P}{\bar{\rho}\bar{g}C_P\bar{T}[\nabla_{\mathrm{ad}} - \nabla + (\varphi/\delta)\nabla_\mu]} \times \left[\frac{L}{M_*}(E_\Omega + E_\mu) + \frac{C_P}{\delta}\frac{\partial\Theta}{\partial t} \right], \qquad (28)$$

where $M_\star = M\left(1 - \Omega^2/2\pi G\rho_{\mathrm{m}}\right)$ is the reduced mass and the other symbols have the same meaning as in [64] and [36].[3] The driving term in the square brackets in the second member is E_Ω. It behaves mainly like $E_\Omega \simeq 8/3\left[1 - \Omega^2/2\pi G\bar{\rho}\right]\left(\Omega^2 r^3/GM\right)$ The term $\bar{\rho}$ means the average on the considered equipotential. The term with the minus sign in the square bracket is the Gratton–Öpik term, which becomes important in the outer layers when the local density is small. This term produces negative values of $U_2(r)$ (noted $U(r)$ from now), meaning that the circulation is going down along the polar axis and up in the equatorial plane. This makes an outward transport of angular momentum, while a positive $U(r)$ gives an inward transport. At lower Z, the Gratton–Öpik term is negligible, which contributes to make larger Ω-gradients in lower Z stars.

Recently [41] rederived the system of partial differential equations, which govern the transport of angular momentum, heat and chemical elements. They expand the departure from spherical symmetry to higher order and include explicitly the differential rotation in latitude, to first order. Boundary conditions for the surface and at the frontiers between radiative and convective zones are also explicitly given in this paper.

Shellular rotation. The differential rotation which results from the evolution and transport of the angular momentum makes the stellar interior highly turbulent. As explained above, the turbulence is very anisotropic, with a much stronger geostrophic-like transport in the horizontal direction than in the vertical one [64], where stabilization is favored by the stable density gradient. This strong horizontal transport is characterized by a large diffusion coefficient D_{h}. Various expressions have been proposed:

- The usual expression for the coefficient ν_{h} of viscosity due to horizontal turbulence and for the coefficient D_{h} of horizontal diffusion, which is of the

---

[3] U_2 is the same as U in Eq. (51).

same order, is, according to [64],

$$D_{\mathrm{h}} \simeq \nu_{\mathrm{h}} = \frac{1}{c_{\mathrm{h}}} r \, |2V(r) - \alpha U(r)| \, , \tag{29}$$

where r is the appropriately defined Eulerian coordinate of the isobar [44]. $V(r)$ is defined by $u_\theta(r, \theta) = V(r) \mathrm{d} P_2(\cos\theta)/\mathrm{d}r$ where u_θ is the horizontal component of the velocity of the meridional currents,[4] $\alpha = 1/2 \, \mathrm{d}\ln r^2 \Omega / \mathrm{d}\ln r$ and c_{h} is a constant of order of unity or smaller. This equation was derived assuming that the differential rotation on an isobaric surface is small [36].

- Reference [29] has derived an expression for the coefficient D_{h} of diffusion by horizontal turbulence in rotating stars. He has obtained

$$D_{\mathrm{h}} \propto r \left(r \overline{\Omega}(r) \, V \, [2V - \alpha U] \right)^{\frac{1}{3}} \, . \tag{30}$$

This expression can be written in the usual form $\nu_{\mathrm{h}} = 1/3 \, l \cdot v$ for a viscosity, where the appropriate velocity v is a geometric mean of three relevant velocities: a velocity $(2V - \alpha U)$ as in Eq. (29) by [64], the horizontal component V of the meridional circulation, the average local rotational velocity $r \overline{\Omega}(r)$. This rotational velocity is usually much larger than either $U(r)$ or $V(r)$, typically by 6–8 orders of a magnitude in an upper main-sequence star rotating with the average velocity.

- From torque measurements in the classical Couette-Tayler experiment [52], [42] have found the following expression:

$$\nu_{\mathrm{h}} = \left(\frac{\beta}{10} \right)^{1/2} \left(r^2 \overline{\Omega}(r) \, [r \, |2V - \alpha U|] \right)^{\frac{1}{2}} \, , \tag{31}$$

with $\beta \approx 1.5 \times 10^{-5}$ [52].

The horizontal turbulent coupling favors an essentially constant angular velocity Ω on the isobars. This rotation law, constant on shells, applies to fast as well as to slow rotators. As an approximation, it is often represented by a law of the form $\Omega = \Omega(r)$ ([64]; see also [15]). Let us note here that the exact value of the diffusion coefficient D_h is not well known. Indeed the values of the numerical factors intervening in the various expressions shown above may vary to some extent. Since the expression of D_h intervenes in the formulas for U_r, for D_{shear} and D_{eff} (see below), these uncertainties have some impact on the amplitudes of the transport mechanisms.

Shear turbulence and mixing. In a radiative zone, shear due to differential rotation is likely to be a most efficient mixing process. Indeed shear instability grows on a dynamical timescale that is of the order of the rotation period [64]. The usual criterion for shear instability is the Richardson criterion, which compares the balance between the restoring force of the density gradient and the excess energy present in the differentially rotating layers:

---

[4] $V(r)$ can be obtained from $U(r)$; see Eq. 2.10 in [64].

$$Ri = \frac{N_{\mathrm{ad}}^2}{(0.8836\,\Omega\frac{d\ln\Omega}{d\ln r})^2} < \frac{1}{4}, \qquad (32)$$

where we have taken the average over an isobar, r is the radius and N_{ad} the Brunt-Väisälä frequency given by

$$N_{\mathrm{ad}}^2 = \frac{g\delta}{H_P}\left[\frac{\varphi}{\delta}\nabla_\mu + \nabla_{ad} - \nabla_{\mathrm{rad}}\right]. \qquad (33)$$

When thermal dissipation is significant, the restoring force of buoyancy is reduced and the instability occurs more easily. Its timescale is however longer, being the thermal timescale. This case is referred to as "secular shear instability". The criterion for low Peclet numbers Pe (i.e., of large thermal dissipation, see below) has been considered by [63], while the cases of general Peclet numbers Pe have been considered by [27, 32], who give

$$Ri = \frac{g\delta}{(0.8836\,\Omega\frac{d\ln\Omega}{d\ln r})^2 H_P}\left[\frac{\Gamma}{\Gamma+1}(\nabla_{ad} - \nabla) + \frac{\varphi}{\delta}\nabla_\mu\right] < \frac{1}{4}. \qquad (34)$$

The quantity $\Gamma = Pe/6$, where the Peclet number Pe is the ratio of the thermal cooling time to the dynamical time, i.e., $Pe = v\ell/K$ where v and ℓ are the characteristic velocity and length scales, and $K = (4acT^3)/(3C_P\kappa\rho^2)$ is the thermal diffusivity. A discussion of shear-driven turbulence by [5] suggests that the limiting Ri number may be larger than $1/4$.

To account for shear transport and diffusion, we need a diffusion coefficient. Amazingly, a great variety of coefficients $D_{\mathrm{shear}} = 1/3v\ell$ have been derived and applied (see a more extended discussion in [45]):

1. Reference [64] defines the diffusion coefficient corresponding to the eddies which have the largest Pe number so that the Richardson criterion is just marginally satisfied. However, the effects of the vertical μ-gradient are not accounted for and the expression only applies to low Peclet numbers. The same has been done by [32], who consider also the effect of the vertical μ-gradient, the case of general Peclet numbers, and, in addition, they account for the coupling due to the fact that the shear also modifies the local thermal gradient. This coefficient has been used by [44] and [13]. The comparisons of model results and observations of surface abundances have led many authors to conclude that the μ-gradients appear to inhibit the shear mixing too much with respect to what is required by the observations ([7, 19, 44]).

2. Instead of using a gradient ∇_μ in the criterion for shear mixing, [7] and [19] write $f_\mu\nabla_\mu$ with a factor $f_\mu = 0.05$ or even smaller. This procedure is not satisfactory since it only accounts for a small fraction of the existing μ-gradients in stars. The problem is that the models depend at least as much (if not more) on f_μ than on rotation, i.e., a change of f_μ in the allowed range (between 0 and 1) produces as important effects as a change of the initial rotational velocity. This situation has led to two other more physical approaches discussed below. Also [19] introduces another factor f_c to adjust the ratio of the transport of the angular momentum and of the chemical elements like [49].

3. Around the convective core in the region where the μ-gradient inhibits mixing, there is anyway some turbulence due to both the horizontal turbulence and the semiconvective instability, which is generally present in massive stars. This situation has led to the hypothesis [28] that the excess energy in the shear, or a fraction α of it of the order of unity, is degraded by turbulence on the local thermal timescale. This progressively changes the entropy gradient and consequently the μ-gradient. This hypothesis leads to a diffusion coefficient D_{shear} given by

$$D_{\text{shear}} = 4 \frac{K}{N_{\text{ad}}^2} \left[\frac{1}{4}\alpha \left(0.8836\,\Omega \frac{d\ln\Omega}{d\ln r} \right)^2 - (\nabla' - \nabla) \right]. \tag{35}$$

The term $\nabla' - \nabla$ in Eq. (35) expresses either the stabilizing effect of the thermal gradients in radiative zones or its destabilizing effect in semiconvective zones (if any). When the shear is negligible, D_{shear} tends toward the diffusion coefficient for semiconvection by [25] in semiconvective zones. When the thermal losses are large ($\nabla' = \nabla$), it tends toward the value

$$D_{\text{shear}} = \alpha(K/N_{\text{ad}}^2) \left(0.8836\,\Omega \frac{d\ln\Omega}{d\ln r} \right)^2, \tag{36}$$

given by [64]. Equation (35) is completed by the three following equations expressing the thermal effects [28]:

$$D_{\text{shear}} = 2K\Gamma \qquad \nabla = \frac{\nabla_{rad} + (\frac{6\Gamma^2}{1+\Gamma})\nabla_{ad}}{1 + (\frac{6\Gamma^2}{1+\Gamma})}, \tag{37}$$

$$\nabla' - \nabla = \frac{\Gamma}{\Gamma + 1}(\nabla_{ad} - \nabla). \tag{38}$$

The system of four equations given by Eqs. (35), (37) and (38) form a coupled system with four unknown quantities D_{shear}, Γ, ∇ and ∇'. The system is of the third degree in Γ. When it is solved numerically, we find that as a matter of fact the thermal losses in the shears are rather large in massive stars and thus that the Peclet number Pe is very small (of the order of 10^{-3}–10^{-4}). For very low Peclet number $Pe = 6\Gamma$, the differences ($\nabla' - \nabla$) are also very small as shown by Eq. (38). Thus, we conclude that Eq. (35) is essentially equivalent, at least in massive stars, to the original Eq. (36) above, as given by [64]. We may suspect that this is not necessarily true in low and intermediate mass stars since there the Pe number may be larger.

4. Reference [55] found that the diffusion coefficient for the shears is modified by the horizontal turbulence. The change can be an increase or a decrease in the diffusion coefficient depending on the various parameters, as discussed below. Thus, we have

$$D = \frac{(K + D_{\mathrm{h}})}{\left[\frac{\varphi}{\delta}\nabla_\mu(1 + \frac{K}{D_{\mathrm{h}}}) + (\nabla_{\mathrm{ad}} - \nabla_{\mathrm{rad}})\right]} \tag{39}$$

$$\times \frac{H_{\mathrm{p}}}{g\delta}\left[\alpha\left(0.8836\Omega\frac{d\ln\Omega}{d\ln r}\right)^2 - 4(\nabla' - \nabla)\right],$$

where D_{h} is the coefficient of horizontal diffusion (cf. [64]). We ignore here the thermal coupling effects discussed by Maeder ([28]) because they were found to be relatively small and they increase the numerical complexity. Interestingly, we see that in regions where $\nabla_\mu \simeq 0$, Eq. (39) leads us to replace K by $(K + D_{\mathrm{h}})$ in the usual expression (cf. [55]), i.e., it reinforces slightly the diffusion in regions which are close to chemical homogeneity. On the contrary, in regions where ∇_μ dominates with respect to $(\nabla_{\mathrm{ad}} - \nabla_{\mathrm{rad}})$, the transport is proportional to D_{h} rather than to K, which is quite logical since the diffusion is then determined by D_{h} rather than by thermal effects. The above result shows the importance of the treatment for the meridional circulation, since in turn it determines the size of D_{h} and to some extent the diffusion by shears.

Of course, the Reynolds condition $D_{\mathrm{shear}} \geq 1/3\nu Re_c$ must be satisfied in order that the medium is turbulent. The quantity ν is the total viscosity (radiative + molecular) and Re_c the critical Reynolds number estimated to be around 10 (cf. [13, 64]). The numerical results indicate that the conditions for the occurrence of turbulence are satisfied.

Transport of the angular momentum. Let us express the rate of change of the angular momentum, $\mathrm{d}\mathcal{L}/\mathrm{d}t$, of the element of mass in the volume ABCD represented in Fig. 3:

$$\frac{\mathrm{d}\mathcal{L}}{\mathrm{d}t} = \mathbf{M},$$

where $\mathbf{M}$ is the momentum of the forces acting on the volume element. We assume that angular momentum is transported only through advection (by a

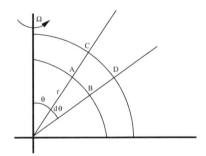

Fig. 3. The momentum of the viscosity forces acting on the element ABCD is derived in the text and the general form of the equation describing the change with time of the angular momentum of this element is deduced. The star rotates around the vertical axis with the angular velocity Ω; r and θ are the radial and colatitude coordinates of point A

velocity field **U**) and through turbulent diffusion, which may be different in the radial (vertical) and tangential (horizontal) directions. The component of the angular momentum aligned with the rotational axis is equal to[5]

$$\underbrace{\rho r^2 \sin\theta d\theta d\varphi dr}_{\text{Mass of ABCD}} \quad \underbrace{r\sin\theta\Omega}_{\text{velocity}} \quad \underbrace{r\sin\theta,}_{\text{distance to axis}}$$

where $\Omega = \dot{\varphi}$. Since the mass of the volume element ABCD does not change, the rate of change of the angular momentum can be written as

$$\rho r^2 \sin\theta d\theta d\varphi dr \frac{d}{dt}(r^2\sin^2\theta\Omega)_{M_r}. \tag{40}$$

Due to shear, forces apply on the surfaces of the volume element. The force on the surface AB is equal to

$$\underbrace{\eta_v}_{\text{vertical viscosity}} \quad \underbrace{r\sin\theta\frac{\partial\Omega}{\partial r}}_{\text{vertical shear}} \quad \underbrace{r^2\sin\theta d\theta d\varphi}_{\text{surface AB}}.$$

The component of the momentum of this force along the rotational axis is

$$\underbrace{\eta_v r^3 \sin^2\theta\frac{\partial\Omega}{\partial r}d\theta d\varphi}_{\text{force on AB}} \quad \underbrace{r\sin\theta.}_{\text{distance to axis}} \quad .$$

The component along the rotational axis of the resultant momentum of the forces acting on AB and CD is equal to

$$\frac{\partial}{\partial r}\left(\eta_v r^4 \sin^3\theta d\theta d\varphi\frac{\partial\Omega}{\partial r}\right)dr. \tag{41}$$

The force on the surface AC due to the tangential shear is equal to

$$\eta_h \quad \underbrace{r\sin\theta\frac{\partial\Omega}{r\partial\theta}}_{\text{tangential shear}} \quad \underbrace{r\sin\theta d\varphi dr,}_{\text{surface AC}}$$

where η_h is the horizontal viscosity. The component along the rotational axis of the resultant momentum of the forces acting on AC and BD is equal to

$$\frac{\partial}{r\partial\theta}\left(\eta_h r^2 \sin^3\theta dr d\varphi\frac{\partial\Omega}{\partial\theta}\right)rd\theta. \tag{42}$$

Using Eqs. (40), (41) and (42), simplifying by $drd\theta d\varphi$, one obtains the equation for the transport of the angular momentum

$$\rho r^2 \sin\theta \frac{d}{dt}(r^2 \sin^2\theta\Omega)_{M_r} = \frac{\partial}{\partial r}\left(\eta_v r^4 \sin^3\theta \frac{\partial\Omega}{\partial r}\right) + \frac{\partial}{\partial\theta}\left(\eta_h r^2 \sin^3\theta \frac{\partial\Omega}{\partial\theta}\right).(43)$$

Setting $\eta_v = \rho D_v$ and $\eta_h = \rho D_h$ and dividing the left and right members by $r^2 \sin\theta$, one obtains

$$\rho\frac{d}{dt}(r^2 \sin^2\theta\Omega)_{M_r} = \frac{\sin^2\theta}{r^2}\frac{\partial}{\partial r}\left(\rho D_v r^4 \frac{\partial\Omega}{\partial r}\right) + \frac{1}{\sin\theta}\frac{\partial}{\partial\theta}\left(\rho D_h \sin^3\theta \frac{\partial\Omega}{\partial\theta}\right).(44)$$

Now, the left-hand side term can be written:

$$\rho\frac{d}{dt}(r^2 \sin^2\theta\Omega)_{M_r} = \frac{d}{dt}(\rho r^2 \sin^2\theta\Omega)_{M_r} - r^2 \sin^2\theta\Omega\frac{d\rho}{dt}|_{M_r}.$$

Using the relation between the Lagrangian and Eulerian derivatives, one has

$$\rho\frac{d}{dt}(r^2 \sin^2\theta\Omega)_{M_r}$$

$$= \frac{\partial}{\partial t}(\rho r^2 \sin^2\theta\Omega)_r + \mathbf{U}\cdot\nabla(\rho r^2 \sin^2\theta\Omega) - r^2 \sin^2\theta\Omega\frac{d\rho}{dt}|_{M_r}. \qquad (45)$$

Using

$$\frac{d\rho}{dt}\bigg|_{M_r} = \frac{\partial\rho}{\partial t}\bigg|_r + \mathbf{U}\cdot\nabla\rho,$$

and the continuity equation

$$\frac{\partial\rho}{\partial t}\bigg|_r = -\mathrm{div}(\rho\mathbf{U}),$$

one obtains $d\rho/dt|_{M_r} + \rho\,\mathrm{div}\mathbf{U} = 0$, which when incorporated in Eq. (45) gives

$$\rho\frac{d}{dt}(r^2 \sin^2\theta\Omega)_{M_r} = \frac{\partial}{\partial t}(\rho r^2 \sin^2\theta\Omega)_r + \nabla(\mathbf{U}\rho r^2 \sin^2\theta\Omega).$$

Developing the divergence in spherical coordinates and using Eq. (44), one finally obtains the equation describing the transport of the angular momentum ([36, 41])

$$\frac{\partial}{\partial t}(\rho r^2 \sin^2\theta\Omega)_r + \frac{1}{r^2}\frac{\partial}{\partial r}(\rho r^4 \sin^2\theta w_r\Omega) + \frac{1}{r\sin\theta}\frac{\partial}{\partial\theta}(\rho r^2 \sin^3\theta w_\theta\Omega)$$

$$= \frac{\sin^2\theta}{r^2}\frac{\partial}{\partial r}\left(\rho D_v r^4 \frac{\partial\Omega}{\partial r}\right) + \frac{1}{\sin\theta}\frac{\partial}{\partial\theta}\left(\rho D_h \sin^3\theta \frac{\partial\Omega}{\partial\theta}\right), \qquad (46)$$

where $w_r = U_r + \dot{r}$ is the sum of the radial component of the meridional circulation velocity and the velocity of expansion/contraction and $w_\theta = U_\theta$, where U_θ is the horizontal component of the meridional circulation velocity. Assuming, as in [64] that the rotation depends little on latitude due to strong horizontal diffusion, we write

$$\Omega(r,\theta) = \bar{\Omega}(r) + \hat{\Omega}(r,\theta),$$

with $\hat{\Omega} \ll \bar{\Omega}$. The horizontal average $\bar{\Omega}$ is defined as being the angular velocity of a shell rotating like a solid body and having the same angular momentum as the considered actual shell. Thus

$$\bar{\Omega} = \frac{\int \Omega \sin^3 \theta d\theta}{\int \sin^3 \theta d\theta}.$$

Any vector field whose Laplacian is nul can be decomposed in spherical harmonics. Thus, the meridional circulation velocity can be written [41]:

$$\mathbf{U} = \underbrace{\sum_{l>0} U_l(r) P_l(\cos\theta)\, \mathbf{e}_r}_{u_r} + \underbrace{\sum_{l>0} V_l(r) \frac{dP_l(\cos\theta)}{d\theta}\, \mathbf{e}_\theta}_{u_\theta},$$

where $\mathbf{e}_r$ and $\mathbf{e}_\theta$ are unit vectors along the radial and colatitude directions, respectively. Multiplying Eq. (46) by $\sin\theta d\theta$ and integrating it over θ from 0 to π, one obtains [36]

$$\frac{\partial}{\partial t}(\rho r^2 \bar{\Omega})_r = \frac{1}{5r^2}\frac{\partial}{\partial r}(\rho r^4 \bar{\Omega}[U_2(r) - 5\dot{r}]) + \frac{1}{r^2}\frac{\partial}{\partial r}\left(\rho D_v r^4 \frac{\partial \bar{\Omega}}{\partial r}\right). \qquad (47)$$

It is interesting to note that only the $l=2$ component of the circulation is able to advect a net amount of angular momentum. As explained in [54] the higher order components do not contribute to the vertical transport of angular momentum. Note also that the change in radius $\dot{r}$ of the given mass shell is included in Eq. (47), which is the Eulerian formulation of the angular momentum transport equation. In its Lagrangian formulation, the variable r is linked to M_r through $dM_r = 4\pi r^2 \rho dr$, and the equation for the transport of the angular momentum can be written:

$$\rho \frac{\partial}{\partial t}(r^2 \bar{\Omega})_{M_r} = \frac{1}{5r^2}\frac{\partial}{\partial r}(\rho r^4 \bar{\Omega} U_2(r)) + \frac{1}{r^2}\frac{\partial}{\partial r}\left(\rho D_v r^4 \frac{\partial \bar{\Omega}}{\partial r}\right). \qquad (48)$$

The characteristic time associated to the transport of Ω by the circulation is [64]

$$t_\Omega \approx t_{KH} \left(\frac{\Omega^2 R}{g_s}\right)^{-1}, \qquad (49)$$

where g_s is the gravity at the surface and t_{KH} the Kelvin–Helmholtz timescale, which is the characteristic timescale for the change of r in hydrostatic models. From Eq. (49), one sees that t_Ω is a few times t_{KH}, which itself is much shorter that the Main-Sequence lifetime.

For shellular rotation, the equation of transport of angular momentum in the vertical direction is in lagrangian coordinates (cf. [36, 64]):

$$\rho \frac{d}{dt}(r^2 \Omega)_{M_r}$$

$$= \frac{1}{5r^2}\frac{\partial}{\partial r}(\rho r^4 \Omega U(r)) + \frac{1}{r^2}\frac{\partial}{\partial r}\left(\rho D r^4 \frac{\partial \Omega}{\partial r}\right). \qquad (50)$$

$\Omega(r)$ is the mean angular velocity at level r. The vertical component $u(r, \theta)$ of the velocity of the meridional circulation at a distance r to the center and at a colatitude θ can be written:

$$u(r, \theta) = U(r)P_2(\cos \theta), \tag{51}$$

where $P_2(\cos \theta)$ is the second Legendre polynomial. Only the radial term $U(r)$ appears in Eq. (50). The quantity D is the total diffusion coefficient representing the various instabilities considered and which transports the angular momentum, namely convection, semiconvection and shear turbulence. As a matter of fact, a very large diffusion coefficient as in convective regions implies a rotation law which is not far from solid body rotation. In this work, we take $D = D_{\text{shear}}$ in radiative zones, since as extra-convective mixing we consider shear mixing and meridional circulation.

In case the outward transport of the angular momentum by the shear is compensated by an inward transport due to the meridional circulation, we obtain the local conservation of the angular momentum. We call this solution the *stationary solution*. In this case, $U(r)$ is given by (cf. [64])

$$U(r) = -\frac{5D}{\Omega}\frac{\partial \Omega}{\partial r} . \tag{52}$$

The full solution of Eq. (50) taking into account $U(r)$ and D gives the *non-stationary solution* of the problem. In this case, $\Omega(r)$ evolves as a result of the various transport processes, according to their appropriate timescales, and in turn differential rotation influences the various above processes. This produces a feedback and, thus, a self-consistent solution for the evolution of $\Omega(r)$ has to be found.

Figure 4 shows the evolution of $U(r)$ in a model of a 20 $M_\odot$ star with $Z = 0.004$ and an initial rotation velocity $v_{\text{ini}} = 300$ km s^{-1} [33]. $U(r)$ is initially positive in the interior, but progressively the fraction of the star where $U(r)$ is negative is growing. This is due to the Gratton–Öpik term in Eq. (2), which favors a negative $U(r)$ in the outer layers, when the density decreases. This negative velocity causes an outward transport of the angular momentum, as well as the shears.[6]

The transport of angular momentum by circulation has often been treated as a diffusion process ([15, 19, 49]). From Eq. (50), we see that the term with U (advection) is functionally not the same as the term with D (diffusion). Physically advection and diffusion are quite different: diffusion brings a quantity from where there is a lot to other places where there is little. This is not necessarily the case for advection. A circulation with a positive value of $U(r)$, i.e., rising along the polar axis and descending at the equator, is as a matter of fact making an inward transport of angular momentum. Thus, we see that when this process

---

[6] When U is negative, the meridional currents turn anticlockwise, i.e., go inward along directions parallel to the rotational axis and go outward in directions parallel to the equatorial plane.

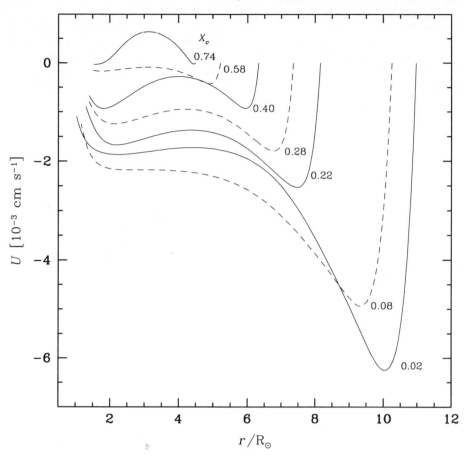

Fig. 4. Evolution of $U(r)$ the radial term of the vertical component of the velocity of meridional circulation in a 20 M$_\odot$ star with $Z = 0.004$ and an initial rotation velocity $v_{\text{ini}} = 300$ km s^{-1}. X_c is the hydrogen mass fraction at the center. Figure taken from [33]

is treated as a diffusion, like a function of $\partial\Omega/\partial r$, even the sign of the effect may be wrong.

The expression of $U(r)$ given above (Eq. (28)) involves derivatives up to the third order; thus Eq. (50) is of the fourth order, which makes the system very difficult to solve numerically. In practice, we have applied a Henyey scheme to make the calculations. Equation 50 also implies four boundary conditions. At the stellar surface, we take (cf. [56])

$$\frac{\partial\Omega}{\partial r} = 0 \quad \text{and} \quad U(r) = 0 \tag{53}$$

and at the edge of the core we have

$$\frac{\partial\Omega}{\partial r} = 0 \quad \text{and} \quad \Omega(r) = \Omega_{\text{core}}. \tag{54}$$

We assume that the mass lost by stellar winds is just embarking its own angular momentum. This means that we ignore any possible magnetic coupling, as it occurs in low-mass stars. It is interesting to mention here that in case of no viscous, nor magnetic coupling at the stellar surface, i.e., with the boundary conditions (53), the integration of Eq. (50) gives for an external shell of mass ΔM [29]

$$\Delta M \frac{d}{dt}(\Omega r^2) = -\frac{4\pi}{5}\rho r^4 \Omega U(r). \tag{55}$$

This equation is valid provided the stellar winds are spherically symmetric. When the surface velocity approaches the critical velocity, it is likely that there are anisotropies of the mass loss rates (polar ejection or formation of an equatorial ring) and thus the surface condition should be modified according to the prescriptions of [29].

Mixing and transport of the chemical elements. A diffusion–advection equation like Eq. (50) should normally be used to express the transport of chemical elements. However, if the horizontal component of the turbulent diffusion D_{h} is large, the vertical advection of the elements can be treated as a simple diffusion [6] with a diffusion coefficient D_{eff}. As emphasized by [6], this does not apply to the transport of the angular momentum. D_{eff} is given by

$$D_{\mathrm{eff}} = \frac{|rU(r)|^2}{30D_h}, \tag{56}$$

where D_{h} is the coefficient of horizontal turbulence. Equation (56) expresses that the vertical advection of chemical elements is severely inhibited by the strong horizontal turbulence characterized by D_{h}. Thus, the change of the mass fraction X_i of the chemical species i is simply

$$\left(\frac{dX_i}{dt}\right)_{M_r} = \left(\frac{\partial}{\partial M_r}\right)_t \left[(4\pi r^2 \rho)^2 D_{\mathrm{mix}} \left(\frac{\partial X_i}{\partial M_r}\right)_t\right] + \left(\frac{dX_i}{dt}\right)_{\mathrm{nucl}}. \tag{57}$$

The second term on the right accounts for composition changes due to nuclear reactions. The coefficient D_{mix} is the sum $D_{\mathrm{mix}} = D_{\mathrm{shear}} + D_{\mathrm{eff}}$ and D_{eff} is given by Eq. (56). The characteristic time for the mixing of chemical elements is therefore $t_{\mathrm{mix}} \simeq R^2/D_{\mathrm{mix}}$ and is not given by $t_{\mathrm{circ}} \simeq R/U$, as has been generally considered [53]. This makes the mixing of the chemical elements much slower, since D_{eff} is very much reduced. In this context, we recall that several authors have reduced by large factors, up to 30 or 100, the coefficient for the transport of the chemical elements, with respect to the transport of the angular momentum, in order to better fit the observed surface compositions (cf. [19]). This reduction of the diffusion of the chemical elements is no longer necessary with the more appropriate expression of D_{eff} given here.

Surface enrichments due to rotation are illustrated in Fig. 5. The tracks are plotted in the plane $(\mathrm{N/C})/(\mathrm{N/C})_{\mathrm{ini}}$ versus P where P is the rotational period in hours. During the evolution the surface is progressively enriched in CNO burning

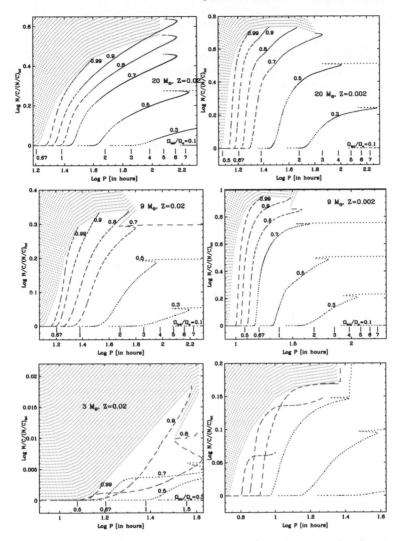

Fig. 5. Evolutionary tracks in the plane surface N/C ratio, normalized to its initial value, versus the rotational period in hours for different initial mass stars, various initial velocities and the metallicities Z=0.02 and 0.002. Positions of some periods in days are indicated at the *bottom* of the figure. The *dotted tracks* never reach the critical limit during the MS phase. The short *dashed tracks* reach the critical limit during the MS phase. The dividing line between the *shaded* and *non-shaded areas* corresponds to the entrance into the phase when the star is at the critical limit during the MS phase. If Be stars are stars rotating at or very near the critical limit, present models would predict that they would lie in the vicinity of this dividing line or above it. Note the different vertical scales used when comparing similar masses at different metallicities. Figure taken from [16]

products, i.e., enriched in nitrogen and depleted in carbon. At the same time, the rotational period increases.

When the effects of the shear and the meridional circulation compensate each other for the transport of the angular momentum (*stationary solution*), the value of U entering the expression for D_{eff} is given by Eq. (52).

1.3 Rotation and Mass Loss

We can classify the effects of rotation on mass loss into three categories.

1. The structural effects of rotation.
2. The changes brought by rotation on the radiation-driven stellar winds.
3. The mass loss induced by rotation at the critical limit.

Let us now consider in turn these various processes.

Structural effects of rotation on mass loss. Rotation, by changing the chemical structure of the star, modifies its evolution. For instance, moderate rotation at metallicities of the Small Magellanic Cloud (SMC) favors redward evolution in the Hertzsprung–Russel diagram. This behavior can account for the high number of red supergiants observed in the SMC [33], an observational fact which is not at all reproduced by non-rotating stellar models.

Now it is well known that the mass loss rates are greater when the star evolves into the red part of the HR diagram; thus in this case, rotation modifies the mass loss indirectly, by changing the evolutionary tracks. The $v_{\text{ini}} = 0, 200,$ 300 and 400 km s^{-1} models lose, respectively, 0.14, 1.40, 1.71 and 1.93 M$_\odot$ during the core He-burning phase (see Table 1 in [33]). The enhancement of the mass lost reflects the longer lifetimes of the red supergiant phase when velocity increases. Note that these numbers were obtained assuming that the same scaling law between mass loss and metallicity as in the MS phase applies during the red supergiant phase. If, during this phase, mass loss comes from continuum-opacity-driven wind then the mass loss rate will not depend on metallicity (see the review by [58]). In that case, the redward evolution favored by rotation would have a greater impact on mass loss than that shown by the computations shown above.

Of course, such a trend cannot continue forever. For instance, at very high rotation, the star will have a homogeneous evolution and will never become a red supergiant [26]. In this case, the mass loss will be reduced, although this effect will be somewhat compensated by other processes: first by the fact that the Main-Sequence lifetime will last longer; second, by the fact that the star will enter the Wolf–Rayet phase (a phase with high mass loss rates) at an earlier stage of its evolution; and third by the fact that the star may encounter the Ω-limit.

Radiation-driven stellar winds with rotation. The effects of rotation on the radiation-driven stellar winds result from the changes brought by rotation to the stellar surface. They induce changes of the morphologies of the stellar winds and increase their intensities.

1.3.1 Stellar Wind Anisotropies

Naively we would first guess that a rotating star would lose mass preferentially from the equator, where the effective gravity (gravity decreased by the effect of the centrifugal force) is lower. This is probably true when the star reaches the Ω-limit (i.e., when the equatorial surface velocity is such that the centrifugal acceleration exactly compensates the gravity), but this is not correct when the star is not at the critical limit. Indeed as recalled above, a rotating star has a non-uniform surface brightness, and the polar regions are those which have the most powerful radiative flux. Thus one expects that in case the opacity does not vary at the surface, the star will lose mass preferentially along the rotational axis. This is correct for hot stars, for which the dominant source of opacity is electron scattering. In that case the opacity only depend on the mass fraction of hydrogen and does not depend on other physical quantities such as temperature. In that way, rotation induces anisotropies of the winds ([14, 31]). This is illustrated in the left panel of Fig. 6. Wind anisotropies have consequences for the angular momentum that a star retains in its interior. Indeed, when mass is lost preferentially along the polar axis, little angular momentum is lost. This process allows loss of mass without too much loss of angular momentum, a process which might be important in the context of the evolutionary scenarios leading to gamma ray bursts. Indeed in the framework of the collapsar scenario ([62]), one has to accommodate two contradictory requirements: on one side, the progenitor needs to lose mass in order to have its H and He-rich envelope removed at the time of its explosion, and on the other hand it must have retained sufficient angular momentum in its central region to give birth to a fast-rotating black-hole.

1.3.2 Intensities of the Stellar Winds

The quantity of mass lost through radiatively driven stellar winds is enhanced by rotation. This enhancement can occur through two channels: by reducing the effective gravity at the surface of the star and by increasing the opacity of the outer layers through surface metallicity enhancements due to rotational mixing.

- *Reduction of the effective gravity:* The ratio of the mass loss rate of a star with a surface angular velocity Ω to that of a non-rotating star of the same initial mass, metallicity and lying at the same position in the HR diagram is given by [39]

$$\frac{\dot{M}(\Omega)}{\dot{M}(0)} \simeq \frac{(1-\Gamma)^{\frac{1}{\alpha}-1}}{\left[1 - \frac{4}{9}(\frac{v}{v_{\text{crit},1}})^2 - \Gamma\right]^{\frac{1}{\alpha}-1}} , \qquad (58)$$

  where Γ is the electron scattering opacity for a non-rotating star with the same mass and luminosity and α is a force multiplier [24]. The enhancement factor remains modest for stars with luminosity sufficiently far away from the Eddington limit [39]. Typically, $\dot{M}(\Omega)/\dot{M}(0) \simeq 1.5$ for main-sequence B-stars. In that case, when the surface velocity approaches the critical limit, the effective gravity decreases and the radiative flux also decreases. Thus the matter becomes less bound when, at the same time, the radiative forces become also weaker. When the stellar luminosity approaches the Eddington limit, the mass loss increases can be much greater, reaching orders of magnitude. This comes from the fact that rotation lowers the maximum luminosity or the Eddington luminosity of a star. Thus it may happen that for a velocity still far from the classical critical limit, the rotationally decreased maximum luminosity becomes equal to the actual luminosity of the star. In that case, strong mass loss ensues and the star is said to have reached the $\Omega\Gamma$ limit [39].

- *Effects due to rotational mixing:* During the core He-burning phase, at low metallicity, the surface may be strongly enriched in both H-burning and He-burning products, i.e., mainly in nitrogen, carbon and oxygen. Nitrogen is produced by transformation of the carbon and oxygen produced in the He-burning core and which have diffused by rotational mixing in the H-burning shell [46]. Part of the carbon and oxygen produced in the He core also diffuses up to the surface. Thus at the surface, one obtains very high value of the CNO elements. For instance a 60 $M_\odot$ with $Z = 10^{-8}$ and $v_{\mathrm{ini}} = 800$ km s^{-1} has, at the end of its evolution, a CNO content at the surface equivalent to 1 million times its initial metallicity! In case the usual scaling laws linking the surface metallicity to the mass loss rates are applied, such a star would lose due to this process more than half of its initial mass.

Mass loss induced by rotation. As recalled above, during the Main-Sequence phase the core contracts and the envelope expands. In case of local conservation of the angular momentum, the core would thus spin faster and faster while the envelope would slow down. In that case, it can be easily shown that the surface velocity would evolve away from the critical velocity (see, e.g., [47]). In models with shellular rotation however an important coupling between the core and the envelope is established through the action of the meridional currents. As a net result, angular momentum is brought from the inner regions to the outer ones. Thus, should the star lose no mass by radiation driven-stellar winds (as is the case at low Z), then one expects that the surface velocity would increase with time and would approach the critical limit. In contrast, when radiation-driven stellar winds are important, the timescale for removing mass and angular momentum at the surface is shorter than the timescale for accelerating the outer layers by the above process and the surface velocity decreases as a function of time. It evolves away from the critical limit. Thus, an interesting situation occurs: when the star loses little mass by radiation-driven stellar winds, it has more chance to lose mass by reaching the critical limit. On the other hand, when the star loses mass at a high rate by radiation-driven mass loss, it has no chance to reach the critical limit and thus to undergo a mechanical mass loss. This is illustrated in the right panel of Fig. 6.

Discussion. At this point it is interesting to discuss three aspects of the various effects described above. First, what are the main uncertainties affecting them? Second, what are their relative importance? And finally, what are their consequences for the interstellar medium enrichment?

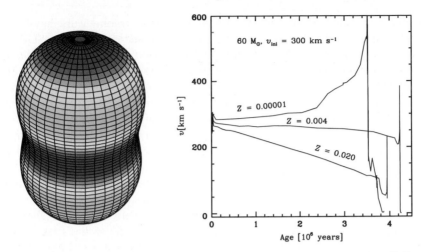

Fig. 6. *Left panel:* Iso-mass loss distribution for a 120 M$_\odot$ star with Log L/L$_\odot$=6.0 and $T_{\rm eff}$ = 30,000 K rotating at a fraction 0.8 of critical velocity (figure from [31]). *Right panel:* Evolution of the surface velocities for a 60 M$_\odot$ star with three different initial metallicities

1.3.3 Uncertainties

In addition to the usual uncertainties affecting the radiation-driven mass loss rates, the above processes pose three additional problems:

1. *What does happen when the CNO content of the surface increases by six orders of magnitude as was obtained in the 60 M$_\odot$ model described above?* Can we apply the usual scaling law between Z and the mass losses? This is what we have done in our models, but of course this should be studied in more detail by stellar wind models. For instance, for WR stars, [59] have shown that at $Z = Z_\odot/30$, 60% of the driving is due to CNO elements and only 10% to Fe. Here the high CNO surface enhancements result from rotational mixing which enrich the radiative outer region of the star in these elements, but also from the fact that the star evolves to the red part of the HR diagram, making an outer convective zone to appear. This convective zone plays an essential role in dredging up the CNO elements at the surface. Thus what is needed here are the effects on the stellar winds of CNO enhancements in a somewhat red part of the HR diagram (typical effective temperatures of the order of Log $T_{\rm eff}$ ~3.8).

2. *Do stars reach the critical limit?* For instance, [2] obtain that during pre-main sequence evolution of rapidly rotating massive stars, "equatorial mass loss" or "rotational mass ejection" never occurs (see also [3]). In these models the condition of zero effective gravity is never reached. However, these authors studied premain sequence evolution and made different hypotheses on the transport mechanisms than in the present work. Since they were interested in the radiative contraction phase, they correctly supposed that "the various

instabilities and currents which transport angular momentum have characteristic times much longer than the radiative-contraction time". This is no
longer the case for the Main-Sequence phase. In our models, we consistently
accounted for the transport of the angular momentum by the meridional currents and the shear instabilities. A detailed account of the transport mechanisms shows that they are never able to prevent the star from reaching the
critical velocity. Another difference between the approach in the work of [2]
and ours is that [2] consider another distribution of the angular velocity than
in our models. They supposed constant Ω on cylindrical surface, while here
we adopted, as imposed by the theory of [64], a "shellular rotation law". They
resolved the Poisson equation for the gravitational potential, while here we
adopted the Roche model. Let us note that the Roche approximation appears
justified in the present case, since only the outer layers, containing little mass,
are approaching the critical limit. The majority of the stellar mass has a rotation rate much below the critical limit and is thus not strongly deformed
by rotation. Thus these differences probably explain why in our models we
reach situations where the effective gravity becomes zero.

3. *What does happen when the surface velocity reaches the critical limit?* Let
us first note that when the surface reaches the critical velocity, the energy
which is still needed to make equatorial matter to escape from the potential
well of the star is still important. This is because the gravity of the system
continues of course to be effective all along the path from the surface to the
infinity and needs to be overcome. If one estimates the escape velocity from
the usual equation energy for a piece of material of mass m at the equator of
a body of mass M, radius R and rotating at the critical,

$$\frac{1}{2}mv_{\mathrm{crit}}^2 + \frac{1}{2}mv_{\mathrm{esc}}^2 - \frac{GMm}{R} = 0, \tag{59}$$

one obtains using $v_{\mathrm{crit}}^2 = GM/R$ that the escape velocity is simply reduced by
a factor $1/\sqrt{2} = 0.71$ with respect to the escape velocity from a non-rotating
body.[7] Thus the reduction is rather limited and one can wonder if matter
will be really lost. A way to overcome this difficulty is to consider the fact
that, at the critical limit, the matter will be launched into a keplerian orbit
around the star. Thus, probably, when the star reaches the critical limit an
equatorial disk is formed like for instance around Be stars. Here we suppose
that this disk will eventually dissipate by radiative effects and thus that the
material will be lost by the star.

Practically, in the present models, we remove the supercritical layers. This
removal of material allows the outer layers to become again subcritical at least
until secular evolution will bring again the surface near the critical limit (see [48]
for more details in this process). Secular evolution during the Main-Sequence
phase triggers two counteracting effects: on one side, the stellar surface expands.
Local conservation of the angular momentum makes the surface to slow down

---

[7] We suppose here that the vector v_{esc} is normal to the direction of the vector v_{crit}.

and the surface velocity to evolve away from the critical limit. On the other hand, meridional circulation continuously brings angular momentum to the surface and accelerates the outer layers. This last effect in general overcomes the first one and the star rapidly reach again the critical limit. How much mass is lost by this process? As seen above, the two above processes will maintain the star near the critical limit for most of the time. In the models, we adopt the mass loss rate required to maintain the star at about 95–98% of the critical limit. Such a mass loss rate is imposed as long as the secular evolution brings back the star near the critical limit. In general, during the Main-Sequence phase, once the critical limit is reached, the star remains near this limit for the rest of the Main-Sequence phase. At the end of the Main-Sequence phase, evolution speeds up and the local conservation of the angular momentum overcomes the effects due to meridional currents; the star evolves away from the critical limit and the imposed "critical" mass loss is turned off.

2 "Spinstars" at Very Low Metallicities?

Let us call "spinstars" those stars with a sufficiently high initial rotation in order to have their evolution significantly affected by rotation. In this section, we present some arguments supporting the view according to which spinstars might have been more common in the first generations of stars in the Universe. A direct way to test this hypothesis would be to obtain measures of surface velocity of very metal poor massive stars and to see whether their rotation is superior to those measured at solar metallicity. At the moment, such measures can be performed only for a narrow range of metallicities for Z between 0.002 and 0.020. Interestingly already some effects can be seen. For instance [22] presents measurements of the projected rotational velocities of a sample of 100 early B-type main-sequence stars in the Large Magellanic Cloud (LMC). He obtains that the stars of the LMC are more rapid rotators than their galactic counterparts and that, in both galaxies, the cluster population exhibits significantly more rapid rotation than that seen in the field (a point also recently obtained by [21]). More recently [40] obtain that the angular velocities of B (and Be stars) are higher in the SMC than in the LMC and MW.

For B-type stars, the higher values obtained at lower Z can be the result of two processes: (1) the process of star formation produces more rapid rotators at low metallicity; (2) the mass loss being weaker at low Z, less angular momentum is removed from the surface and thus starting from the same initial velocity, the low Z star would be less slowed down by the winds. In the case of B-type stars, the mass loss rates are however quite modest and we incline to favor the first hypothesis, i.e., a greater fraction of fast rotators at birth at low metallicity. Another piece of argument supporting this view is the following: in case the mass loss rates are weak (which is the case on the MS phase for B-type stars), the surface velocity is mainly determined by two processes, the initial value on the ZAMS and the efficiency of the angular momentum transport from the core to the envelope. In case of very efficient transport, the surface will receive significant amount of angular momentum transported from the core to the envelope. The main mechanism responsible for the transport of the angular momentum is meridional. The velocities of the meridional currents in the outer layers are smaller when the density is higher thus in more metal-poor stars. Therefore, starting from the same initial velocity on the ZAMS, one would expect that B-type stars at solar metallicity (with weak mass loss) would have higher surface velocities than the corresponding stars at low Z. The opposite trend is observed. Thus, in order to account for the higher velocities of B-type stars in the SMC and LMC, in the frame of the present rotating stellar models, one has to suppose that stars on the ZAMS have higher velocities at low Z. Very interestingly, the fraction of Be stars (stars rotating near the critical velocity) with respect to the total number of B stars is higher at low metallicity ([37, 61]). This confirms the trend discussed above favoring a higher fraction of fast rotators at low Z.

There are at least four other striking observational facts which might receive an explanation based on massive fast-rotating models. *First*, indirect observations indicate the presence of very helium-rich stars in the globular cluster ωCen [50]. Stars with a mass fraction of helium, Y, equal to 0.4 seem to exist, together with a population of normal helium stars with $Y = 0.25$. Other globular clusters appear to host helium-rich stars [4]; thus the case of ωCen is the most spectacular but not the only one. There is no way for these very low mass stars to enrich their surface in such large amounts of helium and thus they must have formed from protostellar cloud having such a high amount of helium. Where does this helium come from? We proposed that it was shed away by the winds of metal-poor fast-rotating stars [34].

Second, in globular clusters, stars made of material only enriched in H-burning products have been observed (see the review by [18]). Probably these stars are also enriched in helium and thus this observation is related to the one reported just above. The difference is that proper abundance studies can be performed for carbon, nitrogen, oxygen, sodium, magnesium, lithium, fluorine, etc., while for helium only indirect inferences based on the photometry can be made. [11] propose that the matter from which the stars rich in H-burning products are formed has been released by slow winds of fast-rotating massive stars. Of course, part of the needed material can also be released by AGB stars. The massive star origin presents however some advantages: first a massive star can induce star formation in its surrounding; thus two effects, the enrichment and the star formation, can be triggered by the same cause. Second, the massive star scenario allows to use a less flat IMF than the scenario invoking AGB stars [51]. The slope of the IMF might be even a Salpeter's one in case the globular cluster lost a great part of its first-generation stars by tidal stripping (see [12]).

Third, the recent observations of the surface abundances of very metal-poor halo stars[8] show the need of a very efficient mechanism for the production of primary nitrogen [8]. As explained in [9], a very nice way to explain this very efficient primary nitrogen production is to invoke fast-rotating massive stars. Very interestingly, fast-rotating massive stars help in explaining the behavior of not only the N/O ratio at low metallicity but also those of the C/O. Predictions for the behavior of the $^{12}C/^{13}C$ ratios at the surface of very metal-poor non-evolved stars have also been obtained [10].

Fourth, below about [Fe/H] <-2.5, a significant fraction of very iron-poor stars are C-rich (see the review by [1]). Some of these stars show no evidence of s-process enrichments by AGB stars and are thus likely formed from the ejecta of massive stars. The problem is how to explain the very high abundances with respect to iron of CNO elements. References [48] and [20] proposed that these stars might be formed from the winds of very metal-poor fast-rotating stars. It is likely that rotation also affects the composition of the ejecta of intermediate mass stars. Reference [48] predict the chemical composition of the envelope of a 7 $M_\odot$ E-AGB star which has been enriched by rotational mixing. The composition presents striking similarities with the abundance patterns observed at the surface of CRUMPS. The presence of overabundances of fluorine and of s-process elements might be used to discriminate between massive and intermediate mass stars.

All the above observations seem to point toward the same direction, an important population of spinstars at low Z. How many? What is the origin of the fast rotation? What are the consequences for the Gamma ray Burst progenitors? All these questions still have to be addressed in a quantitative way and offer nice perspective for future works.

Acknowledgement

My warm thanks to André Maeder whose enlightened theoretical developments allowed to explore the effects of rotation in stellar models.

References

1. Beers, T.C., Christlieb, N.: ARAA, **43**, 531 (2005)
2. Bodenheimer, P., Ostriker, J.P.: Rapidly rotating stars. VI. Pre-main – evolution of massive stars. ApJ **161**, 1101 (1970)
3. Bodenheimer, P., Ostriker, J.P.: Rapidly rotating stars. VIII. Zero-viscosity polytropic sequences. ApJ **180**, 159 (1973)
4. Caloi, V., D'Antona, F.: A&A **463**, 949 (2007)
5. Canuto, V.M.: ApJ **508**, 767 (1998)
6. Chaboyer, B., Zahn, J.-P.: A&A **253**, 173 (1992)

---

[8] These stars are in the field and present [Fe/H] as low as -4, thus well below the metallicities of the globular clusters.

7. Chaboyer, B., Demarque, P., Pinsonneault, M.H.: ApJ **441**, 865 (1995a)
8. Chiappini, C., Matteucci, F., Ballero, S.K.: A&A **437**, 429 (2005)
9. Chiappini, C., Hirschi, R., Meynet, G., Ekström, S., Maeder, A., Matteucci, F.: A&A Lett. **449**, 27 (2006)
10. Chiappini, C., Ekstroem, S., Hirschi, R., Meynet, G., Maeder, A., Charbonnel, C.: A&A Lett. **479**, 9 (2008)
11. Decressin, T., Charbonnel, C., Meynet, G.: A&A **475**, 859 (2007)
12. Decressin, T., Meynet, G., Charbonnel, C., Prantzos, N., Ekström, S.: A&A **464**, 1029 (2007)
13. Denissenkov, P.A., Ivanova, N.S., Weiss, A.: A&A **341**, 181 (1999)
14. Dwarkadas, V.V., Owocki, S.P.: ApJ **581**, 1337 (2002)
15. Endal, A.S., Sofia, S.: ApJ **210**, 184 (1976)
16. Ekström, S., Meynet, G., Maeder, A.: A&A **478**, 467 (2008)
17. Glatzel, W.: A&A **339**, L5 (1998)
18. Gratton, R., Sneden, C., Carretta, E.: ARAA **42**, 385 (2004)
19. Heger, A., Langer, N., Woosley, S.E.: ApJ **528**, 368 (2000)
20. Hirschi, R.: A&A **461**, 571 (2007)
21. Huang, W., Gies, D.R.: ApJ **648**, 591 (2006)
22. Keller, S.C.: PASP **21**, 310 (2004)
23. Kippenhahn, R., Thomas, H.C.: A simple method for the solution of the stellar structure equations including rotation and tidal forces. In: Slettebak, A. (ed.) Proc. IAU Coll. 4, Stellar Rotation, p. 20. Gordon and Breach Science Publishers (1970)
24. Lamers, H.J.G.L.M., Snow, T.P., Lindholm, D.M.: ApJ **455**, 269 (1995)
25. Langer, N., Fricke, K.J., Sugimoto, D.: A&A **126**, 207 (1983)
26. Maeder, A.: A&A **158**, 179 (1987)
27. Maeder, A.: A&A **299**, 84 (1995)
28. Maeder, A.: A&A **321**, 134 (Paper II) (1997)
29. Maeder, A.: A&A **347**, 185 (Paper IV) (1999)
30. Maeder, A.: A&A **399**, 263 (2003)
31. Maeder, A., Desjacques, V.: A&A **372**, L9 (2001)
32. Maeder, A., Meynet, G.: A&A **313**, 140 (1996)
33. Maeder, A., Meynet, G.: A&A **373**, 555, (Paper VII) (2001)
34. Maeder, A., Meynet, G.: A&A **448**, L37 (2006)
35. Maeder, A., Peytremann, E.: A&A **7**, 120 (1970)
36. Maeder, A., Georgy, C., Meynet, G.: A&A Lett. **479**, 37 (2008)
37. Maeder, A., Grebel, E.K., Mermilliod, J.-C.: A&A **346**, 459 (1999)
38. Maeder, A. Meynet, G.: A&A **361**, 159, (Paper VI) (2000)
39. Maeder, A., Zahn, J.P.: A&A **334**, 1000 (Paper III) (1998)
40. Martayan, C., Frémat, Y., Hubert, A.-M., Floquet, M., Zorec, J., Neiner, C., A&A **462**, 683 (2007)
41. Mathis, S., Palacios, A., Zahn, J.-P.: A&A **425**, 243 (2004)
42. Mathis, S., Zahn, J.-P.: A&A **425**, 229 (2004)
43. Mestel, L.: MNRAS **113**, 716 (1953)
44. Meynet, G., Maeder, A.: A&A **321**, 465 (Paper I) (1997)
45. Meynet, G., Maeder, Λ.: A&A **361**, 101, (Paper V) (2000)
46. Meynet, G., Maeder, A.: A&A **390**, 561, (Paper VIII) (2002)
47. Meynet, G., Maeder, A.: ASP Conf. **355**, 27 (2008)
48. Meynet, G., Ekström, S., Maeder, A.: The early star generations: the dominant effect of rotation on the CNO yields. A&A **447**, 623 (2006)
49. Pinsonneault, M.H., Kawaler, S.D., Sofia, S., Demarque, P.: ApJ **338**, 424 (1989)

50. Piotto, G., Villanova, S., Bedin, L.R., Gratton, R., Cassisi, S., Momany, Y., Recio-Blanco, A., Lucatello, S., Anderson, J., King, I.R., Pietrinferni, A., Carraro, G.: ApJ **621**, 777 (2005)
51. Prantzos, N., Charbonnel, C.: A&A **458**, 135 (2006)
52. Richard, D., Zahn, J.-P.: A&A **347**, 734 (1999)
53. Schwarzschild, M.: Structure and Evolution of the Stars, Princeton, Princeton University Press (1958)
54. Spiegel, E., Zahn, J.-P.: A&A **265**, 106 (1992)
55. Talon, S., Zahn, J.P.: A&A **317**, 749 (1997)
56. Talon, S., Zahn, J.P., Maeder, A., Meynet, G.: A&A **322**, 209 (1997)
57. Tassoul, J.L.: The effects of rotation on stellar structure and evolution. In: Willson, L.A., Stalio, R. (eds.) Angular Momentum and Mass Loss for Hot Stars, p. 7. Kluwer Acad. Publ. (1990)
58. van Loon, J.Th.: In: Lamers, H.J.G.L.M., Langer, N., Nugis, T., Annuk, K. (eds.) Stellar Evolution at Low Metallicity: Mass Loss, Explosions, Cosmology. ASP Conf. **353**, 211 (2006)
59. Vink, J.S., de Koter, A.: On the metallicity dependence of Wolf-Rayet winds. A&A **442**, 587 (2005)
60. von Zeipel, H.: MNRAS **84**, 665 (1924)
61. Wisniewski, J.P., Bjorkman, K.S.: ApJ **652**, 458 (2006)
62. Woosley, S.E.: ApJ **405**, 273 (1993)
63. Zahn, J.P.: Rotational instabilities and stellar evolution. In: Ledoux, P. (ed.) Proc. IAU Symp. 59, Stellar Instability and Evolution, p. 185. Reidel, Dordrecht (1974)
64. Zahn, J.-P. A&A **265**, 115 (1992)

Long Baseline Interferometry of Rotating Stars Across the HR Diagram: Flattening, Gravity Darkening, Differential Rotation

A. Domiciano de Souza

Lab. H. Fizeau, CNRS UMR 6525, Univ. de Nice-Sophia Antipolis, Observatoire de la Côte d'Azur, 06108 Nice Cedex 2, France
Armando.Domiciano@unice.fr

Abstract Stellar rotation has been for a long time considered as a second-order effect on theories of stellar structure and evolution. Modern observations proved that stellar rotation is a key parameter to explain many physical mechanisms on stars. In particular optical/IR long baseline interferometry (OLBI) became a key technique to study stellar rotation across the HR diagram. In this contribution we describe the most important OLBI results obtained in the field of stellar rotation since the beginning of the twenty-first century.

1 Introduction

Until a few years ago, rotation has been generally considered only as a second-order effect on theories of stellar structure and evolution. However, a number of serious discrepancies between current models and observations have been noticed (e.g. Maeder & Meynet [36]; Meynet, these proceedings).

In particular, rapid rotation affects stellar shapes and emitted local flux. Fast rotation seems also to be a key parameter to explain the "Be phenomenon" (the existence of epochs where some B stars present hydrogen lines in emission). The presence of differential rotation on stars is also invoked to explain several physical mechanisms, such as the stellar dynamo.

On the other hand, the observational techniques (spectroscopy, polarimetry, etc.) are nowadays very sensitive to the signatures induced by stellar rotation on the observations. In this context optical/IR long baseline interferometry (OLBI) is a technique particularly sensitive to sky-projected shapes and brightness distribution, and it can contribute to significantly improve our understanding of the rotation effects on stars.

Effectively, several amazing results in the field of stellar rotation have been obtained from OLBI observations performed with modern interferometers. These important results could be achieved from the combination of precise OLBI data with a detailed physical modelling work.

Domiciano de Souza, A.: *Long Baseline Interferometry of Rotating Stars Across the HR Diagram: Flattening, Gravity Darkening, Differential Rotation.* Lect. Notes Phys. **765**, 171–194 (2009)
DOI 10.1007/978-3-540-87831-5_7   © Springer-Verlag Berlin Heidelberg 2009

In the next section we present a brief introduction to OLBI, followed by many examples of recent results in the field of stellar rotation across the HR diagram as seen by interferometry.

2 Principles of Optical/IR Long Baseline Interferometry (OLBI)

In this section we briefly present some of the principles of the optical/IR long baseline interferometry (OLBI). A much more extensive and complete description of this modern observing technique can be found in the literature.

Interference fringes obtained from the combination of the light collected by two or more telescopes allow us to measure the *complex visibility*. The Van Cittert–Zernike theorem relates the complex visibility to the Fourier transform of the so-called intensity map of the object I_λ (i.e. the brightness distribution projected onto the sky) normalized by its value at the origin (i.e. the integrated intensity). For this work it is convenient to calculate the complex visibilities in the reference system presented in Fig. 1 so that

$$V(f_y, f_z, \lambda) = |V(f_y, f_z, \lambda)|\, e^{i\phi(f_y, f_z, \lambda)} = \frac{\widetilde{I_\lambda}(f_y, f_z)}{\widetilde{I_\lambda}(0,0)} \tag{1}$$

where $\widetilde{I_\lambda}$ is the Fourier transform of I_λ. Note that I and V are also functions of the wavelength λ. In general, the spatial frequencies f_y and f_z are denoted by u and v so that the Fourier plane is also called uv-plane.

OLBI normally works with diluted apertures (separate telescopes or masked mirrors) where each pair of apertures defines a vector baseline projected onto the sky $B_{\mathbf{proj}}$ along a given direction as indicated in Fig. 1. The spatial frequencies u and v to which we have access are defined by

$$(u, v) = B_{\mathbf{proj}}/\lambda_{\text{eff}} \tag{2}$$

where λ_{eff} is the effective wavelength of the spectral band considered. Ideally one would like to obtain as much information about the object as possible, which means a total coverage of the Fourier plane (at least until the physical limit imposed by the longest baseline available on the interferometer). From Eqs. (1) and (2) we can see that in order to increase the Fourier plane coverage one can (a) observe at several wavelengths and/or (b) observe at different baseline (lengths and orientations). Different baseline orientations can be easily obtained by simply observing at different times to make use of Earth rotation (Earth-rotation synthesis).

As an example we note that the complex visibility amplitude (Eq. (1)) for a spherical star with an angular diameter θ_{UD} and with a uniformly bright intensity map is given by

$$|V(z)| = |2J_1(z)/z| \;; \quad z = \pi\theta_{\text{UD}} B_{\text{proj}}/\lambda_{\text{eff}} \tag{3}$$

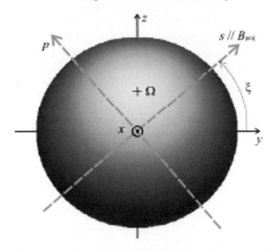

Fig. 1. Adopted reference system for a rotating star. The *cross* indicates the point where the rotation axis crosses the stellar surface. This rotation axis forms an angle i (not represented in the figure) with the observer's direction (x-axis) and its projection onto the sky is parallel to the z-axis. The sky-projected interferometer baseline $\boldsymbol{B}_{\mathrm{proj}}$ forms an angle ξ with the y-axis and defines a new sky-projected coordinate system (s, p) for which the s-direction is parallel to $\boldsymbol{B}_{\mathrm{proj}}$

where J_1 is the Bessel function of the first kind and first order. For the uniform disc model the complex visibility phase is 0 or $\pm 180°$ because the intensity map is centrally symmetric. We note that in the OLBI papers it is usual to adopt the symbol V to represent the complex visibility amplitude instead of the complex visibility.

In addition to the visibilities, other interferometric observables are also commonly used. For example the closure phase, available for simultaneous observations with three or more telescopes, gives direct information about the Fourier transform phase of the target. The closure phase is very useful to determine if the intensity distribution of the studied target is centrally symmetric or not. When a great number of visibilities and closure phases are available it is possible to perform an interferometric image reconstruction as in radio-interferometry (see Sect. 4). We present below another very useful observable accessible from the combination of spectroscopy and interferometry.

2.1 Differential Interferometry

Spectro-interferometry, also called differential interferometry (DI hereafter; Beckers [4]), is a modern technique that combines high spectral and angular resolution, being very sensitive to any physical mechanism inducing chromatic signatures. When considering stellar surfaces such mechanisms can be associated, for example, to stellar spots and large-scale mass motions (e.g. differential and/or rapid rotation, non-radial pulsations, shear currents produced by hydrodynamical instabilities).

DI measures the interference fringe's phase ϕ relative to two different spectral channels (λ and λ_r). For slightly resolved stars, ϕ is proportional to the stellar photocentre ϵ (the barycentre of the sky brightness distribution) difference between the two spectral channels:

$$\phi(\boldsymbol{u}, \lambda, \lambda_r) = -2\pi\boldsymbol{u} \cdot [\epsilon(\lambda) - \epsilon(\lambda_r)] \tag{4}$$

where the vector $\boldsymbol{u}$ is the spatial frequency. The usual procedure in DI is to take λ_r on the adjacent continuum of a spectral line and to put $\epsilon(\lambda_r) = \boldsymbol{0}$, which does not imply any loss of generality. By adopting a reference system as in Fig. 1 we can write the photocentre as

$$\epsilon(\lambda) = \epsilon_y(\lambda)\,\widehat{y} + \epsilon_z(\lambda)\,\widehat{z} \tag{5}$$

and

$$\epsilon_j(\lambda) = \frac{\int\int jI(\lambda, y, z)\,\mathrm{d}y\mathrm{d}z}{\int\int I(\lambda, y, z)\,\mathrm{d}y\mathrm{d}z} = \frac{\int\int jI(\lambda, y, z)\,\mathrm{d}y\mathrm{d}z}{F(\lambda)}\; ;\; j = y, z \tag{6}$$

where $\widehat{y}$ and $\widehat{z}$ are unit vectors, $F(\lambda)$ is the stellar flux spectrum and $I(\lambda, y, z)$ is the intensity map. Note that in OLBI the high angular information is given along the sky-projected baseline (Fig. 1). However, the components y and z can be recovered by data from at least two distinct baseline position angles. The photocentre is measured in a spatial scale and in this chapter it will be given in stellar angular radii units (ρ).

Further details on the DI formalism and its applications are given for example by Chelli and Petrov [7, 8], Vakili et al. [48], and Jankov et al. [27].

2.2 Operating Modern Interferometers

In Sect. 4 we present several recent interferometric results in the field of rapid stellar rotation. These results were obtained with modern interferometers that we very shortly describe here (in alphabetic order), without the pretension of being exhaustive:

- **CHARA** (Center for High Angular Resolution Astronomy; e.g. ten Brummelaar et al. [47]): The Georgia State University's interferometer operates an optical/IR interferometric array on the grounds of Mount Wilson Observatory in the San Gabriel Mountains of southern California. The six light-collecting telescopes of the CHARA Array, each of 1 m aperture, are distributed in a Y-shaped configuration providing 15 baselines ranging from 34.1 to 330.7 m. Several beam combiner instruments exist (from visible to near-IR).
- **NPOI** (Navy Prototype Optical Interferometer; e.g. Armstrong et al. [1]): located near Flagstaff, Arizona, is a long baseline optical interferometer including subarrays for imaging and for astrometry. The imaging subarray consists of six moveable 50 cm siderostats with baseline lengths from 2 to 437 m. The astrometric subarray consists of four fixed 50 cm siderostats with baseline lengths from 19 to 38 m. Spectral coverage ranges from 450 to 850 nm in 32 channels.

- **PTI** (Palomar Testbed Interferometer; e.g. Colavita et al. [11]): Long baseline infrared interferometer installed at Palomar Observatory, California. It was developed by the Jet Propulsion Laboratory, California Institute of Technology, for NASA as a testbed for interferometric techniques applicable to the Keck Interferometer. Baselines up to 110 m are available in the H and K bands.
- **VLTI** (Very Large Telescope Interferometer; e.g. Glindemann et al. [18], Petrov et al. [41]): Located on the summit of Cerro Paranal (Chile), this is the European Southern Observations (ESO) interferometer, which includes large 8 m telescopes called unit telescopes (UTs) and 2 m telescopes called auxiliary telescopes (ATs), but also the optical train that allows the light beam collected by the apertures to be conveyed to the combining instruments. Baseline lengths (AT and UT) range from 16 to 200 m. One mid-IR (MIDI) and one near-IR (AMBER) beam combiner instruments with spectral dispersion are currently operating.

3 Modelling Fast Rotation for Stellar Interferometry: The "Roche–von Zeipel" Model

Theory predicts that rotation can deeply impact the structure and evolution of stars (Meynet, these proceedings; Meynet and Maeder [36]). In particular, rapid rotation affects stellar shapes and emitted local flux. Since OLBI is a technique particularly sensitive to sky-projected shapes and brightness distribution, it can contribute to significantly improve our understanding of the rotation effects on stars.

Johnston and Wareing [29] carried out the first interferometric theoretical study of rotationally distorted stars in the approximation of slow uniform rotation and equator-on stars. Their calculations showed that interferometric measurements of rotational distortion were marginally feasible with the early 1970s interferometers. Modern interferometers, such as those listed in Sect. 2.2, attain the sensitivity and accuracy necessary to directly measure signatures of stellar rotation.

On the other hand, to determine rotational parameters from OLBI data it is also necessary to use physical models adapted to interpret the observations. The development of physically coherent numerical models offers nowadays the complete ingredients to build fast and refined numerical tools allowing a physical interpretation of OLBI observations of rotating stars. The combination of high-quality data and detailed astrophysical models allows a deeper understanding of several physical effects linked to stellar rotation such as geometrical flattening, gravity darkening and differential rotation. We present below a brief description of an astrophysical model developed to interpret OLBI observations on rotating stars.

3.1 The Roche Approximation

Most interferometry-oriented models used nowadays include Roche approximation, gravity darkening and local plane parallel atmospheres. In this section we present the principles and approximations commonly adopted on these models. The formalism and figures presented in this section are based on the work from Domiciano de Souza et al. [12].

The geometrical deformation caused by the rapid rotation of the star is calculated in the Roche approximation (e.g. Roche [45]; Kopal [33]):

(a) Uniform rotation with angular velocity Ω;
(b) All mass M is concentrated in a point at the centre of the star.

The equipotential surfaces for the Roche model are then given by:

$$\Psi\left(\theta\right) = \frac{\Omega^2 R^2\left(\theta\right)\sin^2\theta}{2} + \frac{GM}{R\left(\theta\right)} = \frac{GM}{R_p} \tag{7}$$

where $R(\theta)$ is the stellar radius at colatitude θ, R_p is the polar radius and G is the gravitational constant. In terms of acceleration ($g = -\nabla\Psi$) this potential has two components coming from the gravitational acceleration induced by a point source having the stellar mass and the centrifugal acceleration experienced from a particle at radius R rotating with velocity $v = \Omega R$.

Equation (7) can be rewritten as a cubic equation for $R(\theta)$, where analytical solutions can be found (e.g. Kopal [33]). Even if this analytical solution is used during the modelling, the final results are generally given in terms of parameters described below.

The degree of sphericity D of the star is given by

$$D \equiv \frac{R_p}{R_{eq}} = 1 - \frac{v_{eq}^2 R_p}{2GM} = \left(1 + \frac{v_{eq}^2 R_{eq}}{2GM}\right)^{-1} \tag{8}$$

where R_{eq} and v_{eq} are the equatorial linear radius and rotation velocity, respectively.

The critical or break-up velocity for the Roche model is attained when the centrifugal and gravitational forces are equal. At break-up the material at the stellar equator can easily scape to the circumstellar environment. Some parameters defined at break-up are the critical equatorial radius R_c, the critical equatorial linear v_c and angular Ω_c velocities.

Finally, several useful quantities relating critical and non-critical parameters are listed below:

$$\frac{R_c}{R_p} = \frac{3}{2} \equiv D_c \tag{9}$$

$$v_c = \Omega_c R_c = \sqrt{\frac{GM}{R_c}} = \sqrt{GM\left(\frac{2}{3R_p}\right)} \tag{10}$$

$$\frac{v_{eq}}{v_c} = \sqrt{3\left(1 - D\right)} = \frac{\Omega}{\Omega_c}\frac{R_{eq}}{R_c} \equiv \omega\frac{R_{eq}}{R_c} \tag{11}$$

$$\Omega_{\mathrm{c}} = \frac{v_{\mathrm{c}}}{R_{\mathrm{c}}} = \sqrt{\frac{8GM}{27R_{\mathrm{p}}^3}} \tag{12}$$

$$\frac{\Omega}{\Omega_{\mathrm{c}}} = \omega = \sqrt{3(1-D)}\left(\frac{3}{2}D\right) \tag{13}$$

3.2 The Gravity-Darkening Effect

In order to complete the description of the physical model we consider that the stellar atmosphere may be approximated locally by a plane parallel model with adequate effective temperature $(T_{\mathrm{eff}}(\theta))$ and effective surface gravity $(g(\theta) = |\nabla\Psi(\theta)|)$. We remind that $T_{\mathrm{eff}}(\theta)$ is related to the local bolometric radiative flux $F(\theta)$ by $F(\theta) = \sigma T_{\mathrm{eff}}^4(\theta)$, where σ is the Stefan–Boltzmann constant. For rotating stars the von Zeipel's [52] theorem says that the local flux is proportional to g, or alternatively, $T_{\mathrm{eff}} \propto g^{0.25}$.

This equation for the gravity darkening is strictly valid only for conservative rotation laws (centrifugal force derivable from a potential) and radiative flux in the diffusion approximation. For stars with convective envelopes, Lucy [34] showed that $T_{\mathrm{eff}} \propto g^{0.08}$. More generally, conservative rotation laws result in $T_{\mathrm{eff}} \propto g^\beta$ where the value of β depends on the different approximations chosen for the radiative transfer, opacity laws, model atmospheres, etc. (Claret [10]). The local effective temperature can thus be written as

$$T_{\mathrm{eff}}(\theta) = T_{\mathrm{p}}\left(\frac{g(\theta)}{g_{\mathrm{p}}}\right)^\beta \tag{14}$$

where T_{p} and g_{p} are the polar effective temperature and gravity, respectively.

The expression *gravity darkening* originates from the fact that, near the equator, the local effective gravity g decreases[1] leading to lower T_{eff} with a corresponding decrease in brightness. Figure 2 shows theoretical temperature maps of a star flattened by fast rotation. Due to this combined effect of geometrical deformation and gravity darkening, the star presents a variable aspect for different inclinations.

Stellar interferometry is sensitive to these different aspects induced by fast rotation. Figure 3 shows visibility curves in the region of the first and the second lobes calculated from the model of a fast-rotating star with $v_{\mathrm{eq}}/v_{\mathrm{c}} = 0.81$. Visibility curves along three different baseline directions are shown for models with flattening alone (upper rows) and models with flattening and gravity darkening (lower rows). Clearly, these two effects induce important signatures on the visibility curves: variation of the second lobe height, existence or not of a first zero in $|V|$, dependence of $|V|$ with baseline direction in the first lobe. In particular there is a partial cancellation of the geometrical deformation signature induced by the presence of gravity darkening. For stars rotating fast enough $(v_{\mathrm{eq}}/v_{\mathrm{c}} \geq 0.6 - 0.7)$ these signatures are strong enough to be measured by OLBI (see Sect. 4).

---

[1] At break-up velocity $g = 0$ at the equator.

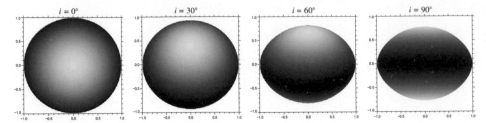

Fig. 2. Effective temperature maps for $R_{eq}/R_p = 1.28$ ($v_{eq}/v_c = 0.81$), $\beta = 0.25$ and selected inclinations i. The polar (maximum) and equatorial (minimum) effective temperatures are $T_p = 35,000\,\text{K}$ and $T_{eq} = 25,100\,\text{K}$, respectively. Axes are normalized by the equatorial radius. The projected geometrical deformation increases with higher inclinations but the stellar size in the horizontal (equatorial) direction is constant (Figure from Domiciano de Souza et al. [12])

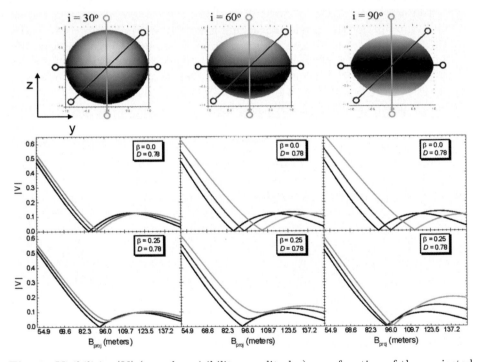

Fig. 3. Visibilities $|V|$ (complex visibility amplitudes) as a function of the projected baseline B_{proj} and inclination i for a modelled fast-rotating star with $v_{eq}/v_c = 0.81$ ($D = 0.78$; same star as in Fig. 2). Part of the first and the second $|V|$ lobes (attained by modern interferometers) are shown. The *upper* (*lower*) row shows curves of $|V|$ for models without (with) gravity darkening $\beta = 0$ ($\beta = 0.25$). The curves cover part of the first and the second $|V|$ lobes for three different projected baseline directions. For clarity, the corresponding temperature maps and baseline directions are illustrated on the *top* of the figure. An angular equatorial radius (ρ_{eq}) of 1 milliarcsec (mas) and an effective wavelength of $\lambda_{eff} = 665.4\,\text{nm}$ were adopted for the plots (Figure adapted from Domiciano de Souza et al. [12])

4 Rapid Rotation Across the HR Diagram: Results from Interferometry

The first attempt to measure the geometrical deformation of a rapidly rotating star was carried out with the Narrabri intensity interferometer on the bright star Altair (Hanbury Brown [19]). Indeed those observations remained too marginal to critically decide between different models of Altair. Today's operating stellar interferometers attain the required accuracies allowing us to constrain rotational parameters from OLBI data.

Since the beginning of the twenty-first century, rapid rotation effects have been directly measured by OLBI on five stars, namely (from hottest to coolest), Achernar, Regulus, Vega, Alderamin, and Altair. In the following we describe the main results obtained on these fast rotators with modern stellar interferometers. A list of physical parameters derived for each from several works is given in Table 1.

Table 1. Summary list with the main results obtained by several recent works based in interferometric observations of five rapidly rotating stars. Physical parameters are derived from the model fitting procedure and some are adopted based on previous works. We do not intend to be exhaustive in this table, but to illustrate how OLBI has been greatly contributing to the study of stellar rotation across the HR diagram

Star	SpType	T_p (K)	T_{eq} (K)	β	R_{eq}/R_p	R_{eq} ($R_\odot$)	v_{eq} (km/s)	i (°)	v_{eq}/v_c	Ref.
Achernar	B3Vpe	20,000	9,500	0.25	1.450	12.0	292	50.0	0.96	[44][3]
Regulus	B7V	15,400	10,314	0.25[1]	1.325	4.16	317	90.0	0.86	[30]
Vega	A0V	10,150	7,900	0.25[2]	1.230	2.78	270	4.7	0.75	[33]
		9,988	7,557	0.25[2]	1.246	2.87	270	4.5	0.77	[34]
Alderamin	A7IV-V	8,440	7,486	0.084[1]	1.298	2.82	283	88.2	0.83	[38]
Altair	A7IV-V	8,500	6,509	0.25[1]	1.237	2.12	277	55.0	0.76	[40]
		8,740	6,890	0.25[2]	1.215	1.99	273	63.9	0.73	[45]
		8,710	6,850	0.25[2]	1.217	2.02	271	62.7	0.73	[42]
		8,450	6,860	0.19[1]	1.221	2.03	286	57.2	0.75	[42]

[1] Derived value;
[2] Fixed value;
[3] Even a gravity-darkened Roche model at break-up cannot reproduce the observed flattening.

4.1 Achernar

The southern star Achernar (α Eri, HD 10144) is the brightest of all Be stars (V = 0.46 mag). A Be star is defined as a non-supergiant B-type star that has presented episodic Balmer lines in emission (Jaschek et al. [28]), whose origin is attributed to a circumstellar envelope (CSE) ejected by the star itself. Physical mechanisms like non-radial pulsations (NRP), magnetic activity, or binarity are invoked to explain the CSE formation of Be stars in conjunction with their fundamental property of rapid rotation. Achernar presents NRP (Vinicius et al. [51]) and also a companion (Kervella and Domiciano de Souza [32]).

Depending on the author (and the technique used) the spectral type of Achernar ranges from B3-B4IIIe to B4Ve (e.g. Slettebak [46]; Balona et al. [3]). The estimated projected rotation velocity $v_{eq} \sin i$ ranges from 220 to 270 km/s and the effective temperature T_{eff} from 15 000 to 20 000 K (e.g. Chauville et al. [7]; Vinicius et al. [51]). The difficulty in deriving more precisely these parameters is a direct consequence of the rapid rotation of Achernar.

Domiciano de Souza et al. [14] measured the apparent rotational flattening of Achernar using the VINCI instrument at the Very Large Telescope Interferometer (VLTI; see Sect. 2.2). They showed that the flattening ratio measured on this star is significantly higher than the limit imposed by the commonly adopted Roche approximation ($R_{eq}/R_p = 1.5$). These conclusions are mainly based on the following:

- Achernar spectra showing an Hα profile in absorption during the VLTI/VINCI observations. This absorption profile shows the absence of an important equatorial CSE that could be responsible for the most part of the measured deformation.
- a physical modelling taking into account both the geometrical deformation of Achernar and the von Zeipel effect, which mimics a less deformed star. This implies that the true flattening is found to be higher when the gravity darkening is included in the model (see Sect. 3).

These first VLTI/VINCI observations of Achernar combined with a physical modelling shed doubts on the classical assumption of Roche approximation (Fig. 4). Deviations from the Roche gravitational potential and the presence of differential rotation, both intimately related to the internal angular momentum distribution, are a promising explanation for such strong rotation deformation measured. Indeed, several differential rotation theories predict surface deformations stronger than that of uniform rotation by considering that the angular velocity increases towards the stellar centre. In particular, Jackson et al. [26] have shown that stellar models of Achernar including differential internal rotation result in a better agreement with the interferometric profile obtained by Domiciano de Souza et al. [14].

Based on a temporal analysis of Hα line profiles, Vinicius et al. [51] investigated more deeply the possibility of a CSE influence on the measured flattening. From a semi-quantitative modelling they concluded that a relatively important equatorial CSE was present (a disc with a radius of $R_{disc} \simeq 3R_\star$ contributing to 27% of the stellar flux) even if the Hα profile was mainly in absorption at the epoch of the VLTI/VINCI campaign. More recently, Kanaan et al. [30] and Carciofi and Domiciano de Souza (priv. comm.) have used more sophisticated models to show that a disc with a radius of $\simeq 3R_\star$ is not compatible with spectroscopic, polarimetric and interferometric observations. However, a smaller disc having an emitting region size of $\simeq 0.0 - 0.5R_\star$ and contributing to a few percent of the total flux cannot be totally excluded. Such residual disc could also

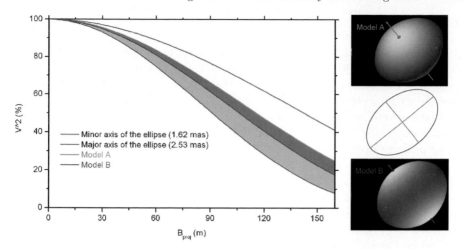

Fig. 4. Comparison of the equivalent uniform disc (UD) squared visibility (V^2) curves corresponding to the axis $2a$ and $2b$ of the best-fit ellipse (major axis $2a = 2.53 \pm 0.06$ milliarcsec (mas), a minor axis $2b = 1.62 \pm 0.01$ mas) with V^2 curves obtained from an interferometry-oriented code for fast rotators (Sect. 3). Based on the physical parameters of Achernar two limit solutions were computed: model A ($i = 50° \Rightarrow v_{eq} = 0.96v_c$) and B ($i = 90° \Rightarrow v_{eq} = 0.79v_c$). These models where calculated for $T_p = 20,000$ K, $M = 6\,M_\odot$ and $R_{eq} = 12.0\,R_\odot$. Clearly, the solutions enclosed by models A (*light grey* plus *dark grey* region) and B (*dark grey* region) cannot reproduce the observed highly oblate ellipse of Achernar (Figures adapted from Domiciano de Souza [13])

contribute to part of the strong flattening observed on Achernar, complementing the photospheric flattening caused by fast rotation.

In addition of being a rotationally deformed Be star, Achernar also presents a polar wind accounting for approximately 5% of the flux of the star in the near-IR (Kervella and Domiciano de Souza [31]). The presence of a polar wind was revealed by a complete analysis of the whole VLTI/VINCI observations available, covering a large range of baseline lengths and position angles. This CSE along the polar direction could be linked to free–free emission from the radiative pressure-driven wind originating from the hot polar caps of the star. Please refer to Stee (these proceedings) for a further discussion about the polar wind and the equatorial CSE of Achernar.

Additional observations with simultaneous and complementary techniques (spectroscopy, polarimetry and interferometry) are necessary to a detailed study of this intriguing and complex object so that we could more precisely evaluate the relative role of the many components present: rotational flattening, equatorial disc, polar wind, companion, gravity darkening and rotation law in the stellar interior, in the surface and in the disc.

4.2 Regulus

Regulus (α Leo) is a bright B7V star also known to be a rapid rotator ($v_{eq} \sin i$ between 250 and 350 km/s). McAlister et al. [35] performed K' band interferometric observations of Regulus using the CHARA interferometer (Sect. 2.2). The observations were performed with baselines ranging from $\simeq 190$ to $\simeq 330$ m spanning several azimuthal angles. The use of these very long baselines was crucial to study the rapid rotation of Regulus since this star has a rather small mean uniform disc angular diameter ($\bar{\theta}_{UD} = 1.47$ mas).

By adopting a model similar to the one presented in Sect. 3. McAlister et al. [35] could derive several physical parameters of Regulus from two distinct approaches. In the first approach a previous grid of models compatible with spectroscopic data was defined followed by a χ^2 minimization procedure applied to the observed visibilities in order to constraint the parameters that were not well defined by the spectroscopic data. The second approach consisted in performing a χ^2 minimization procedure only on the interferometric data (visibilities).

The authors found that both approaches lead to compatible results, except for the von Zeipel parameter β: 0.25 and 0.13 for the first and second approaches, respectively. They argued that the second approach was probably not sensitive enough to the value of β because, although the baselines are already rather long,

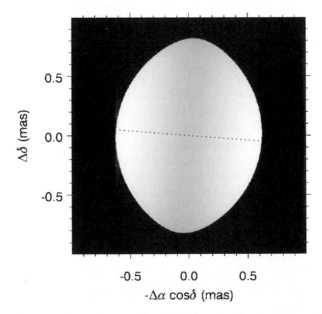

Fig. 5. K' band image (intensity map) of Regulus for the model constrained by the CHARA observations. The inclination is $i = 90°$ and the von Zeipel parameter is $\beta = 0.25$. The *dotted line* indicates the direction of the rotational axis (almost east–west) (Figure from McAlister et al. [35])

they did not reach the second visibility lobe where the signatures of gravity darkening become stronger (see Sect. 3 and Fig. 3). In addition, the CHARA data analysed did not include closure phases and/or spectrally dispersed fringes (differential phases), which are particularly sensitive to the detailed stellar surface intensity distribution profile (as discussed in Sect. 2.1).

Figure 5 shows the intensity map of Regulus in the K' band for the model constrained by the observations. The list of main physical parameters measured for Regulus is given in Table 1.

4.3 Vega

Although Vega (α Lyr), the second brightest star in the northern hemisphere, has a $v_{eq} \sin i$ of only $\simeq$ 15–20 km/s it has been suspected to be a rapidly rotating star seen nearly pole-on since many years (e.g. Gray [23, 24]).

As explained in Sect. 3, an almost pole-on star has a sky-projected intensity distribution that is mostly circular and centrally symmetric, which implies that it will not present a strong visibility amplitude variation as a function of the baseline position angle nor a very smooth transition of the phases (or closure phases) between 0 and $\pm180°$. However, a pole-on rapid rotator will present a strong apparent limb darkening because of the von Zeipel effect. Additionally, unless the star is exactly pole-on, we should expect a weak (but eventually detectable) signature of rotational deformation and gravity darkening on the visibilities and phases. Effectively, modern stellar interferometers are nowadays able to detect rather weak signatures of astrophysical processes on the intensity distributions.

In the particular case of Vega, two recent works based on OLBI observations have directly proved that this star is effectively a distorted, rapidly rotating star seen nearly pole-on (Aufdenberg et al. [2] and Peterson et al. [39, 40]). The results from these two works are based on distinct OLBI observables, namely visibilities and closure phases.

Aufdenberg et al. [2] analysed K' band visibilities measured on Vega with the CHARA array and FLUOR beam combiner (Sect. 2). Figure 6 (top) shows the uv-plane with the CHARA/FLUOR observations and a model of Vega in this Fourier plane. The lower panels show the measured and modelled squared visibilities together with the corresponding residuals from their best-fit model.

Considering the azimuthal coverage seen on the uv-plane, the visibility observations indicate that Vega is apparently a circular star. The fact that Vega does not show an apparent flattening has also been noticed on earlier interferometric studies (Ohishi et al. [38] and Ciardi et al. [9]). On the other hand, the low visibility values measured at the second lobe are a direct indication of an important centre to limb intensity variation. This anomalously high limb darkening measured on an almost circular star strongly suggests that Vega is a rapid rotator seen nearly pole-on.

Peterson et al. [39, 40] analysed interferometric observations of Vega performed in the visible with the NPOI (Sect. 2). The NPOI closure phases show

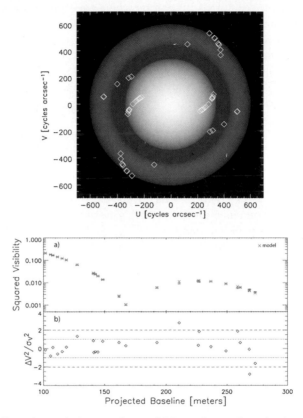

Fig. 6. *Left:* Sampling of the *uv*-plane of Vega obtained with the CHARA array. The *diamonds* represent the monochromatic sampling at 2 mμ within the K' band. The observations span the first and second visibility lobes. *Right:* CHARA V^2 data (and error bars) plotted as a function of projected baseline together with the best-fitting Roche–von Zeipel synthetic V^2 (see Table 1). The second lobe's height is significantly lower than the expected value for a normal limb-darkening effect. Residuals from the best-fit model are also shown in units of σ_{V^2} (Figures from Aufdenberg et al. [2])

a soft transition between values of 0 and $\pm180°$ (Fig. 7 left), which is a strong evidence of an asymmetric intensity distribution. Since Vega is known to have a low $v_{eq} \sin i$ and an apparent circular shape, the asymmetric intensity distribution observed can be interpreted as the signature of the von Zeipel effect (gravity darkening) seen on an almost pole-on star. Note that the star cannot be exactly pole-on, otherwise the phase signatures would be the same as a centrally symmetric object, i.e. 0 or $\pm180°$. The asymmetric intensity distribution derived by Peterson et al. [39, 40] is also shown in Fig. 7.

Peterson et al. [39, 40] and Aufdenberg et al. [2] fitted gravity-darkened Roche models (similar to the one presented in Sect. 3) to their data in order

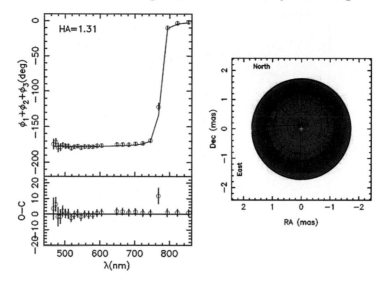

Fig. 7. *Left:* Example of closure phase observations of Vega (*open circles*) and errors, obtained with the NPOI facility in the visible. The *solid line* corresponds to the best-fit model (figure on the *right*). The residuals (observed, O, minus calculated, C) are plotted below. The fact that the closure phases do not show an abrupt transition between 0 and ±180° is a clear signal of the asymmetry in the intensity distribution. *Right:* Intensity distribution corresponding to the best-fit model of Vega to the NPOI visible data. *Blue* is bright, *red* is faint and the *orange* "+" is the subsolar point. The temperature drops more than 2400 K from pole to equator, creating an 18× drop in intensity compared to a 5× drop from limb darkening alone. Although the projected outline is almost perfectly circular, the polar diameter is only 80% of the equator (Figures from Peterson et al. [39, 40])

to derive several physical parameters for Vega (see Table 1). In particular they estimated that Vega rotates at $\omega = \Omega/\Omega_c \simeq 92\%$ of the angular break-up rate ($v_{eq}/v_c = 75\%$) and that the polar inclination angle is $i \simeq 4.8°$.

The fact that Vega is a gravity-darkened star has important implications on the use of this star as a fundamental standard calibrator [25].

4.4 Alderamin

The A7 IV-V star Alderamin (α Cep) was shown to be an oblate and gravity-darkened fast-rotating star by van Belle et al. [50] from Ks band observations performed with the CHARA array.

The flattening of Alderamin can be directly seen from the fit of an ellipse to the individual visibilities converted to uniform diameters (see Fig. 8). The physical parameters of Alderamin were derived with a model similar to the one presented in Sect. 3. The results from van Belle et al. [50] are summarized in Table 1. Although it is angularly smaller, the physical characteristics of Alderamin are similar to Altair (see next section).

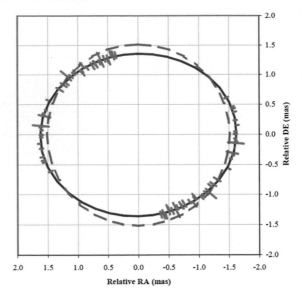

Fig. 8. Data points along the limb of Alderamin for a simple ellipsoidal fit to the uniform disc diameters calculated from each visibility data (Eq. (3)). The fitted ellipse has a major and minor axes of $2a = 1.625 \pm 0.056$ mas and $2b = 1.355 \pm 0.080$ mas. The *dashed line* represents a circular fit for the same data (Figure from van Belle et al. [50])

4.5 Altair

Altair (α Aql; spectral type A7IV-V) is the fast-rotating star that has been most extensively observed by modern interferometers, resulting in many interesting works dedicated to it. In particular, the first interferometric reconstructed image of a main sequence star was obtained for Altair. We describe here the main results from OLBI concerning Altair, one of the brightest stars in the northern sky (V = 0.77). The physical parameters related to stellar rotation derived in all these different works on Altair are summarized in Table 1.

The first successful interferometric observation of the rotational flattening of a fast rotator (Altair) was obtained with PTI in the K band (van Belle et al. [49]). By adopting an equivalent limb-darkened ellipse model these authors showed that Altair's apparent intensity distribution is $\simeq 14\%$ larger in one dimension than the other. After this first result, more extensive OLBI observations were carried out on Altair giving access to larger wavelengths ranges, baselines (lengths and position angles) and interferometric observables.

Altair was observed on May 2001 with three baselines (lengths of 37.5, 29.5 and 64.4 m) of NPOI over 32 spectral channels covering wavelengths from ~450 to 850 nm. Such observations gave rise to high SNR visibilities, closure phases and triple amplitudes around the first visibility minimum of Altair. Ohishi et al. [38] performed a first analysis of these data using a geometrical model to show that Altair not only is oblate but also has a non-centrally symmetric intensity distribution probably caused by a gravity-darkened effect induced by fast rotation.

Next, these NPOI data were analysed by two distinct groups (Domiciano de Souza et al. [16] and Peterson et al. [39, 40]) who fitted the data with physical models of fast rotators similar to the one presented in Sect. 3. Domiciano de Souza et al. [16] also included in their analysis the previously published K band data from PTI and new H and K bands data from VLTI/VINCI. Both works confirmed, from a physical modelling, that Altair is indeed a flattened and gravity-darkened fast-rotating star; the physical models derived by them are in agreement as shown in Table 1. The closure phases were crucial to constraint the gravity-darkening and inclination parameters of the models (see Fig. 9).

It is interesting to note that, because of gravity darkening, the derived effective temperatures indicate that Altair could present external layers which are radiative at the poles and convective at the equator. The hypothesis of a convective equatorial region is supported by other works showing that Altair has a chromosphere and a corona, possibly linked to sub-photospheric convective zones (e.g. Ferrero et al. [17]).

More recently, Monnier et al. [37] observed Altair in the H band using simultaneously four telescopes of the CHARA array (MIRC instrument) to synthesize an elliptical telescope with dimensions 265×195 m (Fig. 10 left). From the mea-

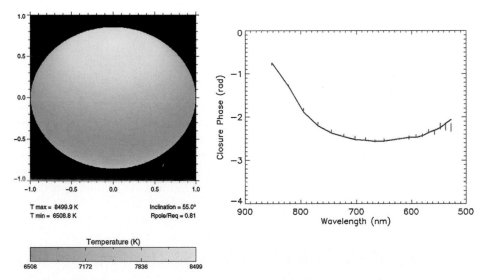

Fig. 9. *Left:* Effective temperature map from the best-fit model obtained for Altair. *Right:* Observed NPOI closure phases (one scan only is shown) and corresponding errors together with the closure phases from the best-fit model (*solid line*). The smooth variation of the closure phase with wavelength (and thus with the spatial frequency) is a strong signature of a non-centrally symmetric intensity distribution (Figures adapted from Domiciano de Souza et al. [16])

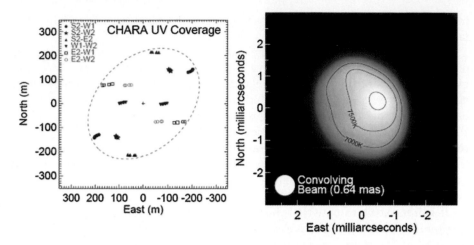

Fig. 10. *Left:* Fourier *uv*-coverage for the Altair observations with the CHARA array, where each point represents the sky-projected separation between one pair of telescopes. The *dashed ellipse* shows the equivalent coverage for an elliptical aperture of 265×195 m, representing the aperture of a telescope that would have an angular resolution equivalent (~ 0.64 mas) to these interferometric observations. *Right:* Intensity distribution at the surface of Altair obtained from image reconstruction using interferometric data recorded with the CHARA array in the H band. Image reconstruction is equivalent to performing an inverse Fourier transform of the interferometric data. In practice, because of the presence of noise and because only a very limited fraction of the *uv*-plane is observed, the image reconstruction is performed using dedicated algorithms such as the maximum entropy method (MEM) in this case. Contours indicate the blackbody temperatures derived from the measured intensities (Figures from Monnier et al. [37])

sured visibilities and closure phases they could reconstruct an image of Altair with an unprecedent angular resolution: ~ 0.64 mas (Fig. 10 right). The interferometric image of Altair shows the signatures of a fast rotator independently of any modelling: an elongated object with an non-centrally symmetric intensity distribution. To estimate physical parameters of Altair related to fast rotation (Table 1) the authors used, nevertheless, a model similar to the one presented in Sect. 3.

5 Stellar Differential Rotation from Spectro-interferometry

In this section, we show how spectro-inteferometry (briefly described in Sect. 2.1) can be used to measure differential rotation on stellar surfaces. The results and formalism presented here are based on the work from Domiciano de Souza et al. [15].

Let us consider a spherical star with radius R presenting a solar-like latitudinal angular differential rotation law on its surface:

$$\Omega(l) = \Omega_{\text{eq}}(1 - \alpha \sin^2 l) = \frac{v_{\text{eq}}}{R}(1 - \alpha \sin^2 l) \qquad (15)$$

where l is the latitude and Ω_{eq} and v_{eq} are the angular and linear velocities at the equator, respectively. The differential rotation parameter $\alpha = (\Omega_{\text{eq}} - \Omega_{\text{p}})/\Omega_{\text{eq}}$ can be negative (acceleration of Ω towards the pole) or positive (acceleration of Ω towards the equator).

Because of this differential rotation the maps of rotational velocity projected onto the observer's direction (v_{proj}) take the form of curved lines instead of the usual vertical straight lines found on uniform rotation. Some examples of these radial velocity maps are shown in Fig. 11 where one can see that they are functions of both α and of the polar inclination i.

These curved velocity maps generated by differential rotation can significantly distort, either deepening or shallowing, the spectral line profiles in comparison with those coming from uniformly rotating stars.

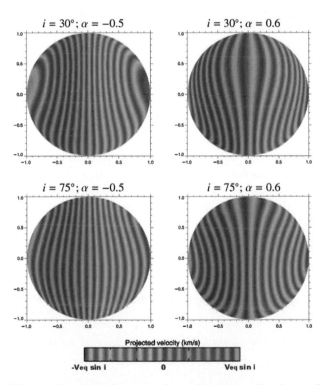

Fig. 11. Radial velocity maps for several combinations of polar inclination i and differential rotation parameter $\alpha = (\Omega_{\text{eq}} - \Omega_{\text{p}})/\Omega_{\text{eq}}$ (assuming a solar-like differential rotation law). A positive velocity corresponds to displacements towards the observer (blue shifts). For differential rotation, regions of constant projected velocity (equal velocity strips) are not straight vertical ones as it is the case for rigid rotation ($\alpha = 0$). A direct consequence of these curved equal velocity strips is that the radial velocity maps become functions of i

By using a method based on the Fourier transform (FT) of photospheric line profiles it is possible to infer (from the zeroes in the FT) several stellar parameters such as the projected rotational velocity and the differential rotation parameter α (e.g. Gray [20, 21]). Recently Reiners and Schmitt [42, 43], Reiners and Royer [44] and others detected differential rotation on several A- and F-type stars using this FT-based technique for spectroscopic observations. For a deeper description of this technique and corresponding results refer to Royer (these proceedings).

When it comes to differential rotation, several authors have shown that this FT technique applied to spectroscopic data alone has difficulties in disentangling distinct effects such as differential rotation, inclination angle and spots [5, 22, 42, 43]. In the following we show how spectro-interferometry allows us to disentangle these effects and, at the same time, allows the use of all Fourier values instead of only the zeroes.

The curved radial velocity maps induced by differential rotation influence the photospheric lines as well as the photocentre position (defined in Sect. 2.1) as a function of the wavelength. Figure 12 shows some examples of intensity maps at selected wavelengths inside a photospheric line for a differentially rotating star together with the spectral line and the photocentre components. Differential rotation changes the spectral profile and the photocentre components ϵ_y and ϵ_z (given in the same reference system of Fig. 1). In particular we have $\epsilon_z(\lambda) \neq 0$, contrary to the uniform rotation case where this quantity is always equal to zero (because the radial velocity maps take the form of vertical strips).

In order to extract the information about differential rotation from the DI observables (spectra and photocentres) one can apply a similar approach used for spectroscopic data alone, i.e. the Fourier transform method. Like for the spectra $F(\lambda)$ (see Royer, these proceedings) the photocentre components can also be expressed as the convolution of a corresponding rotational profile ($G_y(\lambda)$ and $G_z(\lambda)$) and a local intrinsic line profile $H(\lambda)$ (assumed constant over the stellar surface):

$$
\begin{pmatrix} F(\lambda) \\ \epsilon_y(\lambda) \\ \epsilon_z(\lambda) \end{pmatrix} = \begin{pmatrix} 1 \\ F^{-1}(\lambda) \\ F^{-1}(\lambda) \end{pmatrix} H(\lambda) * \begin{pmatrix} G(\lambda) \\ G_y(\lambda) \\ G_z(\lambda) \end{pmatrix} \tag{16}
$$

By evaluating the FT of the quantities above and by using the fact that a convolution becomes a multiplication in the Fourier space we obtain

$$
FT \begin{pmatrix} F(\lambda) \\ \epsilon_y(\lambda) F(\lambda) \\ \epsilon_z(\lambda) F(\lambda) \end{pmatrix} = \widetilde{H}(\sigma) \times \begin{pmatrix} \widetilde{G}(\sigma) \\ \widetilde{G}_y(\sigma) \\ \widetilde{G}_z(\sigma) \end{pmatrix} \tag{17}
$$

where $FT()$ is the Fourier transform operator and the tilded quantities represent the respective Fourier transforms. One can thus define two new quantities independent of H:

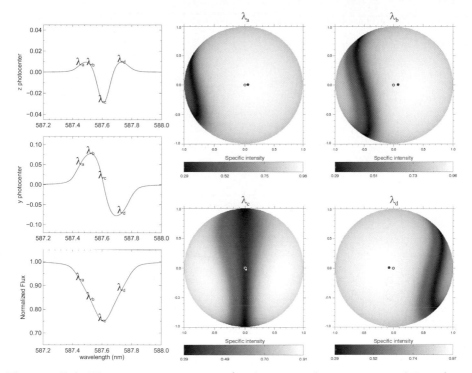

Fig. 12. *Left:* Photocentre components (in the same reference system of Fig. 1), ϵ_z (*top*) and ϵ_y (*middle*), and the normalized spectral flux (*bottom*) across the asymmetric He I $\lambda5876$ line. The calculations were performed for $v_{eq} \sin i = 100$ km/s, $\log g = 4.0$, $T_{eff} = 20{,}000$ K, $i = 45°$ and $\alpha = 0.6$. Photocentre components are given in units of angular stellar radius ρ. The *letters* indicate selected wavelengths corresponding to a region in the *blue line* wing (λ_a), the highest ϵ_y value (λ_b), the central wavelength (λ_c) and the most positive value of ϵ_z (λ_d). *Right:* Intensity maps associated to the DI observables on the *left* side at the four selected wavelengths. The *curved dark* patterns correspond to Doppler shifts of the local line profile caused by differential rotation. These non-symmetrical intensity maps result in a displacement of the stellar photometric barycentre, i.e. the photocentre (*filled circles*), relative to the geometrical centre (*opened circles*)

$$R_y(\sigma) \equiv \frac{|FT(\epsilon_y(\lambda) F(\lambda))|}{|FT(F(\lambda))|} = \frac{|\widetilde{G}_y(\sigma)|}{|\widetilde{G}(\sigma)|} \tag{18}$$

and

$$R_z(\sigma) \equiv \frac{|FT(\epsilon_z(\lambda) F(\lambda))|}{|FT(F(\lambda))|} = \frac{|\widetilde{G}_z(\sigma)|}{|\widetilde{G}(\sigma)|} \tag{19}$$

where the ratios $R_y(\sigma)$ and $R_z(\sigma)$ are valid for $\widetilde{G}(\sigma) \neq 0$. These two quantities are independent from H at all Fourier frequencies σ. Figure 13 shows contour plots of $R_y(\sigma)$ and $R_z(\sigma)$ in the $\alpha - i$ plane for a given Fourier frequency. The dependence of $R_y(\sigma)$ with α and i is similar to what is obtained from

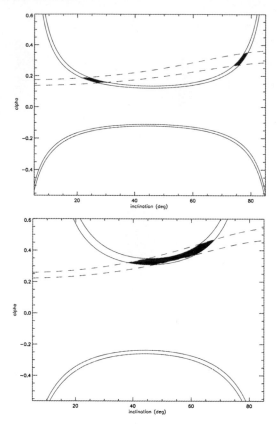

Fig. 13. Two examples of the intersection between R_y (*dashes*) and R_z (*solid*). The finite width for these quantities corresponds to uncertainties of 1% for R_y and 5% for R_z. These values are compatible with the expected uncertainties for modern interferometers and with the relative signal amplitude of the photocentres ϵ_y and ϵ_z. Note that α and i are well constrained (*dark regions*) only when R_y and R_z are considered simultaneously

spectroscopy. Thus, the parameters α (solar-like differential rotation parameter) and i (inclination) can be measured since the quantities $R_y(\sigma)$ and $R_z(\sigma)$ do not depend on those parameters in same manner.

These results show that DI combined with the Fourier transform method is an important tool to study stellar surface differential rotation. However, rather high instrumental performances are needed to determine consistently the required free parameters, notably α and i. In particular, precise differential phase measures as well as high angular and spectral resolution are required. Two modern spectro-interferometers have the required precision, spectral and spatial resolution, and sensitivity required to the study of stellar surfaces structures from DI: VLTI/AMBER (already operating) and CHARA/VEGA (expected to be fully operational in 2008).

6 Conclusions and Future Prospects

We presented the main results in the field of rapid stellar rotation recently obtained from OLBI observations. OLBI from several modern interferometers allowed the determination of many physical parameters of five rapidly rotating stars across the HR diagram (Table 1): Achernar, Regulus, Vega, Alderamin and Altair.

These results proved that to constrain the physical parameters of the stars it is crucial to use physically coherent models oriented to stellar interferometry. Moreover, the scientific output from OLBI data is often greatly improved when one combines several observing techniques such as interferometry, spectroscopy, photometry and polarimetry.

A technique allowing us to measure differential rotation from operating or near-to-operating spectro-interferometers has also been described.

Finally, we would like to note that the results presented here prove that modern OLBI has attained a very mature age. In particular this observing technique now allows us to reconstruct images of stars with ≤ 1 mas resolution. Such images allow us to directly resolve and study large-to-medium spatial structures on the surface of nearby stars (main sequence and/or evolved ones).

In the near future one can expect to obtain very precise and high angular resolution images of stars from the next generation of stellar interferometers. These future imaging interferometers are already under construction and/or study. As examples we mention CHARA/VEGA (2008; PI: D. Mourard), VLTI/MATISSE (around 2009–2010; PI: B. Lopez) and KEOPS: Kiloparsec Explorer for Optical Planet Search (around 2015; PI: F. Vakili). Images with resolutions $\simeq 0.1$ mas can be expected for these next generation of stellar interferometers, which implies *a few hundred resolution elements* over the visible area of bright stars.

Acknowledgment

The author is very thankful to J.-P. Rozelot and C. Neiner for inviting him to participate in the CNRS school "La rotation du Soleil et des étoiles".

References

1. Armstrong, J.T., Mozurkewich, D., Rickard, L.J.: ApJ **496**, 550 (1998)
2. Aufdenberg, J.P., Mérand, A., Coudé du Foresto, V., et al.: ApJ **645**, 664 (2006)
3. Balona, L.A., Engelbrecht, C.A., Marang, F.: MNRAS **227**, 123 (1987)
4. Beckers, J.M.: Opt. Acta **29**, 361 (1982)
5. Bruning, D.H.: ApJ **248**, 274 (1981)
6. Chauville, J., Zorec, J., Ballereau, D., et al.: A&A **378**, 861 (2001)
7. Chelli, A., Petrov, R.G.: A&AS **109**, 389 (1995a)
8. Chelli, A., Petrov, R.G.: A&AS **109**, 401 (1995b)
9. Ciardi, D.R., van Belle, G.T., Akeson, R.L., et al.: ApJ **559**, 1147 (2001)
10. Claret, A.: A&A **131**, 395 (1998)

11. Colavita, M.M., Wallace, J.K., Hines, B.E.: ApJ **510**, 505 (1999)
12. Domiciano de Souza, A., Vakili, F., Jankov, S., Janot-Pacheco, E., Abe, L.: A&A **393**, 345 (2002)
13. Domiciano de Souza, A.: PhD Thesis, Université de Nice-Sophia Antipolis (2003)
14. Domiciano de Souza, A., Kervella, P., Jankov, S., et al.: A&A **407**, L47 (2003)
15. Domiciano de Souza, A., Zorec, J., Jankov, S., et al.: A&A **418**, 781 (2004)
16. Domiciano de Souza, A., Kervella, P., Jankov, S., et al.: A&A **442**, 567 (2005)
17. Ferrero, R.F., Gouttebroze, P., Catalano, S., et al.: ApJ **439**, 1011 (1995)
18. Glindemann, A., Albertsen, M., Andolfato, L.: SPIE **5491**, 447 (2004)
19. Hanbury Brown, R.: In: The Intensity Interferometer, p. 151. Taylor & Francis LTD, London (1974)
20. Gray, D.F.: ApJ **184**, 461 (1973)
21. Gray, D.F.: ApJ **202**, 148 (1975)
22. Gray, D.F.: ApJ **211**, 198 (1977)
23. Gray, R.O.: JRASC **79**, 237 (1985)
24. Gray, R.O.: JRASC **82**, 336 (1988)
25. Gray, R.O.: ASPC **364**, 305 (2007)
26. Jackson, S., MacGregor, K.B., Skumanich, A.: ApJ **606**, 1196 (2004)
27. Jankov, S., Vakili, F., Domiciano de Souza, A., Janot-Pacheco, E.: A&A **377**, 721 (2001)
28. Jaschek, M., Slettebak, A., Jaschek, C.: Be Star Newsletter **4**, 9 (1981)
29. Johnston, I.D.: Wareing, N.C.: MNRAS **147**, 47 (1970)
30. Kanaan, S., Meilland, A., Stee, Ph., Zorec, J., F., Domiciano de Souza, A., et al.: A&A **486**, 785 (2008)
31. Kervella, P., Domiciano de Souza, A.: A&A **453**, 1059 (2006)
32. Kervella, P., Domiciano de Souza, A.: A&A **474**, L49 (2007)
33. Kopal, Z.: Ap&SS **133**, 157 (1987)
34. Lucy, L.B.: Z. Astrophys. **65**, 89 (1967)
35. McAlister, H.A., ten Brummelaar, T.A., Gies, D.R., et al.: ApJ **628**, 439 (2005)
36. Maeder, A., Meynet, J.: Annu. Rev. Astron. Astrophys. **38**, 143 (2000)
37. Monnier, J.D., Zhao, M., Pedretti, E., et al.: Science **317**, 342 (2007)
38. Ohishi, N., Nordgren, T.E., Hutter, D.J.: ApJ **612**, 463 (2004)
39. Peterson, D.M., Hummel, C.A., Pauls, T.A., et al.: Nature **440**, 896 (2006a)
40. Peterson, D.M., Hummel, C.A., Pauls, T.A., et al.: ApJ **636**, 108 (2006b)
41. Petrov, R.G., Malbet, F., Weigelt, G., et al.: A&A **464**, 1 (2007)
42. Reiners, A., Schmitt, J.H.M.M.: A&A **384**, L155 (2002)
43. Reiners, A., Schmitt, J.H.M.M.: A&A **412**, 813 (2003)
44. Reiners, A., Royer, F.: A&A **415**, 325 (2004)
45. Roche, E.A.: Mém. de l'Acad. de Montpellier (Section des Sciences) **8**, 235 (1837)
46. Slettebak, A.: ApJS **50**, 55 (1982)
47. ten Brummelaar, T.A., McAlister, H.A., Ridgway, S.T.: ApJ **628**, 453 (2005)
48. Vakili, F., Mourard, D., Bonneau, D., Morand, F., Stee, P.: A&A **323**, 183 (1997)
49. van Belle, G.T., Ciardi, D.R., Thompson, R.R., et al.: ApJ **559**, 1155 (2001)
50. van Belle, G.T., Ciardi, D.R., ten Brummelaar, T., et al.: ApJ **637**, 494 (2006)
51. Vinicius, M.M.F., Zorec, J., Leister, N.V., Levenhagen, R.S.: A&A **446**, 643 (2006)
52. von Zeipel, H.: MNRAS **84**, 665 (1924)

Is the Critical Rotation of Be Stars Really Critical for the Be Phenomenon?

Ph. Stee and A. Meilland

Université Nice Sophia-Antipolis (UNSA), Observatoire de la Côte d'Azur,
Département FIZEAU – CNRS Avenue Copernic, 06130 Grasse, France
Philippe.Stee@obs-azur.fr

Abstract We aim to study the effect of the fast rotation, stellar wind and circumstellar disks around active hot stars and their effects on the formation and evolution of these massive stars. For that purpose, we obtained, for the first time, interferometric measurements of three active hot stars, namely α Arae, κ CMa and Achernar, using the VLTI /AMBER and VLTI/MIDI instruments which allow us to study the kinematics of the central star and its surrounding circumstellar matter. These data coupled with our numerical code SIMECA (SIMulation pour Etoiles Chaudes Actives) seem to indicate that the presence of equatorial disks and polar stellar wind around Be stars are not correlated. A polar stellar wind was detected for α Arae and Achernar whereas κ CMa seems to exhibit no stellar wind. On the other hand, these two first Be stars are certainly nearly critical rotators whereas the last one seems to be far from the critical rotation. Thus a polar stellar wind may be due to the nearly critical rotation which induces a local effective temperature change following the von Zeipel theorem, producing a hotter polar region triggering a polar stellar wind. This critical rotation may also explain the formation of a circumstellar disk which is formed by the centrifugal force balancing the equatorial effective gravity of the central star. Following these results we try to investigate if critical rotation may be the clue for the Be phenomenon.

1 Introduction

Stellar rotation is certainly a key in our understanding of stellar formation and evolution of massive stars. The fast rotation of active hot ($\sim$20,000 K), massive (10 $M_\odot$) and luminous (L $\sim 10^6$ $L_\odot$) stars, often accompanied by a stellar wind, which leads to the formation of a circumstellar disk, seems to be a major physical effect to finally "slow down" the very high rotational rate of these stars. This very fast rotation may also leads to a situation where the star reaches its "critical" rotation or "breakup" velocity where the centrifugal force counterbalances the gravitational force at the equator. Thus, the matter is no more bounded to the central star and can form a dense equatorial circumstellar disk.

Moreover, it is now well established that at low metallicity, stellar winds are less efficient and thus cannot evacuate enough angular momentum from those fast rotators, which implies that the first generation of massive stars (at zero

Stee, Ph., Meilland, A.: *Is the Critical Rotation of Be Stars Really Critical for the Be Phenomenon?*. Lect. Notes Phys. **765**, 195–205 (2009)
DOI 10.1007/978-3-540-87831-5_8

metallicity) may be all critical rotators! More recently, it seems that active hot stars form as very rapid rotators when they arrived on the main sequence, so that only very few "classical" hot stars may present what is called the "Be phenomenon" through spin up. A detail study of these effects can be found in Martayan et al. [1, 2].

This "Be phenomenon" is finally related to hot stars that have at least exhibited once Balmer lines in emission with infrared excess produced by free–free and free–bound processes in an extended circumstellar disk. There is now a strong evidence that the disk around the Be star α Arae is Keplerian (Meilland et al. [3]) and that this dense equatorial disk is slowly expanding. On the other side there are also clear pieces of evidence for a polar-enhanced wind. This was already predicted for almost critically rotating stars as for a large fraction of Be stars. In 2006, Kervella and Domiciano de Souza [4] showed an enhanced polar wind for the Be star Achernar whereas this Be star presented no hydrogen lines in strong emission during the observations. Thus, it seems that a significant polar wind may be present even if the star is still in a normal B phase, signifying this enhanced polar wind would not be related to the existence of a dense equatorial envelope. However many issues remain unsolved on the actual structure of the circumstellar envelopes in Be stars which probably depends on the dominant mass ejection mechanisms from the central star and on the way the ejected mass is redistributed in the near circumstellar environment. In the following, we present our results for three Be stars, α Arae, κ CMa and Achernar, and try to draw a general scheme regarding the Be phenomenon and the disk formation.

2 α Arae

The star α Arae (HD 158 427, HR 6510, B3 Ve), one of the closest (d=74 pc, Hipparcos, Perryman et al. [5]) Be stars, was observed with the VLTI/MIDI instrument at 10 μm in June 2003 and its circumstellar environment was unresolved even with the 102 m baseline by Chesneau et al. [6]. α Arae was a natural choice as first target due to its proximity but also its large mid-IR flux and its high infrared excess among other Be stars, e.g., E(V-L)~1.8 and E(V-12 μm)~2.23. These first IR interferometric measurements indicated that the size of the circumstellar environment was smaller than predicted by Stee [7] for the K band. The fact that α Arae remains unresolved, but at the same time had strong Balmer emission, has put very strong constraints on the parameters of its circumstellar disk. Independently of the model, they have obtained an upper limit of the envelope size in the N band of $\phi_{max}=$ 4 mas, i.e., 14 R$_\star$ if the star is at 74 pc according to Hipparcos parallax or 20 R$_\star$ if the star is at 105 pc as suggested by the model presented in Chesneau et al. [6].

They finally propose a scenario where the circumstellar environment remains unresolved due to an outer truncation of the disc by an unseen companion. Nevertheless, this companion would be too small and too far away to have any influence on the Be phenomenon itself.

We have observed again α Arae with the VLTI/AMBER instrument during a Science Demonstration Time (SDT) run, on the nights of February 23 and 24, 2005, in medium-resolution mode (R=1500). On the night of 24 February, the observations were made with two UT telescopes, i.e., one interferometric baseline only, and consist of six exposure files. On the following night, three telescopes were used and three baseline data were taken in a series of three exposure files and another series of two. The data have been reduced using the "ammyorick" package developed by the AMBER consortium.[1]

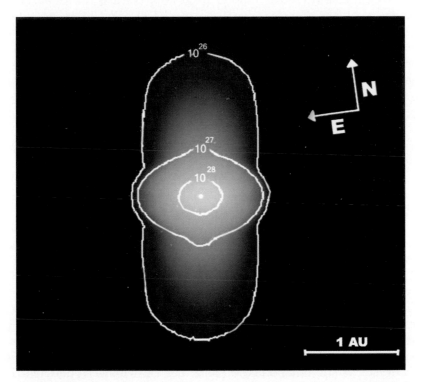

Fig. 1. Intensity map in the continuum at 2.15 μm obtained with SIMECA for our best model parameters. The inclination angle is 55°, the central *bright* region is the flux contribution from the thin equatorial disk whereas the smoother regions originate from the stellar wind. The *brightness* contrast between the disk and the wind is globally ∼30 but can reach 100 if you compare the inner region of the disk with the outer parts of the wind

2.1 Stellar Rotation

We found that α Arae must be rotating very close to its critical velocity since we obtain v_{rot}/v_{crit} ∼97%. This value is far above the conservative estimates

[1] See http: //www−laog.obs.ujf−grenoble.fr/heberges/amber
/article.php3?id_article=81

of ∼75% usually found in the literature for Be stars. The fact that Be stars may be rotating much closer to their critical velocities than it is generally supposed was already outlined by Townsend et al. [8] and Owocki [9]. This nearly critical rotation has quite profound implications for dynamical models of Be disk formation and may be the clue for the Be phenomenon. It may bring sufficient energy to levitate material in a strong gravitational field or at least help other physical processes such as pulsation or gas pressure to provide sufficient energy and angular momentum to create a circumstellar disk. Moreover, observational evidences of this nearly critical rotation are growing, such as the results obtained by Domiciano et al. [10] using interferometric VLTI/VINCI data of Achernar. They measured a rotationally distorted photosphere with an apparent oblateness of 1.56 which cannot be explained using the classical Roche approximation. This scenario follows the original picture by Struve [11] of a critically rotating star, ejecting material from its equatorial regions.

2.2 Polar Wind Enhancement

Our interferometric measurements are also evidencing a polar wind enhancement (see Fig. 1) which was already predicted for almost critically rotating stars. For instance, Cranmer and Owocki [12] and Owocki and Gayley [13] studied the effects of limb darkening, gravity darkening and oblateness on the radiation driving mechanism and found that the tendency for the higher flux from the bright poles to drive material toward the darker equatorial region is outweighed by the opposite tendency for the oblateness of the stellar surface to direct the radiative flux to higher latitudes, i.e., away from the equator. The paper review by Porter and Rivinius [14] also outlines the effect of the inclusion of nonradial line-driving force which reduces the effect of the wind compression to zero and, taking into account the gravity darkening, results in a polar wind enhancement. In a 2006 paper, Kervella and Domiciano de Souza [4] have evidenced an enhanced polar wind for the Be star Achernar whereas this Be star presents no hydrogen lines in strong emission. Thus, it seems that a significant polar wind may be present even if the star is still in a normal B phase, i.e., this enhanced polar wind does not seem to be related to the existence of a dense equatorial envelope, as already outlined by Kervella and Domiciano de Souza [4].

2.3 New Results for α Arae

1. We were able to propose a possible scenario for its circumstellar environment which consists of a thin equatorial disk + polar-enhanced winds that are successfully modeled with our SIMECA code.
2. We found that the disk around α is compatible with a dense equatorial matter confined in the central region whereas a polar wind is contributing along the rotational axis of the central star. Between these two regions the density must be low enough to reproduce the large visibility modulus (small extension) obtained for two of the four VLTI baselines. This new scenario is also

compatible with the previous MIDI measurements and the fact that the outer part of the disk may be truncated by an unseen companion at 32 $R_\star$.

3. We obtain for the first time the clear evidence that the disk is in Keplerian rotation.

4. We found that that α Arae must be rotating very close to its critical velocity.

More details can be found in Meilland et al. [3].

3 κ CMa

κ CMa (HD 50013, HR 2538) is one the brightest Be star of the southern hemisphere (V=3.8, K=3.6). It is classified as a B2IVe star, and the distance deduced from Hipparcos parallax is 230 ±30 pc. The measured vsini values range from 220 km s^{-1} (Dachs et al. [15]; Mennickent et al. [16]; Okazaki [17]; Prinja [18]) to 243 km s^{-1} (Zorec et al.[19]), its radius is 6 $R_\odot$ (Dachs et al. [15]; Prinja [18]) and its mass is 10 $M_\odot$ (Prinja [18]).

Dedicated observations of κ CMa using medium spectral resolution (1500) were carried out during the night of December 26, 2004, with the three VLTI 8m ESO telescopes UT2, UT3 and UT4. The data were reduced using the amdlib (v1.15)/ammyorick (v0.54) software package developed by the AMBER consortium.

3.1 Is κ CMa Really Similar to α Arae?

Following the previous section we concluded that α Arae fits very well within the classical scenario for the "Be phenomenon", i.e., a fast-rotating B star close to its breakup velocity surrounded by a Keplerian circumstellar disk with an enhanced polar wind. This scenario was also confirmed for the Be star Achernar by Kervella and Domiciano de Souza [4] using VLTI/VINCI data, even if, for this latter case, the star was not in its active Be phase, i.e., without any strong emission line and no circumstellar disk (see next section). Nevertheless, Achernar was still a nearly critical rotator and was still exhibiting an enhanced polar stellar wind. Our study of κ CMa shows that this star does not fit very well within this classical scenario.

3.2 κ CMa: The Black Sheep of the Family?

1. We clearly found that κ CMa is not a critical rotator. If this star was rotating close to its breakup velocity, i.e., V_c=463 km s^{-1}, the inclination angle would be around 27° in order to obtain a measured vsini=210 km s^{-1}. With this inclination angle the maximum flattening corresponding to a geometrically very thin disk is 1.12. Since we measure a flattening of about 2±0.7 this inclination angle can be ruled out. In our best SIMECA model the star is rotating at only 52% of its critical velocity (Fig. 2).

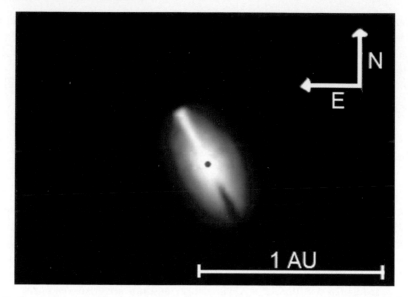

Fig. 2. Intensity map in the continuum at 2.15 µm obtained with SIMECA for our best model parameters. The inclination angle is 60°, the central *black dot* represents the κ CMa photosphere (0.25 mas); the bright part in the equatorial disk is produced by the over-density which is oriented along the B₁ baseline. This over-density is also responsible for a 30% emission excess in the asymmetric V part of the Brγ line

2. The rotation law within the disk appears to be non-Keplerian. A Keplerian rotating law will produce a narrower double-peaks separation in the Paβ line profile. Using a simple axi-symmetric Keplerian disk model the double-peaks separation would be about $90 \, \mathrm{km \, s^{-1}}$ whereas we measure an asymmetric double-peaks separation of about $160 \, \mathrm{km \, s^{-1}}$. Even if we subtract the emission of the over-density producing a larger double-peaks separation by contributing to the V peak of the emitting Paβ line, we still obtain a double-peaks separation of about $120 \, \mathrm{km \, s^{-1}}$. The exponent of the rotation law used for our best SIMECA model is 0.32 whereas it should be 0.5 for a purely Keplerian disk.

3. Finally, the asymmetry presently detected in the disk of κ CMa is very hard to explain within the "one-armed" viscous disk framework. Following the viscous disk models by Okazaki [17] and the observational detection of "one-armed" oscillations in the disk of ζ Tau by Vakili et al. [20] and γ Cas by Berio et al. [21], the precessing period of such oscillations should be confined within a few years up to about 20 years for the longer ones. On the other side, compiling all the observational data available, we obtain a period of about 80 years for κ CMa which is too long compared to theoretical predictions. Considering a 80 years period and assuming a mass for κ CMa of about $10 M_\odot$, we could suspect the presence of an undetected (spectroscopically and photometrically speaking) companion to confine the over-density. The expected distance of such a companion should be 40 AU, i.e., $\sim$1500 $R_\star$. This is unrealistically

large. Moreover, the fact that this over-density remains confined along the major axis of the disk is a very nice coincidence.

More details can be found in Meilland et al. [22].

4 Achernar

The star Achernar (α Eridani, HD 10144) is the nearest (d = 44 pc, Hipparcos) and brightest (V = 0.46 mag) Be star. Depending on the author and the technique used, the spectral type of Achernar ranges from B3-B4IIIe to B4Ve (e.g., Slettebak [23], Balona et al. [24]). The estimated projected rotational velocity vsini ranges from 220 to 270 km s^{-1} and the effective temperature $T_{\rm eff}$ from 15,000 to 20,000 K (Chauville et al. [25], Vinicius et al. [26]). Achernar rapid rotation (>80% of its critical velocity) induces mainly two effects on the stellar structure: a rotational flattening and a gravity darkening described by the von Zeipel effect (von Zeipel [27]) (Fig. 3).

Fig. 3. Artist's view of Achernar with its distorted photosphere due to its nearly critical rotation and its intermittent circumstellar disk whereas the polar wind seems to be permanent. Copyright Meilland 2007

4.1 Achernar: An "Extreme" Rotator?

Achernar rotational flattening was measured for the first time by Domiciano de Souza et al. [10] using the VLTI and its test bed VINCI instrument. The measured flattening ratio is $R_{eq}/R_{pole} > 1.5$.

On the other hand, following the Hα line profiles variations, Vinicius et al. [26] suggested the presence of a circumstellar disk contribution instead of a purely flattened photosphere without any circumstellar contribution. However, Kervella and Domiciano de Souza [4] reprocessed the whole VINCI visibility data set with a rather complete (u,v) plane coverage in order to evidence whether there is an equatorial or a polar gaseous extension, or both. Finally they concluded that there might be a polar extension.

4.2 Achernar: A Not-so-Critical Rotator with a Small Disk?

Our main objective was to use a physical model of Achernar circumstellar envelope using the whole VLTI/VINCI data set obtained by Kervella & Domiciano de Souza [4] in order to investigate whether the flattening ratio is smaller than the one they have estimated and can rather be accounted by a small circumstellar contribution. The second objective was to compare our results with the finding of a circumstellar environment around Achernar by Vinicius et al. [26] and study its possible disk evolution between 1991 and 2002.

Again, we have used the SIMECA code to interpret the VINCI data from Kervella and Domiciano de Souza [4] who rather used a simple model by assuming a 2D elliptical Gaussian envelope superimposed on a uniform ellipse for the central distorted star. We have evidenced a clear polar wind with a 1–5% contribution to the total flux and a spatial extension greater than 10 $R_\star$ in agreement with the results obtained by Kervella and Domiciano de Souza [4]. We also conclude that at the time of the VINCI observations Achernar seems to have no equatorial disk but rather a confined polar stellar wind with a poorly constrained ~20° opening angle.

The stellar and disk parameters used in this study were compatible with those obtained by Vinicius et al. [26], i.e., same photospheric density of 6.3 × 10^{11} hydrogen atoms per cm^3 and roughly the same effective temperature, i.e., 15,000 K for Vinicius, whereas we have adopted a latitude-dependent effective temperature ranging from 8500 K (equator) to 20,000 K (pole). The plot in Vinicius et al. [26] showing the specific intensity at the equator as a function of the disk radial extension < 3 $R_\star$ is also compatible with our finding of a wind+disk scenario with a disk < 5 $R_\star$ since we are lacking interferometric data at small baselines to strongly constrain the disk size. Nevertheless, our results strongly disagree with the disk/star flux ratio of 27% in Vinicius whereas we obtain a value < 5%. Using a value of 27% would have produced visibilities out of the VINCI measurements.

On the other hand the study of Achernar's spectroscopic data between 1991 and 2002 made by Vinicius et al. [26] evidenced an evolving equatorial disk. Thus, we have used these Hα line profiles to study their variations and found a clear signature of the formation dissipation of the equatorial disk. The disk evolution follows three phases; the first two were reproduced with an outburst scenario whereas we need another (unknown) physical effect to explain the third phase corresponding to the final contraction of the disk. As already outlined in Kervella and Domiciano de Souza [4], it seems that the polar wind may be present independently of the phase of the central star (B or Be phase), i.e., the polar stellar wind does not seem to be linked to the presence of a disk or a ring around the star.

More recently, using VLT/VISIR data Kervella & Domiciano de Souza [28] found that Achernar may have a A spectral type companion located at 12.3 AU at the distance of Achernar. If the companion orbit is very eccentric (not yet determined) it may be responsible for the disk formation by tidal effects when coming close to the periaster which may occur in 2013 (date estimated from the pseudo-periodic line profiles variation of 13 years). On the other side, the VLTI/VINCI measurements were made in 2000, i.e., close to the apoaster and thus without triggering the disk formation which was not detected in the Kervella and Domiciano de Souza [4] data.

More details can be found in Kanaan et al. [29].

5 A Possible Clue for the Be Phenomenon?

Following these three examples, we may wonder if classical Be stars are really an homogeneous group or if there are sub-groups where the Be phenomenon may be triggered by different physical processes?

These studies seem to indicate that we may have at least two sub-groups. The first one, with α Arae and Achernar, may be dominated by the nearly critical rotation of the central star. This nearly critical rotation might be responsible for the equatorial disk formation. Nevertheless, for α Arae, the disk seems to be permanent whereas for Achernar it seems to form and dissipate with a timescale on the order of a decade of years. To interpret this variability, binarity may be the clue for this phenomenon: as in the case of δ Scorpii, a companion with an eccentric orbit coming close to the periaster may be responsible for the disk formation triggered by tidal effects. For α Arae a putative companion was also detected by Chesneau et al. [6], but if its orbit is more circular, the disk may be permanent and confined within the Roche lobe of the system.

In the second group (i.e., κ CMa type), the central star may be a non-critical rotator, without a Keplerian disk and no detectable polar wind. In this case, the rapid stellar rotation (0.52 $V_{critical}$) is not sufficient to compensate the local gravity more than 15–20% and thus cannot be responsible for the disk formation. Other physical mechanisms must be advocate. For κ CMa, its higher effective temperature (22,500 K) and its sub-giant class (IV) may be responsible for a disk formation by radiative pressure on the gas. Thus, only the more luminous objects may form a disk only by radiative pressure as for γ Cas (B0IV) and ζ Tau (B2IV) without advocating for a rotationally supported disk. Note that for these three objects we have detected a precessing "over-density" structure within the disk which becomes no more axi-symmetrical. Our results are summarized in Table 1 and a complete discussion can be found (in french) in Meilland [30].

Table 1. Comparison of the stellar rotation, geometry and kinematics of the circumstellar disk of the 3 studied Be stars α Arae, κ CMa and Achernar (see Meilland [30])

	α Arae	κ CMa	Achernar
Equatorial disk			
Stellar Rotation	97 % V_c	52% V_c	$\sim V_c$
Radius (K band)	32 $R_\star$	23 $R_\star$ independent of λ	Intermittent $R_{max} \sim 4.8$ $R_\star$ if Keplerian
Flux (K band)	40%	50%	<5% (2002)
Expansion	negligible	negligible	0.2 kms^{-1}
Disk rotation (β parameter)	0.48 (quasi Keplerian)	0.3 (sub Keplerian)	no disk detected in 2000
Polar wind			
Extension	>10 $R_\star$	not detected	>10 $R_\star$
Flux (K band)	1–5%	X	3–4% (2002) (H band)
Opening angle	50 $\pm10^o$	X	5–40°

Acknowledgement

We thank Alain Spang our data reduction Guru and Damien Mattei for the SIMECA code developments support.

The AMBER project has benefited from funding from the French Centre National de la Recherche Scientifique (CNRS) through the Institut National des Sciences de l'Univers (INSU) and its "Programmes Nationaux" (ASHRA, PNPS).

This research has also made use of the ASPRO observation preparation tool from the JMMC in France, the SIMBAD database at CDS, Strasbourg (France) and the Smithsonian/NASA Astrophysics Data System (ADS). This publication makes use of data products from the Two Micron All Sky Survey.

References

1. Martayan, C., Hubert, A.-M., Floquet, M., et al.: A&A **445**, 931 (2006a)
2. Martayan, C., Frémat, Y., Hubert, A.-M., et al.: A&A **452**, 273 (2006b)
3. Meilland, A., Stee, Ph., Vannier, M., et al.: A&A **464**, 59 (2007)
4. Kervella, P., Domiciano de Souza, A.: A&A **453**, 1059 (2006)
5. Perryman, M.A.C., Lindegren, L., Kovalevsky, J., et al.: A&A **323**, 49 (1997)
6. Chesneau, O., Meilland, A., Rivinius, T., et al.: A&A **435**, 275 (paper I) (2005)
7. Stee, Ph.: A&A **403**, 1023 (2003)
8. Townsend, R.H., Owocki, S.P., Howarth, I.D.: MNRAS **350**, 189 (2004)

9. Owocki, S.P.: ASPC **337**, 101 (2005)
10. Domiciano de Souza, A., Kervella, P., Jankov, S., et al.: A&A **407**, L47 (2003)
11. Struve, O.: ApJ **73**, 94 (1931)
12. Cranmer, S.R., Owocki, S.P.: ApJ **440**, 308 (1995)
13. Owocki, S.P., Gayley, K.G.: ASPC **131**, 237 (1998)
14. Porter, J.M., Rivinius, Th.: PASP **115**, 1153 (2003)
15. Dachs, J., Poetzel, R., Kaiser, D. A&AS **78**, 487 (1989)
16. Mennickent, R.E., Vogt, N., Barrera, L.H.: A&AS **106**, 427 (1994)
17. Okazaki, A.: A&A **318**, 548 (1997)
18. Prinja, R.K.: MNRAS **241**, 721 (1989)
19. Zorec, J., Frémat, Y., Cidale, L.: A&A **441**, 235 (2005)
20. Vakili, F., Mourard, D., Stee, Ph., et al.: A&A **335**, 261 (1998)
21. Berio, Ph., Stee, Ph., Vakili, F., et al.: A&A **345**, 203 (1999)
22. Meilland, A., Millour, F., Stee, Ph., et al.: A&A **464**, 73 (2007)
23. Slettebak, A.: ApJ **50**, 55 (1982)
24. Balona, L.A., Engelbrecht, C.A., Marang, F.: MNRAS **227**, 123 (1987)
25. Chauville, J., Zorec, J., Ballereau, D., et al.: A&A **378**, 861 (2001)
26. Vinicius, M.M.F., Zorec, J., Leister, N.V., et al.: A&A **446**, 643 (2006)
27. von Zeipel, H.: MNRAS **84**, 665 (1924)
28. Kervella, P., Domiciano de Souza, A.: A&A **474**, L49 (2007)
29. Kanaan, S., Meilland, A., Stee, Ph., et al.: A&A **486**, 785 (2008)
30. Meilland, A.: Evolution, gèomètrie et cinèmatique des enveloppes circumstellaires des ètoiles chaudes: Apport des instruments AMBER et MIDI, Thesis, 2007, Université Nice Sophia Antipolis-UFR Science

On the Rotation of A-Type Stars

F. Royer

Observatoire de Paris, GEPI – bâtiment 11, 5 place Jules Janssen, 92195 Meudon,
France
frederic.royer@obspm.fr

Abstract We discuss the derivation of the projected rotational velocities ($v \sin i$) in
general, detail the different methods and emphasize the Fourier transform methods.
The effects of gravity darkening and differential rotation are detailed and reviewed in
the case of A-type stars. Finally the distributions of rotational velocities in the range
of A-type stars are discussed.

List of Abbreviations and Symbols

FT   Fourier transform
FWHM Full width at half maximum
PDF   Probability density function

1 Introduction

Rotation is one of the fundamental stellar parameters and it affects stellar formation, evolution and death. Since work began on the subject [58], it has been
observed that rotation rate strongly depends on the spectral type, and A-type
stars are known to be fast rotators on average. A review of the characteristics
of A-type stars can be found in [4].

Figure 1 shows the variation of the rotation rate as a function of the stellar
spectral type. The rotation rate is estimated in this case by the projected rotational velocity ($v \sin i$, see Sect. 2). The striking feature in the plot is the clear
dichotomy between hot (earlier than about F5) and cool (later than about F5)
stars. Whereas the cool stars rotate slowly (typically $v \sin i < 10 \, \mathrm{km \, s^{-1}}$), the hot
stars rotate on the average at velocities higher than $100 \, \mathrm{km \, s^{-1}}$. The difference
between these two subgroups is that the cool stars harbor convective envelopes,
and the generally accepted hypothesis is the generation of a magnetic brake by
interaction of the convective envelope and rotation in a dynamo process.

In the Hertzsprung–Russell diagram, A-type stars are located close to the
limit where the convective zone appears in the stellar photosphere and where

Royer, F.: *On the Rotation of A-Type Stars*. Lect. Notes Phys. **765**, 207–230 (2009)
DOI 10.1007/978-3-540-87831-5_9   © Springer-Verlag Berlin Heidelberg 2009

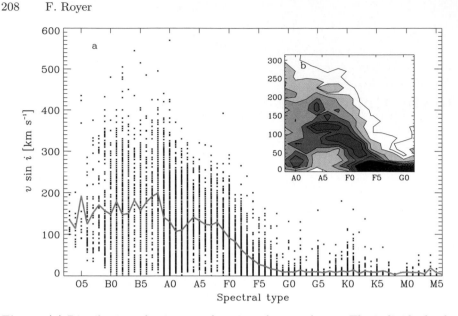

Fig. 1. (a) Distribution of $v \sin i$ as a function of spectral type. The individual values (*crosses*) are taken from the compilation in [20] and completed by $v \sin i$ data from [2] and [50]. The variation of the average $v \sin i$ per spectral type (*solid thick line*) is also over-plotted. (b) The sub-panel in the *upper-right* corner represents the color-coded scale density of points in the same diagram $v \sin i$ – spectral type; both axes have the same scale as in (a). *Darker tones* stand for higher densities

the average rotation rate decreases with later spectral type. In Fig. 1a, one can also notice the drop in average $v \sin i$ around A0–A1, which corresponds to the over-density of slow rotators in sub-panel (Fig. 1b). These aspects will be addressed in the following sections.

Different methods exist to measure stellar rotation, depending on the observed signature of rotation:

(i) the spectral line profile broadening, which gives the projected equatorial velocity: $v \sin i$,

(ii) the photometric modulation of starlight, due to the presence of spots on the surface, which allows the derivation of the rotational period,

(iii) the Rossiter effect which distorts the radial velocity curve in some eclipsing binary systems,

(iv) the shape of the stellar disk, measured by interferometric observations, to derive the oblateness of the star caused by fast rotation.

The first three methods are overviewed in [54], and the last and most recent one is detailed in [59]. Methods (ii), (iii) and (iv) are limited to specific objects: stars with spots (ii), eclipsing binary systems (iii), nearby and bright stars (iv), whereas the first method can be applied to a very wide range of objects.

This chapter focuses first on the rotational broadening and the methods for measuring the $v \sin i$ parameter, with special emphasis on the Fourier analysis.

Sections 3 and 4 overview the measurement of differential rotation and gravity darkening in A-type stars. Section 5 details the analysis of the distributions of rotational velocities for A-type stars.

2 Measuring Rotational Broadening

2.1 Line Broadening

In Fig. 2a, the star is represented by a sphere rotating as a solid body. The Cartesian coordinates are chosen so that the line of sight is along the z-axis and that the rotation axis lies in the plane y–z. The inclination i is the angle from the z-axis to the rotation axis. $\mathbf{\Omega}$ is the angular velocity vector, and for any point at the surface of the star defined by a radius vector $\mathbf{R}$, the resulting velocity $\boldsymbol{v}$ is the cross product:

$$\boldsymbol{v} = \mathbf{\Omega} \times \mathbf{R}. \tag{1}$$

As $\mathbf{\Omega}$ lies in the y–z plane, its coordinates can be written as $(0, \Omega_y, \Omega_z)$. Let (R_x, R_y, R_z) be the coordinates of the radius vector $\mathbf{R}$. The Doppler shift is due to the component of the velocity along the line of sight, i.e., the z-axis, namely v_z, and

$$v_z = -R_x\, \Omega\, \sin i. \tag{2}$$

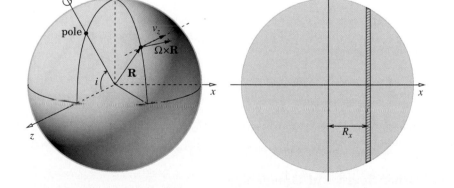

Fig. 2. Solid body rotating star. (**a**) Three-dimensional sketch. The rotation axis $\mathbf{\Omega}$ is inclined at an angle i with respect to the line of sight along the z-axis. The vector $\mathbf{\Omega}$ lies in the y–z plane. For an arbitrary point at the surface, defined by the radius vector $\mathbf{R}$, the velocity is $\boldsymbol{v} = \mathbf{\Omega} \times \mathbf{R}$. The z-component of this velocity (v_z) gives the Doppler shift. (**b**) The apparent disk of the star is the projection on the x–y plane. The rotation axis is projected onto the y-axis. According to (2), the apparent disk can be divided in strips parallel to the y-axis (*hatched area*) having a Doppler shift proportional to R_x

The maximum shifts occur at the stellar limbs, where $|R_x| = R$:

$$v_{\text{limb}} = R\,\Omega\,\sin i, \tag{3}$$
$$= v_e\,\sin i, \tag{4}$$

where v_e is the equatorial velocity. The datum v_{limb} is usually called $v\sin i$.

The Doppler broadening is the result of the integration of these components v_z over the apparent disk. The broadening profile b is defined as the observed fraction of light at each Doppler shift. The integrated light from the strips shown in Fig. 2b is proportional to $\sqrt{1 - v_z^2/v_{\text{limb}}^2}$; thus axial rotation of a homogeneous spherical rigid body results in a semi-elliptic rotation profile. When taking into account the center-to-limb darkening, and assuming a linear darkening law,[1] a parabolic term $1 - v_z^2/v_{\text{limb}}^2$ appears in the integrated light from the strips shown in Fig. 2b.

The rotation profile can be written:

$$b(\lambda') = \underbrace{c_1\sqrt{1 - \lambda'^2}}_{\text{elliptic term}} + \underbrace{c_2\,(1 - \lambda'^2)}_{\text{parabolic term}}\;, \tag{5}$$

with

$$\Delta\lambda = \lambda_0\,v\sin i/c, \tag{6}$$
$$\lambda' = (\lambda - \lambda_0)/\Delta\lambda, \tag{7}$$
$$c_1 = \frac{2\,(1 - \epsilon)}{\pi\,v\sin i\,(1 - \epsilon/3)}, \tag{8}$$
$$c_2 = \frac{\epsilon}{2\,v\sin i\,(1 - \epsilon/3)}. \tag{9}$$

As it is detailed in [24], an observed spectrum $s(\lambda)$ can be approximated by the convolution product:

$$s(\lambda) = h(\lambda) * b(\lambda) * p(\lambda)\,, \tag{10}$$

where $h(\lambda)$ is "true" spectrum of the star, $b(\lambda)$ the broadening function and $p(\lambda)$ the instrumental profile.

2.1.1 Other Sources of Broadening

In addition to the rotation, different mechanisms induce other types of broadening.

---

[1] Linear limb darkening: let θ be the angle between the surface normal and the line of sight. The local intensity is a function of θ and using a linear approximation it can be written $I(\theta) = I_0\,(1 - \epsilon(1 - \cos\theta))$, where ϵ is the limb-darkening coefficient. In Fig. 2, the angle θ can be expressed as $\cos\theta = R_Z/R$.

Pressure Broadening

In a star, the atoms absorbing the light continuously collide with other particles. The other particles can be electrons, ions and atoms of the same element or another as the absorber, or in cool stars, molecules. This shortens the lifetime of an atomic energy level, which broadens the absorption lines. This effect is well described by a Lorentzian profile.

Thermal Broadening

Due to its thermal motion, each absorbing atom has a component of velocity along the line of sight. This results in a shift of the observed absorption line. Since this shift is proportional to the particle velocity, and this velocity is given by a Maxwellian distribution, this broadening is well described by a Gaussian profile.

Microturbulence

Motions of the photospheric gas as a whole also introduce Doppler shifts. If the characteristic dimensions of the moving material are small compared to unit optical depth, it broadens the lines in the same way the particle velocity distribution does. Because microturbulence velocities are generally very small compared to other broadeners, they are usually assumed to have an isotropic Gaussian shape.

Macroturbulence

When the photospheric turbulent cells are large enough to capture incoming photons, the cells produce their own spectra. Each macro-cell gives a spectrum shifted by the velocity of the cell. As the individual spectra of many macro-cells are usually viewed simultaneously, this results in a broadening.

Instrumental Effects

Instrumental broadening can be measured by fitting emissions lines of calibration lamps. It depends on the instrument, but it is often assumed Gaussian.

2.2 Techniques

The projected rotational velocities that can be found in the literature are derived with different techniques. As pointed out in [25], there is no "standard" technique for measuring projected rotational velocity. The most widely used is based on the measurement of the width of spectral lines. One can also artificially broaden a synthetic spectrum and find the best match with the observed spectrum. Finally the $v \sin i$ can be derived from the analysis of the broadened profiles in the frequency space. These different techniques are detailed below.

2.2.1 FWHM Calibration

As a simple and straightforward technique, the full width at half maximum (FWHM) of spectral lines is often used as a measurement of the broadening mechanisms. The effect of increasing rotational velocity on the spectrum is illustrated in Fig. 3. The lines become wider and shallower with increasing $v \sin i$. One can also see an example of the effect of blends of neighboring lines which makes the lines around 4490 Å impossible to measure individually.

A set of $v \sin i$ standard stars is given in [55]. It provides $v \sin i$ and FWHM for stars with various spectral types (from late O to late F) and allows a calibration of $v \sin i$ against FWHM. The data are shown in Fig. 4 for the three different lines used according to the spectral type. These standard stars, known as "the Slettebak system", have been widely used, but several authors question the $v \sin i$ scale defined by this system [6, 50, 59], arguing that it underestimates $v \sin i$. The largest homogeneous catalog of $v \sin i$ of A-type stars derived using the "Slettebak system" [3] has been since statistically corrected for an offset with respect to a scale defined using Fourier transform [50].

When the rotational broadening is not the dominant mechanism, one has to carefully remove the broadening due to the instrumental profile or the intrinsic line profile. Authors sometimes build their own calibration (e.g., [18]) and take into account instrumental effects. The instrumental width is usually quadratically subtracted from the observed width of the spectral lines.

Cross-Correlation

The FWHM can also be measured on the cross-correlation function (CCF) of the spectrum with a box-shaped template. This technique gives reliable results as long as the rotational broadening is small, $\lesssim 30 \, \mathrm{km \, s^{-1}}$ [38], and therefore the other broadening mechanisms are part of the calibration. This method is mostly used for cool stars. The measured width of the CCF (σ_{obs}) is usually represented as the quadratic sum of the width due to rotational broadening (σ_{rot}) and the width due to other broadening mechanisms (σ_0):

$$\sigma_{\mathrm{obs}}^2 = \sigma_{\mathrm{rot}}^2 + \sigma_0^2 . \tag{11}$$

σ_0 is thus the width of the CCF of the non-rotating star; it depends on the spectral resolution of the observed spectra and the intrinsic width of the stellar absorption lines. The projected rotational velocity is calibrated against the width of the CCF using this formula:

$$v \sin i = A \sqrt{\sigma_{\mathrm{obs}}^2 - \sigma_0^2} . \tag{12}$$

The coupling constant A is determined using artificially broadened spectra and is derived from the slope of the relation $v \sin i^2 - \sigma_{\mathrm{obs}}^2$. Since the broadening

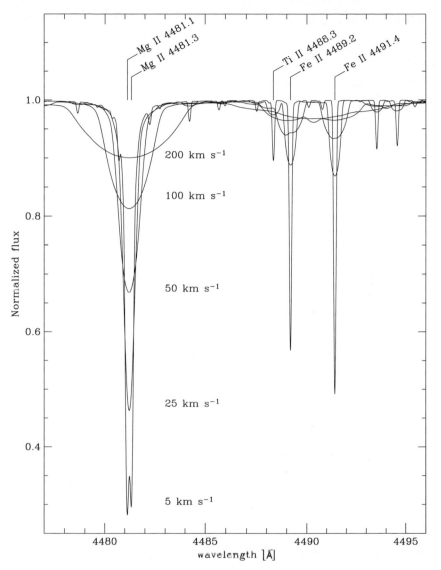

Fig. 3. Synthetic spectra for different rotational broadenings. They correspond to $T_{\mathrm{eff}} = 10,000\,\mathrm{K}$ and $\log g = 4.0\,\mathrm{dex}$. The $v \sin i$ values are given next to the Mg II doublet at the level of its depth for the corresponding velocity. The spectral lines of interest are identified in the *upper part* of the panel

mechanisms are a function of the temperature and gravity, a smooth dependence for σ_0 as a function of the stellar color and luminosity class is expected. The width for non-rotating objects is thus modeled as a function of $B - V$. Figure 5 shows the relation obtained for non-rotating calibrators observed with FEROS [36], with a spectral resolution $R = 48,000$.

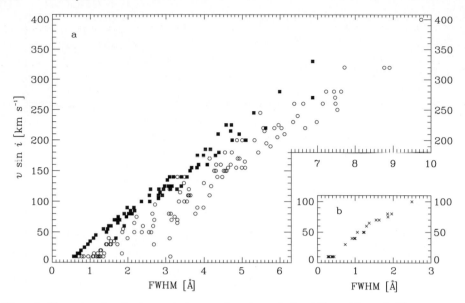

Fig. 4. Rotational velocity $v \sin i$ as a function of the full width at half maximum from the standard stars in [55]. **(a)** FWHM derived from the He I 4471 line (O9- to B8-type stars, *open circles*) and Mg II 4481 line (B8- to F0-type stars, *filled squares*). **(b)** FWHM derived from the Fe I 4476 line (F0- to F8-type stars, *crosses*)

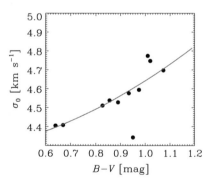

Fig. 5. Variation of width of the CCF σ_0 for a non-rotating star as a function of $B - V$ for the calibrators observed with the FEROS spectrograph [36]. The fitted polynomial relation is shown as a *continuous line*

2.2.2 Spectral Synthesis

This method consists in the computation of a complete spectral range including all the observed lines. It requires a grid of model atmospheres and atmospheric parameters (T_{eff}, $\log g$, micro-/macroturbulent velocities, metallicity) in order to interpolate in the grid for a given star. As far as the spectral range to be computed is concerned, the atomic data are mandatory: line list and corresponding oscillator strengths. Spectral synthesis is a trial-and-error method and allows the adjustment of parameters such as element abundances, Doppler broadening and Doppler shift, until the best fit of the observed spectrum is achieved. Many examples can be found in the literature [16, 27, 28, 30, 32].

This technique is time consuming and is often used to derive abundances primarily. However, $v \sin i$ is a by-product of this technique and it offers a powerful way to derive rotational velocities in case of severe blending. Moreover, rotation can be more accurately accounted for than a simple convolution with a rotation profile, the static intensity can be integrated over the rotating disk, and effects such as differential rotation or gravity darkening can be included [27].

2.3 Fourier Analysis

The first application of Fourier analysis in the determination of stellar rotational velocities was undertaken some 75 years ago [7]. In Fourier space, the convolution product (10) becomes

$$S(\nu) = H(\nu) B(\nu) P(\nu) , \tag{13}$$

where S, H, B and P are, respectively, the Fourier transforms (hereafter FT) of s, h, b and p.

The FT of a semi-elliptic profile is dominated by a Bessel function of the first kind, and a rotation profile in Fourier space will display typical lobes of decreasing amplitude with higher frequency. Figures 6a, 6b and 6c illustrate the shape of the rotation profile in wavelength and frequency space. Since the earliest work [7], it is known that the loci of the zeros ν_n of the FT of a rotation profile depend on the $v \sin i$ value:

$$\nu_n \propto (\lambda_0 \, v \sin i)^{-1}, \tag{14}$$

where λ_0 is the wavelength of a measured spectral line.

From (13), the zeros of the observed profile $S(\nu)$ will be the union of the zeros of H, B and P. If the intrinsic profile of the line $h(\lambda)$ is sharp enough and close to a Dirac impulse function, then in the frequency domain, $H(\nu)$ is a constant function and thus negligible. Assuming that the instrumental profile $p(\lambda)$ is Gaussian, $P(\nu)$ is also Gaussian (as illustrated in Figs. 6a and 6b) and does not add zeros. But the combined influence of spectral resolution and noise is shown in Fig. 6g.

Measuring the loci of the zeros of the observed profile in Fourier space offers a way to derive the $v \sin i$. Usually only the first zero is measured, which makes the $v \sin i$ value based on a single parameter, as when the FWHM is measured.

A dimensionless value q_n of the locus of the nth zero can be defined, depending only on the limb-darkening coefficient ϵ, and

$$\nu_n = q_n \, c \, (\lambda_0 \, v \sin i)^{-1}, \tag{15}$$

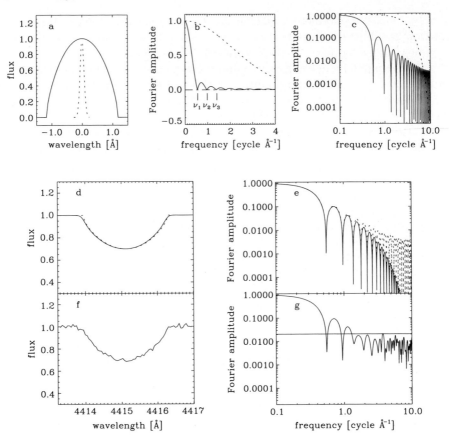

Fig. 6. Simulated line profiles in the wavelength space and in the Fourier space. (**a**) Theoretical rotational profile (*solid line*) corresponding to $v \sin i = 80 \, \text{km s}^{-1}$ computed according to (5) for $\lambda_0 = 4415.122$ Å and an instrumental Gaussian profile (*dashed line*) corresponding to a resolution of 25,000. (**b**) Fourier transforms of the profiles from panel (**a**) in linear scale. The loci of the first three zeros of the rotational profile are indicated (ν_1, ν_2 and ν_3). (**c**) Same as (**b**) in log–log scale. (**d**) Simulated spectral line: pure rotational profile (*dashed line*) corresponding to $v \sin i = 80 \, \text{km s}^{-1}$ computed according to (5) and its convolution with an instrumental profile (*solid line*) corresponding to a resolution of 25,000. (**e**) Fourier transforms of the profiles from panel (**d**). The FT of the convolved profile (*solid line*) is the product of the two profiles from panel (**c**). (**f**) Theoretical profile with additional noise (corresponding to a signal-to-noise ratio of about 80). (**g**) Fourier transform of the profile in panel (**f**). The *solid horizontal line* stands for the noise level derived using the formula given in [57]

where c is the speed of light. The loci q_n have been approximated by polynomial expressions of ϵ [14]:

$$q_1 = 0.610 + 0.062\,\epsilon + 0.027\,\epsilon^2 + 0.012\,\epsilon^3 + 0.004\,\epsilon^4, \tag{16}$$

$$q_2 = 1.117 + 0.048\,\epsilon + 0.029\,\epsilon^2 + 0.024\,\epsilon^3 + 0.012\,\epsilon^4, \tag{17}$$

$$q_3 = 1.619 + 0.039\,\epsilon + 0.026\,\epsilon^2 + 0.031\,\epsilon^3 + 0.020\,\epsilon^4. \tag{18}$$

The $v \sin i$ of some 800 B8- to F2-type stars has been determined using the first zero of the Fourier profile [49, 50]. The essential step in this analysis is the search for suitable spectral lines to measure the $v \sin i$. The candidate lines should not be blended and should offer a good access to the continuum. The list of their candidate lines is given in Table 1. The validity of a line varies in the whole range of spectral type covered by their sample, and they used two criteria to reject the lines:

- a priori, if the skewness of the line exceeds 0.15 in a synthetic spectrum with corresponding $T_{\rm eff}$ and $v \sin i$ (*hatched* area in Fig. 7),
- a posteriori, if the shape of the observed FT profile is considered as spurious by comparison with a theoretical profile (*gray* area in Fig. 7).

Figure 7 summarizes this selection in a synthetic plot of the validity of each line as a function of spectral type and $v \sin i$.

The larger the rotational broadening, the wider the wavelength range covered by a given line. The assumption that there is no additional zero due to the intrinsic profile H may be wrong. This can be checked by comparing the observed FT with the FT of the intrinsic *synthetic* profile, on the same wavelength range [37]. The zeros due to rotation are the ones in the observed FT that are not present in the FT of the intrinsic profile.

Table 1. List of the 23 spectral lines used for the $v \sin i$ measurement [49, 50]

Wavelength [Å]	Element	Wavelength [Å]	Element
4215.519	Sr II	4488.331	Ti II
4219.360	Fe I	4489.183	Fe II
4226.728	Ca I	4491.405	Fe II
4227.426	Fe I	4501.273	Ti II
4235.936	Fe I	4508.288	Fe II
4242.364	Cr II	4515.339	Fe II
4261.913	Cr II	4520.224	Fe II
4404.750	Fe I	4522.634	Fe II
4415.122	Fe I	4563.761	Ti II
4466.551	Fe I	4571.968	Ti II
4468.507	Ti II	4576.340	Fe II
4481.126	Mg II†		
4481.325	Mg II†		

† Wavelength of both components are indicated for the magnesium doublet line.

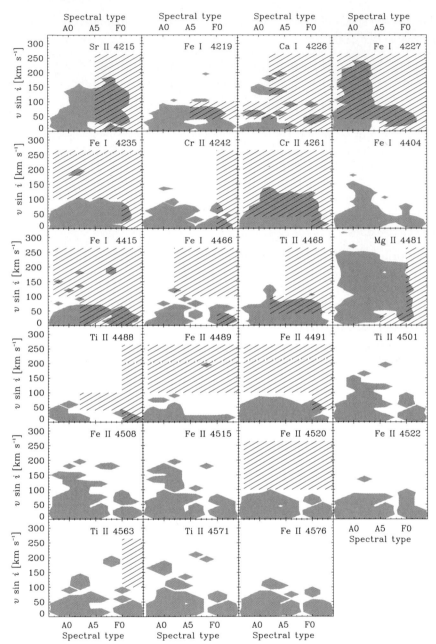

Fig. 7. Validity of the candidate lines for $v \sin i$ measurement. The *gray contour* represents the coverage in spectral type and rotational broadening where the observed Fourier profile of the line is considered as valid and rotation dominated. The *hatched boxes* stand for the areas where the skewness of the line in a synthetic spectrum (with given $v \sin i$ and a T_{eff} corresponding to the spectral type) is higher than 0.15. A line is then considered as valid in the *gray*, non-hatched, zones

2.3.1 Measuring the Broadening Function

The uses of the FT mentioned above are applied to individual spectral lines. As it is illustrated in Fig. 7, the number of suitable lines at high $v \sin i$ and/or late spectral type becomes scarcer and scarcer, due to the occurrence of blends. In order to overcome this problem, it is possible to deconvolve the observed spectrum by a synthetic non-rotating template and extract the broadening function $b(\lambda)$. It can be done using least square deconvolution methods (LSD, [47]). The broadening function can be analyzed in the Fourier space, and the resulting signal-to-noise ratio being increased compared to single lines, it easily allows the measurement of the second zero on the Fourier profile.

When both first and second zeros can be measured, the ratio q_2/q_1 can be used as a signature of the discrepancy with respect to the rotation of a spherical rigid body. Assuming a star to be a sphere rotating as a solid body is indeed only an approximation. Three different mechanisms make a rotating star differ from the simple model presented in Fig. 2:

(i) fast rotation diminishes the gravitational potential at the equator and alters the sphericity of the star,
(ii) gravity darkening changes the flux distribution on the stellar surface, due to temperature gradient,
(iii) the rotation law may depend on the latitude on the stellar surface.

It has been shown that the ratio q_2/q_1 is a direct indicator for solar-like latitudinal differential rotation [46] and gravity darkening [41]. If the ratio q_2/q_1 is due to gravity darkening, there is a monotonic dependence on the equatorial

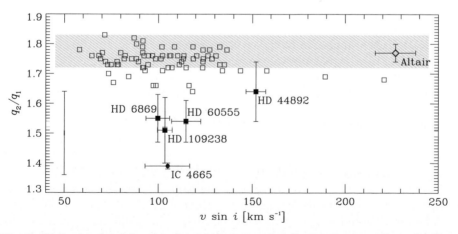

Fig. 8. Derived values of q_2/q_1 plotted against $v \sin i$ for A-type stars, from the literature: *squares* [45], *circle* [43], *diamond* [44]. The *gray region* is consistent with solid body rotation for arbitrary limb darkening (19): $1.72 < q_2/q_1 < 1.83$ for $0 < \epsilon < 1$. Identified differential rotators are indicated as *filled symbols* whereas solid rotators are represented by *open symbols*. The error bar on the *left* side is the median error for *open squares* [45]

velocity v of the star, for a given stellar model (T_{eff}, mass) and darkening law. If it is due to latitudinal differential rotation, the parameter of the rotation law can directly be derived from q_2/q_1.

In the case of solid rotation, the ratio q_2/q_1 has been approximated by a polynomial expression of ϵ [14]:

$$q_2/q_1 = 1.831 - 0.108\,\epsilon - 0.022\,\epsilon^2 + 0.009\,\epsilon^3 + 0.009\,\epsilon^4. \tag{19}$$

Figure 8 displays the ratios q_2/q_1 computed for A-type stars and taken from the literature [43–45] as a function of $v \sin i$. These objects are detailed in the following sections.

3 Differential Rotation

The Sun is known to rotate differentially, and its angular velocity can be approximated by

$$\Omega(\ell) = \Omega_{\text{equator}}\left(1 - \alpha\,\sin^2\ell\right), \tag{20}$$

where ℓ is the latitude on the stellar surface and $\alpha_\odot = 0.2$, i.e., the solar equator rotates 20% faster than the pole.

As mentioned previously, the main difference between photospheres of the Sun and A-type stars is the convective envelope. There is no or very thin convective envelope in A-type stars. The onset of convection has been directly searched [26], using line bisectors. They found that the granulation boundary , which delimits convection in the H-R diagram, goes through spectral types F0V, F2.5IV and F5III. This granulation boundary is represented in Fig. 9. The observations of convection in A-type stars are summarized in [56].

Differential rotation is detected in F-type stars, as early as F0 [47]. Whereas no indication of differential rotation has been found in A-type stars in early works [22], a few objects have recently been reported [43, 44].

Figure 8 displays the sample of 76 A0–A1 stars [44], with $60 < v \sin i < 150\,\text{km s}^{-1}$, studied using the Fourier analysis of the broadening profile (*squares*). Out of this sample, three stars show a low ratio q_2/q_1 signature of differential rotation   (HD 6869, HD 60555, HD 109238). For one object (HD 44892) the ratio q_2/q_1 could also be due to gravity darkening, whereas the ratio was not compatible with gravity darkening for the three previous candidates.

The rotation and temperature dependence of latitudinal differential rotation has been reviewed [42] using recent results, and it is noticed that the A-type differential rotators are located close to the granulation boundary. As it can be seen in Fig. 9 it is the case on the main sequence, but for the most evolved differential rotators the location is significantly bluer than the granulation boundary. It could nevertheless be due to uncertainties related to the location of the granulation boundary for the luminosity class IV.

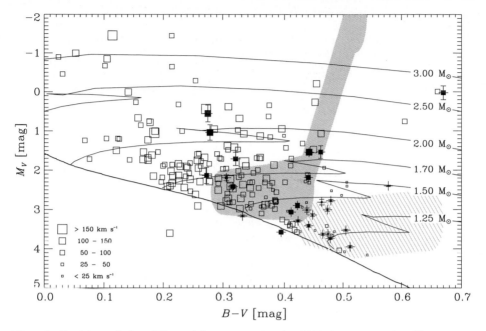

Fig. 9. Position of the differential rotators in the H-R diagram, using HIPPARCOS data [17]. The *plotted* data come from [45] (for the A-type stars) and from [42] for the later types. Identified differential rotators are indicated as *filled squares* whereas solid rotators are represented by *open squares*. The symbol size is representative of the $v \sin i$, and error bars in M_V are given for differential rotators only. Evolutionary tracks and ZAMS [31] are over-plotted as *solid lines*. The "granulation boundary" [26] (*thick solid gray line*) and the "rotation boundary" [23] (*thick dashed gray line*) are over-plotted, transformed into M_V absolute magnitude [21] and $B - V$ color indices [52]

4 Gravity Darkening

When a star is oblate, it has a larger radius at its equator than it does at its poles. As a result, the poles have a higher surface gravity, and thus temperature and brightness. Thus, the poles are gravity brightened, and the equator gravity darkened. This is the so-called von Zeipel theorem [60]. The gravity dependence of the surface temperature is described by

$$T_{\mathrm{eff}} \propto g^{\beta} \, . \tag{21}$$

The values of the parameter β is discussed in many publications:

- $\beta = 0.25$, the "classical" value [60], valid for conservative rotation law in stars without convective envelope,
- $\beta = 0.08$, derived for stars with convective envelopes [33].

Tabulations of β as a function of mass and evolutionary stages are given by [10]. Figure 10 shows this variation of the parameter β as a function of effective

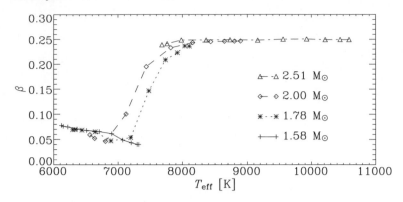

Fig. 10. Variation of the gravity-darkening parameter β as a function of $T_{\rm eff}$ from model calculations [10]. The different symbols stand for different initial masses and are restricted to the main-sequence evolutionary phase

temperature is the range covered by A-type stars, and one can see the effect of the transition from convective to radiative envelopes.

4.1 Vega

Vega (α Lyr) is the second brightest star in the northern celestial hemisphere and an extensively studied "standard" object. It has been qualified as "arguably the next most important star in the sky after the Sun" [27]. Vega is an A0V star, with $T_{\rm eff} = 9550\,\rm K$ [8]. Its narrow line spectrum gives a projected rotational velocity $v \sin i = 25 \pm 2\,\rm km\,s^{-1}$ [50].

This low $v \sin i$ star has been discovered to be a rapidly rotating star seen nearly pole-on. It has been first detected by the accurate measurements of two line profiles, Fe I 4528 Å and Ti II 4529 Å, which display a flat-bottomed profile instead of the expected cupsy shape [27]. This flat-bottomed profile is due to the temperature gradient over the stellar photosphere. Their spectral synthesis gave $v = 245 \pm 15\,\rm km\,s^{-1}$ and $i = 5.1° \pm 0.3°$. More recently, the same authors [29] published a new analysis with a revised value of $\log g$ for Vega which led to an inclination angle closer to 8°. Their new results are $v = 160 \pm 10\,\rm km\,s^{-1}$ and $i = 7.9° \pm 0.5°$.

Vega has been since studied with interferometric measurements [40]. Their results are closer to the initial results from spectroscopy [27]: $v = 274 \pm 14\,\rm km\,s^{-1}$ and $i = 4.54° \pm 0.33°$. Their model of the apparent disk of Vega is displayed in Fig. 12b. The temperature gradient over the photosphere goes from $9988 \pm 61\,\rm K$ at the poles down to $7557 \pm 261\,\rm K$ at the equator.

4.2 Altair

Altair (α Aql) is an A7IV-V main sequence star, with $T_{\rm eff} = 7550\,\rm K$ [16]. It is known to be a fast rotator with a projected rotational velocity $v \sin i = 217\,\rm km\,s^{-1}$

[50]. As the projection effect is much lower than in Vega, the oblateness of Altair due to rapid rotation is measurable. Altair is indeed the first star whose oblateness has been measured using interferometric observations [59]. They derived an axial ratio $a/b = 1.140 \pm 0.029$ and a projected rotational velocity $v \sin i = 210 \pm 13 \,\mathrm{km\,s^{-1}}$. They put some constraint on the inclination angle i, determined to be larger than $30°$ within 1-σ.

In order to put more constraint on the inclination angle, a spectroscopic analysis using Fourier methods has been done [44]. They used the ratio q_2/q_1 derived from the FT of the broadening function to derive the equatorial velocity [41]. Contrary to Vega, the spectroscopic determination of the inclination angle did not focus on a few specific line profiles but took into account a global profile using more than 600 lines. They found that $i > 68°$ on a 1-σ level.

The locus of Altair in v and i from the different determinations in the literature is shown in Fig. 11. The loci follow portions of function $\arcsin(v^{-1})$ because of the observational constraint of the $v \sin i$. The chronological sequence [59, 44, 13, 39] shows the improvement in the constraints on the inclination angle. The latest result [39] gives $v = 273 \pm 13 \,\mathrm{km\,s^{-1}}$ and $i = 63.9° \pm 1.7°$. Their model of the apparent disk of Altair is displayed in Fig. 12a. The temperature gradient over the photosphere goes from $7840 \pm 140 \,\mathrm{K}$ at the poles down to $6890 \pm 60 \,\mathrm{K}$ at the equator.

It is worth noticing that the value of the parameter β in the spectroscopic study [44], chosen to be ≈ 0.09 for Altair based on the tables from [10], is not agreed on in the interferometric analyses [13, 39]. The latter authors chose the "classical" value $\beta = 0.25$, which provides a better fit of the observations by

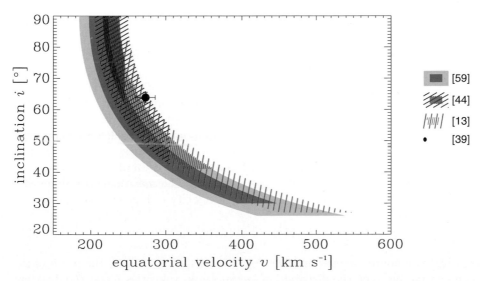

Fig. 11. Loci of Altair in the equatorial velocity–inclination plane. The *shaded/dark and the hatched/light areas* are respectively, 1-σ and 2-σ, determinations from the literature, derived from interferometric [13, 59] and spectroscopic observations [44]. The *filled circle* is the result from interferometric measurement by [39]

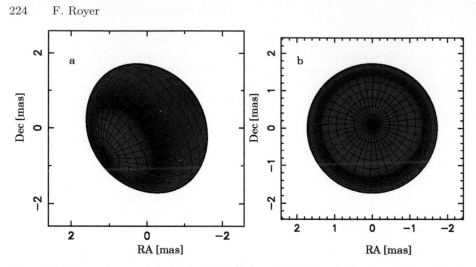

Fig. 12. False color model of (**a**) Altair [39] and (**b**) Vega [40], as seen from Earth. Both models are on the same spatial scale, but the color-coded temperature scale is different (see text). *Blue* is hot and bright, *red* is cool and faint

their model. The choice of the correct gravity-darkening law is crucial and the interpretation of Altair observations (spectroscopic and interferometric) bear on this limitation.

5 Distribution of Rotational Velocities

5.1 Projection Effect

Rotational velocities measured using spectroscopic data are projected along the line of sight. Although the inclination angle i can be measured for some objects, as seen above, its determination is usually not feasible for individual objects. But it is possible to statistically correct a distribution Φ of $v\sin i$ data from the projection effect and recover Υ, the distribution of equatorial velocities. It is generally assumed that stellar rotation axes are randomly oriented. This hypothesis has been tested many times [19, 24] and it is still the most valid. Under this assumption, it ensues that $\langle \sin i \rangle = \pi/4$ and there is a simple relation between moments of the distributions of v and $v\sin i$ [9]. The mean equatorial velocity $\langle v \rangle$ can be derived from the mean projected velocity of a given distribution:

$$\langle v \rangle = \frac{4}{\pi} \langle v\sin i \rangle \tag{22}$$

The probability density function (hereafter PDF) of $v\sin i$ is the result of the convolution between the distribution of equatorial velocities v and the distribution of inclination angles i:

$$\Phi(v\sin i) = \int \Upsilon(v)\, P(v\sin i | v)\, \mathrm{d}v, \tag{23}$$

where $P(v \sin i|v)$ is the conditional probability of $v \sin i \in [v, v + dv]$ and $\varUpsilon(v)$ is the PDF of equatorial velocities.

Under the assumption of randomly oriented rotation axes, the conditional probability $P(v \sin i|v)$ in (23) is

$$P(v \sin i|v) = \begin{cases} \dfrac{v \sin i}{v \sqrt{v^2 - (v \sin i)^2}} & \text{if } v > v \sin i \\ 0 & \text{if } v \leq v \sin i \end{cases} \tag{24}$$

Combining (23) and (24), the PDF of $v \sin i$ becomes

$$\varPhi(v \sin i) = \int_{v \sin i}^{\infty} \varUpsilon(v) \frac{v \sin i}{v \sqrt{v^2 - (v \sin i)^2}} \, dv, \tag{25}$$

which is an Abelian integral. There is an analytical solution:

$$\varUpsilon(v) = -\frac{2 v^2}{\pi} \frac{d}{dv} \int_v^{\infty} \frac{v \, \varPhi(v \sin i)}{v \sin i^2 \sqrt{v \sin i^2 - v^2}} \, d(v \sin i). \tag{26}$$

Details about this Abelian form of the convolution can be found in [9].

Two ways are therefore possible to compute the final distribution of equatorial velocities $\varUpsilon(v)$. Either (26) can be used, or the Lucy–Richardson deconvolution technique [34, 48].

5.2 Distributions of Equatorial Velocities in A-Type Stars

Among the A-type main-sequence stars, several types of chemically peculiar stars are found: the metallic-line (Am) stars and the Ap stars. These peculiar stars rotate more slowly in average than the normal A-type stars [3]. The Am phenomenon is often found in binary stars [11] and slow rotation can be partly due to tidal braking. Ap stars have strong magnetic fields [5], and their slow rotation can be the result of a magnetic braking.

The distribution of $v \sin i$ changes from early A- to late A-type stars, as it can be noticed in Fig. 1b. Using homogeneous $v \sin i$ data, the distributions of rotational velocities in A-type stars have been deconvolved in order to recover the distributions of equatorial velocities [3]. The observed bimodality in the distributions was attributed to the dichotomy between normal and chemically peculiar stars. The resulting distributions are shown in Fig. 13.

More recently, the distributions of rotational velocities for normal stars only (main sequence, non binary, non-peculiar) have been analyzed in the spectral type range B9–F0 [51]. They observe a net lack of slow rotators ($v < 70 \, \text{km s}^{-1}$) in A2- to A9-type stars. On the other hand, they found that the bimodality in the distribution for A0–A1 stars is still present, even when all known binary and chemically peculiar stars are excluded.

The spectral type subsample and the luminosity class selection differ between [3] and [51]. In order to compare the distributions on a uniform basis, the selection criteria used in [3] have been applied to data from [51]: divisions in spectral

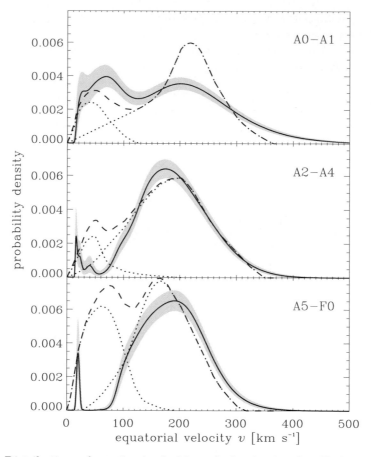

Fig. 13. Distributions of rotational velocities v for luminosity class V objects in three spectral type ranges: A0–A1, A2–A4 and A5–F0. *Solid thick lines* are the distributions of true equatorial velocities $\Upsilon(v)$ and the *gray strips* are their associated variability bands. *Dotted lines* are the distributions found in [3] for peculiar and normal stars (respectively slow and fast rotators), and *dashed lines* represent the sum of both peculiar and normal star PDF for each spectral type range

types A0–A1, A2–A4 and A5–F0, and discarding all stars more evolved than the luminosity class V. Known binary stars and chemically peculiar stars are still discarded from the sample. Details about the subgroups are given in Table 2 as in [51].

The Lucy-Richardson method [34, 48] was applied to correct from the projection effect and recover the distributions of true equatorial velocities. The results (Fig. 13) show that the distributions are well in agreement with the previous distributions of normal stars for the later spectral type subgroups, A2–A4 and A5–F0. On the opposite, for A0–A1 stars, the bimodality is still present.

The excess of slow rotators can be represented as the difference of the observed distribution of equatorial from a Maxwellian distribution [12] fitted to

Table 2. Estimated bandwidth for the kernel method and mean integrated squared error: for each subsample, the size n (number of stars) is given, as well as the estimated $\hat{h}$ according to [53], computed in the logarithmic velocity scale, and the variability band width ε

Subsample	n	$\hat{h}$	ε
A0–A1	244	0.153	0.0435
A2–A4	245	0.147	0.0443
A5–F0	208	0.141	0.0491

the observed fast rotator mode. When the distributions are projected back onto the $v\sin i$ scale, this excess of slow rotators amounts approximately 80% of the stars with $v\sin i < 40\,\mathrm{km\,s^{-1}}$, 70% for $v\sin i < 60\,\mathrm{km\,s^{-1}}$ and 60% for $v\sin i < 100\,\mathrm{km\,s^{-1}}$. This excess is larger to what was found by Dworetsky [15].

It is argued that rotational velocity alone discriminates the normal A-type stars from the peculiar ones [1]. The definition of chemically peculiarity used in [1, 3] differs from the classical one [35], and the nature of the slow A0–A1 rotators is still to be understood.

References

1. Abt, H.A.: Does rotation alone determine whether an A-type star's spectrum is abnormal or normal? ApJ **544**, 933 (2000).
2. Abt, H.A., Levato, H., Grosso, M.: Rotational velocities of B stars. ApJ **573**, 359 (2002).
3. Abt, H.A., Morrell, N.I. The relation between rotational velocities and spectral peculiarities among A-type stars. ApJS **99**, 135 (1995).
4. Adelman, S.J.: The physical properties of normal A stars. In: Zverko, J., Žižňovský, J., Adelman, S.J., Weiss, W.W., (eds.) The A-Star Puzzle, IAU Symp. vol. 224, p. 1 (2004).
5. Babcock, H.W.: Magnetic fields of the A-type stars. ApJ **128**, 228 (1958).
6. Brown, A.G.A., Verschueren, W.: High s/n echelle spectroscopy in young stellar groups. ii. rotational velocities of early-type stars in Sco OB2. A&A **319**, 811 (1997).
7. Carroll, J.A.: The spectroscopic determination of stellar rotation and its effect on line profiles. MNRAS **93**, 478 (1933).
8. Castelli, F., Kurucz, R.L.: Model atmospheres for Vega. A&A **281**, 817 (1994).
9. Chandrasekhar, S., Münch, G.: On the integral equation governing the distribution of the true and the apparent rotational velocities of stars. ApJ **111**, 142 (1950).
10. Claret, A.: Comprehensive tables for the interpretation and modeling of the light curves of eclipsing binaries. A&AS **131**, 395 (1998).
11. Debernardi, Y.: Investigation on Am stars binarity. In: Reipurth, B., Zinnecker, H. (eds.) Birth and Evolution of Binary Stars, IAU Symp. vol. 200, p. 161 (2000).
12. Deutsch, A.J.: Maxwellian distributions for stellar rotations. In: Slettebak, A. (ed.) Stellar Rotation, IAU Colloquium No. 4, p. 207. Reidel, Dordrecht (1970).

13. Domiciano de Souza, A., Kervella, P., Jankov, S., Vakili, F., Ohishi, N., Nordgren, T.E., Abe, L.: Gravitational-darkening of Altair from interferome-try. A&A **442**, 567 (November 2005).
14. Dravins, D., Lindegren, L., Torkelsson, U.: The rotationally broadened line profiles of Sirius. A&A **237**, 137 (1990).
15. Dworetsky, M.M.: Rotational velocities of A0 stars. ApJS **28**, 101 (1974).
16. Erspamer, D., North, P.: Automated spectroscopic abundances of A and F-type stars using echelle spectrographs. ii. abundances of 140 A-F stars from Elodie. A&A **398**, 1121 (2003).
17. ESA. The Hipparcos and Tycho catalogues. ESA-SP 1200 (1997).
18. Fekel, F.C.: Rotational velocities of B, A, and early-F narrow-lined stars. PASP **115**, 807 (2003).
19. Gaigé, Y.: Stellar rotational velocities from the v sin i observations – inversion procedures and applications to open clusters. A&A **269**, 267 (1993).
20. Głębocki, R., Stawikowski, A.: Catalog of projected rotational velocities. Acta Astron. **50**, 509 (2000).
21. Gómez, A.E., Luri, X., Mennessier, M.-O., Torra, J., Figueras, F.: The luminosity calibration of the HR diagram revisited by Hipparcos. In: Battrick, B. (ed.) Hipparcos Venice'97, vol. SP-402, p. 207. ESA Publications Division (1997).
22. Gray, D.F.: A quest for differential stellar rotation in A stars. ApJ **211**, 198 (1977).
23. Gray, D.F.: The rotational break for G giants. ApJ **347**, 1021 (1989).
24. Gray, D.F.: The observation and analysis of stellar photospheres, 3rd edn. Cambridge University Press (2005).
25. Gray, R.O., Garrison, R.F.: The early a type stars – refined MK classification, confrontation with Stroemgren photometry, and the effects of rotation. ApJS **65**, 581 (1987).
26. Gray, D.F., Nagel, T.: The granulation boundary in the H-R diagram. ApJ **341**, 421 (1989).
27. Gulliver, A.F., Hill, G., Adelman, S.J.: Vega: a rapidly rotating pole-on star. ApJ Lett. **429**, L81 (1994).
28. Hill, G.M.: Compositional differences among the A-type stars. 2: Spectrum = up to v sin i = 110 km/s. A&A **294**, 536 (1995).
29. Hill, G., Gulliver, A.F., Adelman, S.J.: A spectral study of Vega: a rapidly rotating pole-on star. In: Zverko, J., Žižňovský, J., Adelman, S.J., Weiss, W.W. (eds.) The A-Star Puzzle, IAU Symp. vol. 224, p. 35 (2004).
30. Holweger, H., Hempel, M., Kamp, I.: A search for circumstellar gas around normal A stars and lambda bootis stars. A&A **350**, 603 (1999).
31. Lejeune, T., Schaerer, D.: Database of Geneva stellar evolution tracks and isochrones for *(ubv)J(ri)C jhkll' m*, hst-wfpc2, Geneva and Washington photometric systems. A&A **366**, 538 (2001).
32. Lemke, M.: Abundance anomalies in main sequence A stars. i – iron and titanium. A&A **225**, 125 (1989).
33. Lucy, L.B.: Gravity-darkening for stars with convective envelopes. Z. Astrophys. **65**, 89 (1967).
34. Lucy, L.B.: An iterative technique for the rectification of observed distributions. AJ **79**, 745 (1974).
35. Mathys, G.: Rotation and properties of A- and B-type chemically peculiar stars. In: Maeder, A., Eenens, P. (eds.) Stellar Rotation, IAU Symp. vol. 215, p. 270 (2004).
36. Melo, C.H.F., Pasquini, L., De Medeiros, J. R.: Accurate vsin i measurements in M 67: the angular momentum evolution of $1.2\,m_\odot$ stars. A&A **375**, 851 (2001).

37. Mora, A., Merín, B., Solano, E., Montesinos, B., de Winter, D., Eiroa, C., Ferlet, R., Grady, C.A., Davies, J.K., Miranda, L.F., Oudmaijer, R.D., Palacios, J., Quirrenbach, A., Harris, A.W., Rauer, H., Cameron, A., Deeg, H.J., Garzón, F., Penny, A., Schneider, J., Tsapras, Y., Wesselius, P.R.: Export: spectral classification and projected rotational velocities of Vega-type and pre-main sequence stars. A&A **378**, 116 (2001).

38. Nordström, B., Latham, D.W., Morse, J.A., Milone, A.A.E., Kurucz, R.L., Andersen, J., Stefanik, R.P.: Cross-correlation radial-velocity techniques for rotating F stars. A&A **287**, 338 (1994).

39. Peterson, D.M., Hummel, C.A., Pauls, T.A., Armstrong, J.T., Benson, J.A., Gilbreath, G.C., Hindsley, R.B., Hutter, D.J., Johnston, K.J., Mozurkewich, D., Schmitt, H.: Resolving the effects of rotation in altair with long-baseline interferometry. ApJ **636**, 1087 (2006a).

40. Peterson, D.M., Hummel, C.A., Pauls, T.A., Armstrong, J.T., Benson, J.A., Gilbreath, G.C., Hindsley, R.B., Hutter, D.J., Johnston, K.J., Mozurkewich, D., Schmitt, H.R.: Vega is a rapidly rotating star. Nat. **440**, 896 (2006b).

41. Reiners, A.: The effects of inclination, gravity darkening and differential rotation on absorption profiles of fast rotators. A&A **408**, 707 (2003).

42. Reiners, A.: Rotation- and temperature-dependence of stellar latitudinal differential rotation. A&A **446**, 267 (2006).

43. Reiners, A., Hünsch, M., Hempel, M., Schmitt, J.H.M.M.: Strong latitudinal shear in the shallow convection zone of a rapidly rotating A-star. A&A **430**, L17 (2005).

44. Reiners, A., Royer, F.: Altair's inclination from line profile analysis. A&A **428**, 199 (2004a).

45. Reiners, A., Royer, F.: First signatures of strong differential rotation in A-type stars. A&A **415**, 325 (2004b).

46. Reiners, A., Schmitt, J.H.M.M.: On the feasibility of the detection of differential rotation in stellar absorption profiles. A&A **384**, 155 (2002).

47. Reiners, A., Schmitt, J.H.M.M.: Rotation and differential rotation in field F- and G-type stars. A&A **398**, 647 (2003).

48. Richardson, W.H.: Bayesian-based iterative method of image restoration. J. Opt. Soc. Am. **62**, 55 (1972).

49. Royer, F., Gerbaldi, M., Faraggiana, R., Gómez, A.E. Rotational velocities of A-type stars. i. measurement of v sin i in the southern hemisphere. A&A **381**, 105 (2002).

50. Royer, F., Grenier, S., Baylac, M.-O., Gómez, A.E., Zorec, J.: Rotational velocities of A-type stars. ii. measurement of v sin i in the northern hemisphere. A&A **393**, 897 (2002).

51. Royer, F., Zorec, J., Gómez, A. E.: Rotational velocities of A-type stars. iii. velocity distributions. A&A **463**, 671 (2007).

52. Schmidt-Kaler, T.: Landolt-Börnstein, Group 6: Astronomy Astrophysics and Space research, vol. VI/2b, chap. 5.2.1 Classification of Stellar Spectra, p. 33. Springer-Verlag (1982).

53. Sheather, S.J., Jones, M.C.: A reliable data-based bandwidth selection method for kernel density estimation. J. R. Statist. Soc B. **53**(3), 683 (1991).

54. Slettebak, A.: Determination of stellar rotational velocities. In: IAU Symp. 111: Calibration of Fundamental Stellar Quantities, p. 163 (1985).

55. Slettebak, A., II Collins, G.W., Boyce, P.B., White, N.M., Parkinson, T.D.: A system of standard stars for rotational velocity determinations. ApJS **29**, 137 (1975).

56. Smalley, B.: Observations of convection in A-type stars. In: Zverko, J., Žižňovský, J., Adelman, S.J., Weiss, W.W. (eds.) The A-Star Puzzle, IAU Symp. vol. 224, p. 131 (2004).
57. Smith, M.A., Gray, D.F.: Fourier analysis of spectral line profiles – a new tool for an old art. PASP **88**, 809 (1976).
58. Struve, O., Elvey, C.T.: Algol and stellar rotation. MNRAS **91**, 663 (1931).
59. van Belle, G.T., Ciardi, D.R., Thompson, R.R., Akeson, R.L., Lada, E.A.: Altair's oblateness and rotation velocity from long-baseline interferometry. ApJ **559**, 1155 (2001).
60. von Zeipel, H.: The radiative equilibrium of a rotating system of gaseous masses. MNRAS **84**, 665 (1924).

The Solar Magnetic Field: Surface and Upper Layers, Network and Internetwork Field

V. Bommier

Laboratoire d'Étude du Rayonnement et de la Matière en Astrophysique, CNRS UMR 8112 – LERMA, Observatoire de Paris, Section de Meudon, 5 place Jules Janssen, 92195 Meudon, France
V.Bommier@obspm.fr

Abstract After having presented the magnetic field effects on the radiation emitted in an atomic or molecular line (Zeeman effect, Hanle effect), we present two applications: (i) the measurement of the magnetic field vector in solar prominences during the ascending phase of Cycle XXI. By transferring the results on synoptic maps of the filaments, a general organization of the large-scale surface field becomes visible; (ii) the measurements of the surface magnetic field made by recent spectropolarimeters like THEMIS. The result is the division of the surface magnetic field into two classes: (a) the network field, nearly vertical and 100 G or more (local average value), located in active regions and at the frontiers of supergranules; (b) the internetwork field, turbulent in direction with an horizontal trend, with a local average value of 20 Gauss or less.

1 Introduction

In astrophysics, the information about the observed object can be retrieved from the analysis of the received radiation only. The information about the magnetic field is transported by the radiation polarization. The polarization may roughly be introduced as follows: if a polarizing device is interposed on the light ray, and if in addition the transmitted radiation changes when the device is rotated, then the light is polarized and its polarization degree can be measured. With the help of the theory of the magnetic field effect on the line polarization, the result of the polarization measurement can be converted in magnetic field information. At the surface of the Sun, the matter state is a plasma, which means a gas globally neutral but containing moving charged particles. Their movement is controlled by the magnetic field. Therefore the forecasting of the solar matter ejections (the so-called space weather) will necessarily make use of magnetic field maps as initial or surface conditions for extrapolation or evolution modelling.

Besides, the magnetic field is the product of the internal dynamo, or of the local dynamos if they exist. Thus, information on the dynamos can be derived from the magnetic field measurement that can be realized only at the surface or

Bommier, V.: *The Solar Magnetic Field: Surface and Upper Layers, Network and Internetwork Field*. Lect. Notes Phys. **765**, 231–259 (2009)
DOI 10.1007/978-3-540-87831-5_10

in the upper-layer. In addition, the upper-layer magnetic field, marked out by the prominences magnetic field, shows a global structure linked to the differential rotation visible in Figs. 19 and 20 of the present course.

In the following course, first the basic ingredients of the magnetic field measurement in astrophysics will be introduced. The tool is the radiation polarization, which is described by the four Stokes parameters. We will concentrate on atomic (or molecular) line polarization. The magnetic field acts on the line polarization via two effects: the Zeeman and Hanle effects, which will be presented and compared. Then, two applications will be introduced: the solar prominences measurement via analysis of the Hanle effect and the photospheric magnetic field measurement via analysis of the Zeeman effect.

2 The Basic Ingredients of the Magnetic Field Measurement in Astrophysics

2.1 The Radiation Polarization: The Stokes Parameters

I,Q,U,V. The four Stokes parameters describe the intensity and polarization of a pencil of radiation propagating along the Oz axis in the positive direction. Considering the Ox and Oy axes perpendicular to Oz, the full definition of the Stokes parameters requires also the consideration of the OX and OY axes obtained by rotating the xOy referential by $+45°$ around the Oz axis, and also the consideration of the rotating basis vectors

$$\begin{cases} \boldsymbol{e}_+ = -\frac{1}{\sqrt{2}}(\boldsymbol{e}_x + i\boldsymbol{e}_y) \\ \boldsymbol{e}_- = +\frac{1}{\sqrt{2}}(\boldsymbol{e}_x - i\boldsymbol{e}_y) \end{cases}, \tag{1}$$

$\boldsymbol{e}_x$ and $\boldsymbol{e}_y$ being the basis vectors of the xOy reference frame (see the definition of the whole set of axes in Fig. 1). If one considers the intensity I_i along one of the axes $(i = x, y, X, Y, +, -)$, it is the average value of the square of the incident radiation electric field vibrating along the i direction:

$$I_i = \left\langle E_i^2 \right\rangle, \tag{2}$$

and it is the intensity that would be transmitted through a polarizing device oriented in this axis direction. The four Stokes parameters are the following combinations

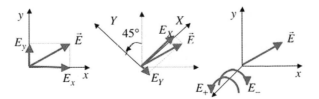

Fig. 1. Referentials for the Stokes parameters definition

$$\begin{cases} I = I_x + I_y = I_X + I_Y = I_+ + I_- \\ Q = I_x - I_y \\ U = I_X - I_Y \\ V = I_+ - I_- \end{cases} \tag{3}$$

I is the intensity. Q and U are related to the linear polarization (see below). V is related to the circular polarization: $p_c = -V/I$ is the circular polarization degree, provided that it is recalled that e_+ is the *left-handed* circular polarization unit vector whereas e_- is the *right-handed* one. One can find other definitions of the Stokes parameters

$$\begin{cases} I = \langle E_x^2(t) \rangle + \langle E_y^2(t) \rangle \\ Q = \langle E_x^2(t) \rangle - \langle E_y^2(t) \rangle \\ U = 2 \langle E_x(t) E_y(t) \rangle \\ V = 2 \langle E_x(t) E_y(t + \pi/2\omega) \rangle \end{cases} \tag{4}$$

or, introducing the complex generalization of the electric field vector

$$E_i = \frac{1}{\sqrt{2}} \left[\mathcal{E}_i(t) e^{-i\omega t} + \mathcal{E}_i^*(t) e^{+i\omega t} \right] , \tag{5}$$

one has also

$$\begin{cases} I = \langle \mathcal{E}_x(t) \mathcal{E}_x^*(t) \rangle + \langle \mathcal{E}_y(t) \mathcal{E}_y^*(t) \rangle \\ Q = \langle \mathcal{E}_x(t) \mathcal{E}_x^*(t) \rangle - \langle \mathcal{E}_y(t) \mathcal{E}_y^*(t) \rangle \\ U = \langle \mathcal{E}_x(t) \mathcal{E}_y^*(t) \rangle + \langle \mathcal{E}_y(t) \mathcal{E}_x^*(t) \rangle \\ V = i \left[\langle \mathcal{E}_x(t) \mathcal{E}_y^*(t) \rangle - \langle \mathcal{E}_y(t) \mathcal{E}_x^*(t) \rangle \right] \end{cases} \tag{6}$$

However, the definition that displays most clearly the measurement method is the one of Eq. (3).

The linear polarization. Let us consider a linearly polarized pencil of radiation. Figure 2 displays the intensity transmitted through a polaroid plate rotating around the Oz axis. The direction where the maximum intensity I_{max} is obtained is referred to with the α_0 angle. The linear polarization degree is defined by

$$p_\ell = \frac{I_{max} - I_{min}}{I_{max} + I_{min}} = \frac{\sqrt{Q^2 + U^2}}{I} \le 1 , \tag{7}$$

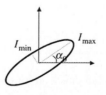

Fig. 2. Linear polarization parameters: intensity transmitted through a polaroid plate rotating around the Oz axis

and the linear polarization direction α_0 is defined (modulo $\pi/2$) by a

$$\tan 2\alpha_0 = \frac{U}{Q} , \tag{8}$$

more precise definition (modulo π) being given by

$$\cos 2\alpha_0 = \frac{Q}{\sqrt{Q^2 + U^2}} , \quad \sin 2\alpha_0 = \frac{U}{\sqrt{Q^2 + U^2}} . \tag{9}$$

The transformation of the linear polarization parameters Q and U in an axis rotation of angle α is as follows:

$$\begin{cases} Q' = Q \cos 2\alpha + U \sin 2\alpha \\ U' = -Q \sin 2\alpha + U \cos 2\alpha \end{cases} , \tag{10}$$

where Q' and U' are the linear polarization Stokes parameters in the new frame. Finally, the four Stokes parameters obey the inequality

$$\sqrt{Q^2 + U^2 + V^2} \leq I , \tag{11}$$

which leads to the definition of the total polarization rate (polarization rate 'in the sense of the Stokes parameters')

$$p = \frac{\sqrt{Q^2 + U^2 + V^2}}{I} \leq 1 . \tag{12}$$

2.2 The Zeeman Effect

The atomic structure. The effect of the magnetic field on an isolated energy level is represented in Fig. 3. Under the magnetic field effect the energy level is split into $2J + 1$ Zeeman sublevels, each sublevel energy behaving linearly with the magnetic field strength. This is due to the fact that the magnetic Hamiltonian is given by

$$H_{\mathrm{magn}} = -\boldsymbol{M} \cdot \boldsymbol{B} , \tag{13}$$

where $\boldsymbol{M}$ is the magnetic dipolar momentum. If the quantization axis Oz is chosen along the magnetic field $\boldsymbol{B}$, $-\boldsymbol{M} \cdot \boldsymbol{B}$ is proportional to J_z (for an isolated level, or in weak field), the kinetic momentum component along Oz. Thus, the

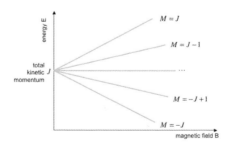

Fig. 3. The Zeeman effect on an isolated energy level

energy of the sublevel characterized by the magnetic quantum number M ($-J \leq M \leq +J$) is

$$E = g_J M \hbar \omega_{\mathrm{L}} , \tag{14}$$

where the proportionality factor g_J is the Landé factor, characteristic of the energy level, and ω_{L} is the Larmor pulsation, related to the Larmor frequency as usual $\omega_{\mathrm{L}} = 2\pi\nu_{\mathrm{L}}$, where $\nu_{\mathrm{L}}/B \simeq 1.4$ MHz/G. The Larmor pulsation depends linearly on the magnetic field strength

$$\omega_{\mathrm{L}} = \frac{\mu_{\mathrm{B}} B}{\hbar} , \tag{15}$$

where μ_{B} is the Bohr magneton (SI units):

$$\mu_{\mathrm{B}} = \frac{q\hbar}{2m_e} , \tag{16}$$

where q and m_e are, respectively, the electron charge and mass.

The Landé factor. In the LS coupling scheme (in most cases for light atoms), let us recall that a level is denoted by $^{2S+1}L_J$ in spectroscopic notation, where S, L and J are, respectively, the spin, orbital and total kinetic momentum quantum numbers, associated to the kinetic momenta $\boldsymbol{S}$ and $\boldsymbol{L}$ that are each the sum of the individual electron momenta, and to the total kinetic momentum $\boldsymbol{J} = \boldsymbol{L} + \boldsymbol{S}$. In the spectroscopic notation, the L value is indicated by a letter: S, P, D, F, G, H for $L = 0, 1, 2, 3, 4, 5$, respectively. For instance, the level labelled $^3D_3$ has $L = 2$, $S = 1$ and $J = 3$. Given these numbers, the Landé factor is given by

$$g_J = \frac{3}{2} + \frac{S(S+1) - L(L+1)}{2J(J+1)} . \tag{17}$$

The upper and lower level Landé factors are given in the Kurucz's line data basis:

$$\text{http://kurucz.harvard.edu/linelists.htm,} \tag{18}$$

from which a compilation of solar and stellar lines can be found in the Tarbes site of the solar data basis BASS2000:

$$\text{http://bass2000.bagn.obs-mip.fr/New2003/Pages/info\_raie.html.} \tag{19}$$

For molecular lines, let us mention the MZL (Molecular Zeeman Library) software for Landé factor calculation by B. Leroy, accessible in the Meudon site of the solar data basis BASS2000:

$$\text{http://bass2000.obspm.fr/software.php.} \tag{20}$$

The end of the Zeeman effect: the Paschen–Back effect. When the magnetic field strength increases, the Zeeman splitting may become non-negligible with respect to the level distance, due for instance to the fine or hyperfine structure. In other words, the magnetic hamiltonian (see Eq. (13)) may become non-negligible with respect to the fine-structure hamiltonian. And whereas for a weak field the magnetic hamiltonian is a perturbation of the fine-structure hamiltonian (Zeeman regime), for a very strong field this is the fine-structure hamiltonian that becomes a perturbation of the magnetic hamiltonian (Paschen–Back regime). Whereas the magnetic hamiltonian is diagonal in the uncoupled (L, M_L, S, M_S) basis due to the fact that the magnetic dipolar momentum is $\boldsymbol{M} = \boldsymbol{L} + 2\boldsymbol{S}$, the fine-structure hamiltonian is diagonal in the coupled (L, S, J, M_J) basis; in strong field J is no more a 'good quantum number', and the field dependence is given by $M_L + 2M_S$ (Paschen–Back effect). As it is different from M_J, the slope of the sublevels is not the same in the Zeeman and Paschen–Back regimes. In the transition regime the levels are curved and the magnetic field effect is non-linear. The levels having the same M_J magnetic quantum number, but coming from levels of different J quantum number, are coupled by the magnetic field and 'anticross'. An example of transition from the Zeeman effect to the Paschen–Back effect in helium is given in Fig. 4.

The conservation of the kinetic momentum and the polarization of the Zeeman components. The various Zeeman components have characteristic intrinsic polarization, resulting from the angular momentum conservation during the photon emission process. The possible polarizations are represented

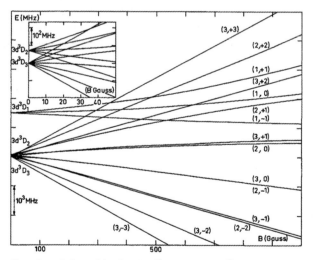

Energies of the sublevels of $3d^3D_{3,2,1}$ and $2p^3P_{2,1,0}$ as functions of the magnetic field B. On each sublevel is reported the corresponding (J^*, M_J) quantum numbers

Fig. 4. Transition from the Zeeman to the Paschen–Back effect in helium. From [1]

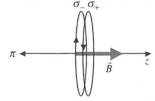

Fig. 5. Polarization of the Zeeman components 1

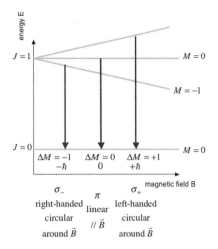

Fig. 6. Polarization of the Zeeman components 2

in Fig. 5: the π polarization is the linear polarization parallel to the magnetic field, whereas the two σ polarizations are the two circular polarizations (right-handed for σ_-, left-handed for σ_+) around the magnetic field. As for the angular momentum component along Oz (which is parallel to the magnetic field) the π polarization has a zero component, whereas the σ_+ (resp. σ_-) polarization has a positive (resp. negative) component. Consider now the various Zeeman components emitted by a normal Zeeman triplet (i.e. having $J = 1$ in the upper level and $J = 0$ in the lower one, see Fig. 6): the upper level has three Zeeman sublevels having $M_J = -1, 0, +1$, whereas the lower level has only one sublevel having $M_J = 0$. Thus, when in the transition (photon emission) $\Delta M = -1$, the emitted photon wears the angular momentum $-\hbar$, so its polarization is σ_- (right-handed, clockwise), whereas if in the transition $\Delta M = +1$, the emitted photon wears the angular momentum $+\hbar$, so its polarization is σ_+ (left-handed, counterclockwise), and if in the transition $\Delta M = 0$, the emitted photon wears the angular momentum 0, so its polarization is π. As, also, the different Zeeman components do not have exactly the same frequency due to the magnetic field dependence of the sublevel energy, it results that *in the Zeeman effect, the polarization varies along the profile of the global line.*

2.3 The Hanle Effect

The scattering polarization. As explained in Fig. 7, right-angle scattering is a linearly polarizing mechanism. This is due to the combination of the two following features: (a) in a classical view, a linearly polarized incident radiation excites an atomic dipole with vibration direction parallel to the incident polarization. This vibrating dipole then radiates; (b) but it cannot radiate along its vibration direction. The result is that an unpolarized incident radiation is scattered into a linearly polarized radiation, the linear polarization direction being perpendicular to the scattering plane.

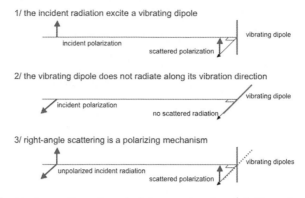

Fig. 7. Scattering: a linearly polarizing mechanism (*right-angle* scattering)

The line polarizability. From above, it is expected that the line would be 100% linearly polarized in right-angle scattering. Actually, this is rigorously true only when the upper level has the kinetic momentum $J' = 1$ and the lower level $J = 0$, which is the so-called normal Zeeman triplet that is quantitatively described by the classical model of the vibrating dipole. In the general case, there is a maximum polarization rate, which is less or equal to unity and which depends on the level kinetic momenta. One may have an idea (only an idea, not the exact value) of this *polarizability* with the W_2 factor listed in Table I of Landi Degl'Innocenti [2]. Looking at the values in this table, it is clear that W_2, and then the line polarizability, ranges between 0 and 1.

The modification of the scattering polarization by the magnetic field. Let us consider the right-angle scattering linear polarization and introduce now a non-zero magnetic field along the line of sight (Fig. 8). The field induces the Larmor precession of the electrons (and then of the vibrating dipole) about it. The combination of the dipole damping and of the Larmor precession induces the *two main features of the Hanle effect*: (a) the depolarization with respect to the zero-field scattering polarization; (b) the rotation of the linear polarization direction.

The Hanle effect: the characteristic quantity. In the presence of a magnetic field, the linear polarization degree depends on the competition between the dipole damping and the Larmor precession. Thus, the characteristic quantity of the Hanle effect is the product $\omega\tau$, where ω is the Larmor pulsation (ignoring now the L subscript) and τ the upper level life time. Two cases can be considered (see Fig. 9):

- $\omega\tau \approx 1$: the depolarization is partial, i.e. the polarization degree is non-zero and the linear polarization direction is rotated. The Hanle effect can be used to measure the quantity $\omega\tau$. Thus, if the magnetic field is known the Hanle effect can be used to measure the upper level life time: this is the main laboratory application of the Hanle effect. In astrophysics, on the contrary one uses lines having a known life time, and the Hanle effect is used

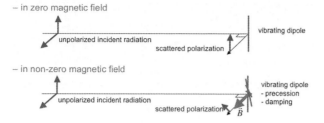

Fig. 8. The Hanle effect in non-zero magnetic field: depolarization, rotation of the polarization direction

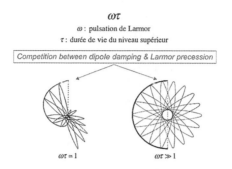

Fig. 9. Competition between the damping and the Larmor precession

to measure the magnetic field strength. The field direction can be recovered also. There is however the fundamental ambiguity, whose physical basis lies in the fact that the atom size is small compared to the incident radiation wavelength, so that the atom is insensitive to the propagation sense of the incident radiation. Thus, from the Hanle effect two fields symmetrical with respect to the line of sight cannot be distinguished: this is the fundamental degeneracy (or ambiguity).

– $\omega\tau \gg 1$: the depolarization is total, i.e. the polarization degree is zero in right-angle scattering and field lying along the line of sight. This is the case of the forbidden lines of the solar corona, due to their very long life times. In this case the Hanle effect is 'saturated' and the field strength cannot be measured, but the field direction remains possible to determine. The fundamental ambiguity exists also, but it has to be complemented with the Van Wleck ambiguity (Van Wleck angle: $3\cos^2\theta = 1$). If the polar angle between the field and the incident radiation is the Van Wleck angle, the polarization degree is zero whatever be the field azimuth. A given polarization may correspond to two field vectors having their polar angles on either side of the Van Wleck angle. Thus, there is at least four magnetic fields possible for a given measurement.

The Hanle effect: a simple quantum description. A quantum description of the Hanle effect can be simply outlined in the case of the 'normal Zeeman triplet' line, analogous to the vibrating dipole of the classical theory. The normal

Fig. 10. Referentials for the quantum description of the Hanle effect

Zeeman triplet line has $J = 1$ in the upper level and $J = 0$ in the lower level. The referential is the one of Fig. 10, and the magnetic fields lie along the quantization axis Oz. The incident radiation propagates along Ox having a linear polarization parallel to e_y, the unit vector of the Oy axis. In terms of the rotating vectors e_+ and e_- defined in Eq. (1), the polarization of the incident radiation is

$$e_y = -\frac{1}{i\sqrt{2}}(e_+ + e_-) \,, \tag{21}$$

so that the photon state before absorption is

$$-\frac{1}{i\sqrt{2}}(|\sigma_+\rangle + |\sigma_-\rangle) \,. \tag{22}$$

Thus, the atomic wavefunction just after absorption is

$$|\psi(0)\rangle = -\frac{1}{i\sqrt{2}}(|+1\rangle + |-1\rangle) \,. \tag{23}$$

Applying the evolution solution of the Schrödinger equation, and introducing an ad hoc damping coefficient to take into account the coupling between the atom and the vacuum leading to the finite life time of the upper level, one gets the wavefunction at time t just before emission:

$$|\psi(t)\rangle = -\frac{1}{i\sqrt{2}}e^{-\frac{t}{2\tau}}\left(e^{-i\omega t}|+1\rangle + e^{+i\omega t}|-1\rangle\right) \,. \tag{24}$$

In this wavefunction, the *coherence* between the two basic states $|+1\rangle$ and $|-1\rangle$ is

$$c_n c_m^* = e^{-\frac{t}{\tau}(1+2i\omega\tau)} \,. \tag{25}$$

The photon state just after emission is the translation of the atomic state

$$-\frac{1}{i\sqrt{2}}e^{-\frac{t}{2\tau}}\left(e^{-i\omega t}|\sigma_+\rangle + e^{+i\omega t}|\sigma_-\rangle\right) \,, \tag{26}$$

so that the polarization of the radiation emitted along Oz is

$$e = \frac{1}{\sqrt{2}}e^{-\frac{t}{2\tau}}\left(e_y \cos\omega t - e_x \sin\omega t\right) \,. \tag{27}$$

Indeed, the emitted polarization results of the above process averaged over a great number of atoms or averaged over time, with various emission time t. As it is visible in the right part of Fig. 10, the result of the averaging process is a decrease in the polarization degree and a rotation of the polarization direction. The average coherence is

$$\langle c_n c_m^* \rangle = \left\langle e^{-\frac{t}{\tau}(1+2i\omega\tau)} \right\rangle , \qquad (28)$$

which is a quantity that decreases with $\omega\tau$. This shows the quantum effect of the magnetic field, which is the *partial destruction of the Zeeman coherences*. The formalism well adapted to describe this phenomenon is the one of the atomic density matrix [3].

The Hanle effect: a non-linear and anisotropic effect. The Hanle effect is not isotropical: Figure 11 represents the effect for the three basic field vector directions, for a scattering point located above the Sun. It is visible that the effect is highly different for these three directions. Besides, the Hanle effect is highly non-linear: the average effect is not the effect of the average field. This is why the Hanle effect can be detected even in unresolved fields, for instance field having various directions in the resolution element (such field is often called a 'turbulent field', turbulence meaning spatial unresolved variation). Thus, the Hanle effect has been used to detect the turbulent unresolved field of the Quiet Sun [4].

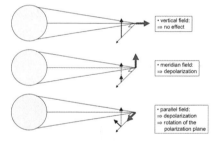

Fig. 11. The Hanle effect is neither isotropical nor linear

TABLE I

Domain of sensitivity to the Hanle effect of selected lines of astrophysical interest. B_{typ} is the 'typical' magnetic field defined by $\omega\tau = 1$ (cf. Section 2), and p_{lim} the maximum theoretical value of the polarization degree obtained for an infinite height above the solar limb

Spectrum	λ (Å)	Transition	B_{typ} (G)	P_{lim}
Fe XIV	5303	$3p\,^2P_{3/2} \to 3p\,^2P_{1/2}$	5×10^{-6}	0.43
C III	1909	$2s2p\,^3P_1 \to 2s^2\,^1S_0$	1.1×10^{-5}	1
He I	10 830	$2p\,^3P_{2,1,0} \to 2s\,^3S_1$	0.83	
He I (D$_3$)	5875	$3d\,^3D_{3,2,1} \to 2p\,^3P_{2,1,0}$	6	
He I (D$_3$)	major comp.	$3d\,^3D_{3,2,1} \to 2p\,^3P_{2,1}$	6	
He I (D$_3$)	minor comp.	$3d\,^3D_1 \to 2p\,^3P_0$	16	1
C IV	1548	$2p\,^2P_{3/2} \to 2s\,^2S_{1/2}$	22.5	0.43
N V	1239	$2p\,^2P_{3/2} \to 2s\,^2S_{1/2}$	28.7	0.43
O VI	1032	$2p\,^2P_{3/2} \to 2s\,^2S_{1/2}$	34.7	0.43
Si IV	1394	$3p\,^2P_{3/2} \to 3s\,^2S_{1/2}$	78.2	0.43
Si III	1206	$3s3p\,^1P_1 \to 3s^2\,^1S_0$	295.	1
Lα	1216		53.2	0.27
Lβ, Hα	1026, 6563		16	
Lγ, Hβ	992, 4861		7	

Fig. 12. Sensitivity of their Hanle effect for a series of lines. From [5]

The line sensitivity. From Hanle effect measurements, the field strength is recoverable in the domain defined by $0.1 \leq \omega\tau \leq 10$, which depends on the line via τ. The 'typical' field B_{typ} being defined by the central value $\omega\tau = 1$, Fig. 12 is a list of typical fields for a series of lines of astrophysical interest (from [5]). It can be seen in this table that this typical field may change a lot from line to line: the line choice must be adapted to the searched for magnetic field strength.

2.4 Comparison Between the Zeeman and Hanle Effects

The Hanle effect domain is defined by $\omega\tau \approx 1$: in the energy diagram, this corresponds to overlapping Zeeman sublevels broadened by their natural width. When the broadened sublevels overlap, the Zeeman coherences may exist (even if partially destroyed) and the Hanle effect takes place. When the broadened sublevels are well separated, the Zeeman coherences are destroyed and one is in the *saturated* Hanle effect regime. As for the Zeeman effect, the question is to know if the levels broadened by the Doppler width overlap or not. When the

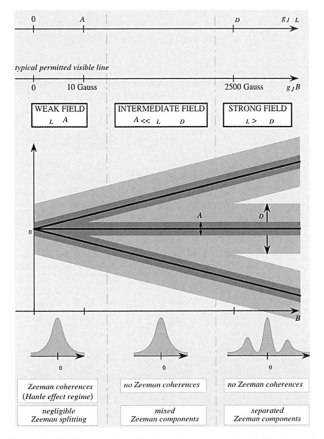

Fig. 13. Different domains for the magnetic field strength

Doppler broadened sublevels overlap, the Zeeman components are unresolved in the line profile. When the Doppler broadened sublevels are well separated, the Zeeman components are well separated, each having its own polarization. As the Doppler width is in general much larger than the natural width (this is the case of the solar atmosphere), one can define three field domains that are represented in Fig. 13. In the intermediate domain, the Zeeman components exist though not separated, so that they induce a polarization variation along the line profile. Thus, we can list the main differences between the Hanle and Zeeman effects:

- The Hanle effect acts on the linear polarization only; the Zeeman effects act on both linear and circular polarizations.
- The Hanle effect is constant along the line profile, because it results from the atomic polarization (existence of Zeeman coherences and imbalance between the Zeeman sublevel populations): thus, a filter is sufficient for the polarization measurement. The Zeeman effect induces a wavelength-dependent polarization: a spectropolarimeter is required for its measurement.
- The Hanle effect is *non-linear* (the average effect is not the effect of the average field), and consequently it is sensitive to unresolved structures (like the 'turbulent' field). The Zeeman effect is a *linear* effect (insensitive to the unresolved structures).
- The Hanle effect requires an initially polarizing mechanism, on which it acts as a depolarizing secondary mechanism. This initially polarizing mechanism may be the scattering of an anisotropic radiation, as described above. The absorption of the anisotropic radiation creates an *atomic (or molecular) polarization*, which is an imbalance between the Zeeman sublevel populations. The emitted polarization is the signature of this induced atomic polarization. As for the Zeeman effect, there is no condition on the incident radiation to observe it.
- The Hanle effect acts on the atomic polarization which is a NLTE phenomenon. The Zeeman effect can be observed even in LTE.

3 The Magnetic Field Observed in Solar Prominences

3.1 What Is a Solar Prominence?

As visible in the example of Fig. 14, a prominence is a concentration of matter colder and denser than the surrounding Corona, probably suspended by magnetic field lines above the underlying solar surface. A prominence is like a wall having feet separating two regions of different photospheric polarity (entering or outgoing magnetic field). When seen from above, this wall appears like a dark elongated absorbent, then called a *filament*. But the Sun rotates and when seen at the limb, the filament appears no more in absorption but in emission (by scattering) in front of the darker sky, thus taking the name of *prominence*. Thus, prominence and filament are two designations of the same solar object.

A description of the physical conditions in prominences can be found in [6].

Fig. 14. Quiescent prominence observed on 15 August 1980 with the Meudon spectroheliograph, in the Hydrogen Hα line

As the prominence radiates by scattering of an anisotropic incident radiation field like schematized in Fig. 11, the Hanle effect may take place if in addition the magnetic field strength and the upper level life time have compatible orders of magnitude. In zero magnetic field the polarization direction would be parallel to the solar limb (scattering polarization); the rotation of the linear polarization direction with respect to the solar limb was first observed by Lyot [7, 8], but its interpretation in terms of Hanle effect was later due to Hyder [9]. However a sign error was remaining and it was necessary to wait for the work by Leroy [10], who observed the prominences during the whole ascending phase of Cycle XXI and who gained one order of magnitude in polarimetric accuracy, on the one hand, and the density matrix approach for the Hanle effect theory [3] on the other hand, to get quantitative results on the prominence magnetic field measurement by interpretation of the Hanle effect.

3.2 Steps of the Magnetic Field Measurement by the Hanle Effect

The prominence magnetic field measurement by interpretation of the Hanle effect has required several steps:

– At Pic-du-Midi, more than 300 prominences have been observed in the He I D$_3$ line 5876 Å, during the years 1974–1982 (ascending phase of Cycle XXI). This line was chosen as absent from the photospheric spectrum, so that vertical motions in the prominence do not alter the incident radiation properties. In addition this line is nearly optically thin in prominences and insensitive to the depolarizing effect of collisions. The line was observed through a filter, so that its two components were not resolved. The linear polarization was measured, which consists of two parameters, the polarization degree and direction. Three magnetic field vector coordinates are searched for; that is why with one single line the information is insufficient; information has to be added to achieve the interpretation. This additional information was the horizontality of the field, which can be derived from statistical considerations about the observations (see next section). Adding this information, horizontal field, the magnetic field strength and azimuth can be derived from the observations, but two solutions remain indiscernible that are symmetrical with respect to the line of sight: this is the fundamental ambiguity. Thus, an average field strength of 6–10 G was obtained.

– At Sacramento Peak, with the Stokes II instrument which was a spectropolarimeter, the two components of the He I D_3 line were resolved. Indeed this line has six fine-structure components (upper level $3d^3 D_{3,2,1}$; lower level $2p^3 P_{2,1,0}$,) and, due to the smallness of the fine structure, five components are unresolved forming a global component, whereas the $3d^3 D_1 \rightarrow 2p^3 P_0$ component is apart and separated from the other one. As visible in Fig. 12, these two components do not have the same B_{typ}, which means that they do not have the same sensitivity to the magnetic field. Thus, by combining the linear polarization data from two lines of different sensitivities, it is possible to determine the three components of the field vector. The fundamental ambiguity remains however, and actually one has two solutions symmetrical with respect to the line of sight. The horizontality of the field vector was thus demonstrated in 18 [11] and 2 [12] quiescent prominences.

– The Pic-du-Midi instrument was then modified to observe a second line. This second line was hydrogen $H\beta$ or $H\alpha$, alternatively. $H\beta$ is optically thin, and with this line the result was the same as with the Sacramento Peak Stokes II observing the resolved He I D_3: the horizontality of the magnetic field was obtained in 18 prominences [13], but the ambiguity remains. However, contrary to the He I D_3 line, the hydrogen lines are sensitive to the depolarizing effect of collisions with protons and electrons. This is due to the smallness of the energy difference between n, l, j and $n, l \pm 1, j'$ levels, which is specific to hydrogen. The corresponding cross-sections have been computed by using a semi-classical formalism [14] and introduced in the density matrix statistical equilibrium [15]. With two lines four linear polarization parameters are measured: two polarization degrees and two polarization directions. Three magnetic field vector coordinates are searched for, and one extra parameter can be determined. Here this extra parameter was then the electron density via the depolarizing effect of collisions: it was found to range between 10^9 and 4×10^{10} cm^{-3} [16]. But the fundamental ambiguity remained unresolved and two ambiguous solutions were derived from each measurement. Solving the fundamental ambiguity was possible by using hydrogen $H\alpha$, which is not optically thin contrary to He I D_3. For modelling the incident radiation, it is then necessary to take into account the prominence internal radiation [16], but this modifies the symmetry of the ambiguous solutions, so that by combining one optically thin line and one optically thick line, each pair of solutions have not the same symmetry so that the field is obtained by selecting the common solution between the two pairs. Thus the fundamental ambiguity was solved in 14 prominences [18], and the average angle between the field vector and the filament long axis was found to be 35° giving a field vector more or less aligned with the filament long axis.

In the case of solar prominences, we have found three methods for solving the fundamental ambiguity:

-   Observing the prominences on two (or more) following days. Due to the solar rotation, the scattering angle and then the solution symmetry are not the same for each of the 2 days, so that the ambiguity can be solved by selecting the common solution between the two pairs [19]. This method was successfully applied to a series of cases observed at the Pic-du-Midi.
-   Statistical study of the symmetry of the solutions on a large number of prominences. This was applied to the 300 prominences observed at the Pic-du-Midi [20].
-   Joined analysis of optically thin and thick line observations, as described above. As the first one, this method is based on a modification of the scattering geometry: in the present case, the incident radiation does not have the same geometry for the two lines, because the prominence internal radiation has to be taken into account for the optically thick line.

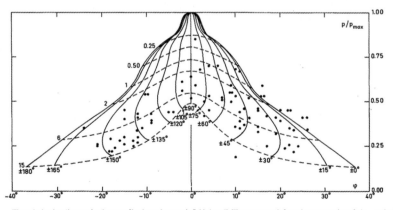

The polarization diagram for the case of horizontal magnetic fields ($\psi = \frac{\pi}{2}$). The curves result from the computation of p/p_{max} and φ for various values of B and θ. Full lines correspond to constant values of θ while dotted lines give the variation of p/p_{max} as a function of φ for B constant and expressed in Gauss. The points refer to the average values of p/p_{max} and φ observed on 82 different quiescent prominences. It is clearly seen that the intensity of magnetic field in quiescent prominences in most cases lies between 1 and 15 Gauss, which is exactly the range of intensities where the Hanle effect has a well visible influence.

Fig. 15. Abacus for the Hanle effect, plotted for an horizontal field. The vertical series of *full lines* are constant azimuth and increasing strength lines. The horizontal series of *dashed lines* are constant strength and varying azimuth lines. The abscissa is the rotation of the polarization direction, and the ordonnae is the polarization degree, divided by the zero-field polarization degree p_{max}, which is the largest one (*right angle scattering*). Case of the He I D$_3$ line of prominences. From [21]

3.3 First Result: The Field Horizontality

A useful tool of the Hanle effect interpretation is the abacus that can be plotted as the one in Fig. 15. An abacus can be plotted for each value of the polar angle ψ of the magnetic fiels vector with respect to the vertical axis (the solar radius). As the observation results in two parameters, the linear polarization degree and direction, these two parameters define a point in the axis system of the abacus, the polarization direction giving the abscissa and the polarization degree giving the ordonnae (note that the zero-field polarization, p_{max}, has to be computed from a model). The field strength and azimuth can be directly read from the abacus system of lines. This gives the structure of a Hanle data inversion code as ours, where a linear interpolation between the abacus lines can even be implemented.

Figure 15 is the abacus for the prominence horizontal field observed with the He I D_3 line. It can be seen that the system of curves has a central vacuum (around the ordonnae axis). Such a vacuum is absent in the abaci of non-horizontal fields (see Fig. 6a and b of [21]). As the points in Fig. 15, which represent the polarization observations, show the same central vacuum, it can be statistically inferred from this figure (from [21]) that the prominence field is horizontal. This statistical result was however later confirmed on individual cases by two-line analysis, in the Sacramento Peak data ([11, 12]), as well as in the modified Pic-du-Midi data ([13, 18]), as stated above.

3.4 Second Result: The Inverse Polarity

As it can be seen in Fig. 16, in most cases the two ambiguous solutions that are symmetrical with respect to the line of sight (right-angle scattering) do not correspond to the same orientation with respect to the neighbouring photospheric polarities. A prominence of *normal polarity* has its field going from outgoing photospheric field $(+)$ towards incoming photospheric field $(-)$. This corresponds to the Kippenhahn–Schlüter model of Fig. 17. On the contrary, a prominence of *inverse polarity* has its field going from incoming photospheric field $(-)$ towards outgoing photospheric field $(+)$. This corresponds to the Kuperus–Raadu model type that includes a neutral point. Three methods have been described above for solving the ambiguity: all three methods have led to the same unexpected result: *the quiescent prominences are of inverse polarity in their biggest majority.*

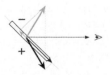

Fig. 16. Most general case for the two ambiguous solutions that are symmetrical with respect to the line of sight: they do not correspond to the same orientation with respect to the photospheric neighbouring polarities

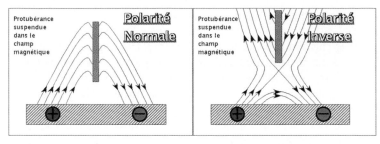

Fig. 17. The two models of prominence field line, associated to the two different orientations of the field with respect to the neighbouring photospheric polarities. The left-hand side model is the Kippenhahn-Schlüter model, the right-hand side model is the Kuperus–Raadu model

3.5 Global Structure Derived from the Measurements in 300 Prominences

Given these two results, we reinvestigated the sample of 300 prominences observed in one single integrated line at the Pic-du-Midi. Assuming the horizontality of the field, it was then possible to determine the field strength and azimuth. We averaged the measurements in each prominence, leading to the determination of one average field vector for each prominence/filament. However this field vector was ambiguous, and we solved the fundamental ambiguity by using the above result of a biggest majority of inverse polarity prominences, following the method described below. As a result, we have one unique field vector, averaged over the prominence, and, when plotted on filament synoptic maps, this draws a large-scale structure of the high-altitude solar magnetic field, which is summarized below.

The fundamental ambiguity solution. For solving the fundamental ambiguity, we followed the scheme outlined in Fig. 18. When the two ambiguous solutions correspond to different prominence polarity, we selected the inverse polarity solution. This is the biggest majority of cases, and we called this method 'the inverse method'. But some cases exist where this method cannot be applied, because the two ambiguous solutions point towards the same prominence side. That kind of cases is much less numerous. This is the case of the side-on seen prominences. It can be remarked that, in this case, the two ambiguous solutions have opposite field components along the prominence long axis. As the filaments are aligned along a solar neutral line, we selected this solution for which the long axis field component is in continuity with the one of other prominences of the same neutral line where the inverse method is applicable. We called this method 'the longitudinal method'.

The prominences, tracer of the solar magnetic field. We then reported the obtained field vectors (one per prominence) on the synoptic maps of spots, active regions and filaments of the Meudon Observatory, on which we added the neutral lines, the photospheric polarities and some Polar Crown prominences

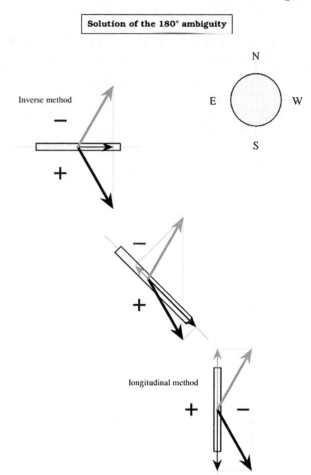

Fig. 18. Solution of the ambiguity applied to the sample of 300 prominences: choice of the inverse solution, when possible (inverse method), and, when impossible (minority of cases), continuity of the long axis field component (longitudinal method). The true solution is in *black*, the ambiguous solution is in *grey*

taken from the MacIntosh maps. In Fig. 19 we give a typical one, Carrington rotation 1697 (July 1980). On this map, the continuity of the filament long axis field component is visible along a neutral line, and in addition its sign changes when crossing successive neutral line from north to south, or from east to west. It appears also that the prominence field is in accordance with alignment along a north–south general line distorted by the differential rotation.

The prominences observed at the Pic-du-Midi cover the ascending phase of Cycle XXI, so that it is possible to schematize the prominence field evolution along the XI years solar cycle, as shown in Fig. 20. The alternation of the long axis field vector component from one neutral line to the neighbouring one is

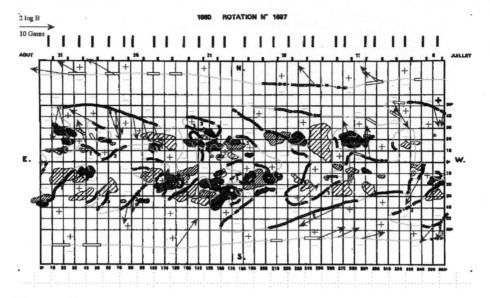

Fig. 19. Carrington rotation 1697. The *red* vectors indicate the inverse method for the ambiguity solution. The *orange* vectors indicate the longitudinal method for the ambiguity solution. The synoptic map is due to the Meudon Observatory, on which we have added the neutral line, the photospheric polarities and some Polar Crown prominences taken from the MacIntosh maps

visible. The alignment of the field along a north–south general line distorted by the differential rotation is also visible. The Polar Crown formation and disparition at the pole just after the activity maximum and the pole polarity reversal are also visible.

4 The Magnetic Field Observed in the Photosphere

In this section we present measurements based on the interpretation of the Zeeman effect. For such measurements a spectropolarimeter is needed, because the polarization varies along the spectral line. The used spectropolarimeter was the one of the THEMIS telescope ('Télescope Héliographique pour l'Étude du Magnétisme et des Instabilités Solaires'). This telescope has the original feature of being 'polarization free'; i.e. the polarization analysis is performed on axis, before any oblique reflection. The second original feature of THEMIS is being able to simultaneously record several spectral windows, in order to probe the solar atmosphere along its depth, because the different lines simultaneously observed are formed at different altitudes. A more detailed description of the THEMIS instrument can be found in Arnaud et al. [22], although it has to be updated with the tip-tilt correction, which has been modified and is now operational, and the polarization analyser quarter-wave plate positions that are now free to take any position needed.

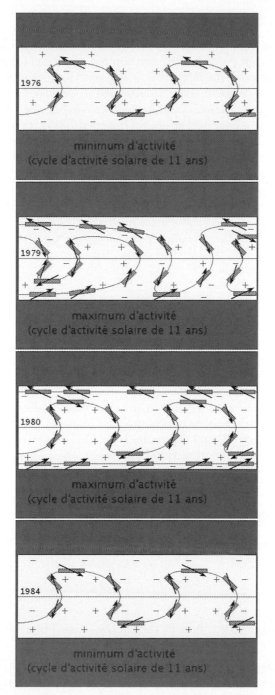

Fig. 20. Scheme of the prominence field evolution along the solar cycle (synoptic maps)

4.1 Need of an Inversion Method

The theory is able to answer the question: given a magnetic field vector, what would be the emergent polarization? and computations can be achieved by applying the theory. However, the interpretation of spectropolarimetric measurements requires the *inverse* process: given an observed polarization, what would be the magnetic field vector(s) which could have created it? As for answering the question one has to come back to the source magnetic field, this back operation is called *inversion*. In the previous case of the Hanle effect in prominences, the inversion method consisted of achieving a linear interpolation in the abacus, which is nothing else than a grid of model results. This is the first kind of inversions. However, in the case of the interpretation of the Zeeman effect of photospheric lines, advantage can be taken of the Levenberg–Marquardt algorithm to build up an inversion code, where the algorithm makes the calculation progresses towards the minimum of the χ^2 (discrepancy between the observed and modelled profiles). This algorithm is described below, together with its application to the UNNOFIT inversion code.

4.2 The UNNOFIT Inversion Method and Code

The theory at the basis of the UNNOFIT code ([23, 24]) is the Unno–Rachkowsky solution of the radiative transfer equation, which gives the polarization of the radiation emerging from a magnetized atmosphere (see, for instance, [25]). This theory was further complemented by Landi Degl'Innocenti and Landolfi [26] who introduced the magneto-optical effects and the damping (Voigt profiles). The atmosphere is modelled by using the Milne–Eddington approximation, which is described below. This approximation concerns an atmosphere in LTE, so that the notions of LTE and departure from LTE are introduced below before. The Unno–Rachkowsky solution being analytical, the Levenberg–Marquardt algorithm can easily be applied, following Auer's idea [27]. This algorithm is finally introduced below.

The local thermodynamical equilibrium (LTE). The local thermodynamical equilibrium (LTE) is defined by the fact that the level populations are given by the Boltzmann law applied to the local temperature T:

$$n_i = \frac{g_i e^{-\frac{E_i - E_0}{kT}}}{\sum_i g_i e^{-\frac{E_i - E_0}{kT}}} \, , \tag{29}$$

where g_i is the degeneration and E_i the energy of the i level (0 is the fundamental level).

The rate equations. Out of LTE (in NLTE), the level populations are only the solutions of the rate equations, or statistical equilibrium equations

$$\frac{dn_i}{dt} = 0 = \sum_j C_{j \to i} n_j - n_i \sum_j C_{i \to j} \, , \tag{30}$$

$C_{i \to j}$ being the transition probability from state i to state j.

The *atomic polarization* is a typical NLTE phenomenon, because it is an imbalance between the Zeeman sublevels populations. Such an imbalance cannot result from the Boltzman law, the Zeeman sublevels having the same (or nearly the same) energy. If the populations of levels $+M$ and $-M$ are different, the average kinetic momentum $\langle J_z \rangle$ is not zero and the medium is said to be *oriented*. For a normal Zeeman triplet the so-called *orientation* is the quantity $n_{+1} - n_{-1}$. Generally, in astrophysics there is no orientation, because the kinetic momentum conservation law would imply that kinetic momentum be brought to the medium, so that a pre-existent kinetic momentum has to be found. However, the populations may vary with M without any difference between $+M$ and $-M$. In this case the medium is said to be *aligned*, because the average kinetic momentum is not the same along the three axes x, y, z: $\langle J^2 \rangle - 3 \langle J_z^2 \rangle$ is non-zero. This quantity is proportional to the so-called *alignment* which is $n_{+1} + n_{-1} - 2n_0$ for a normal Zeeman triplet. The rate equations well adapted to polarization studies, including the Zeeman coherences, are the density matrix statistical equilibrium equations [3].

The rate equations involve

- the radiative processes: absorption, induced emission, spontaneous emission, from and towards each level;
- the collisional processes. If the connected levels i and j have different energies (i.e. are different): the collisions are said to be *inelastic* collisions. If, on the contrary, i and j are two Zeeman sublevels of the same level, the collisions are said to be *elastic* collisions, but two cases have to be considered: (a) $i \neq j$: in this case the collisions are said to be *depolarizing collisions*; (b) $i = j$: in this case the collision is strictly elastic (actually, such collisions do not transfer population so that they are invisible in the rate equations).

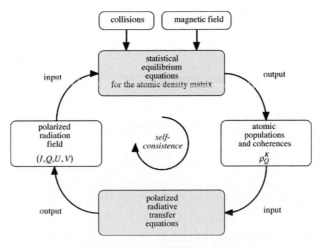

Fig. 21. The NLTE problem

These are the collisions that establish LTE (due to the microreversibility laws of the collisional processes), so that NLTE are the *weakly dense* media.

The NLTE problem. The NLTE problem is the self-consistent solution of the system of equations schematized in Fig. 21, because the radiative processes entering the statistical equilibrium equations depend on the incident radiation which is the solution of the transfer equation, whereas the level populations that are solutions of the statistical equilibrium equations enter the transfer equation. Various techniques, numerical and analytical, have been developed to solve this problem. In the case of LTE, the level populations being given by the Boltzman law, there is no need to solve the statistical equilibrium equations.

The Milne–Eddington atmosphere model. The Milne–Eddington atmosphere model, which is a particular case of LTE, is defined by the following:

(1) There is no macroscopic velocity field;
(2) The η_0 ratio between the line and continuum absorption coefficients does not vary with depth in the atmosphere;
(3) The Planck function (cgs units)

$$B = \frac{2h\nu^3}{c^2} \frac{1}{e^{\frac{h\nu}{kT}} - 1} \tag{31}$$

is a linear function of the optical depth τ along the atmosphere vertical axis

$$B(\tau) = B_0 + B_1\tau \ . \tag{32}$$

This is a model of the temperature variation in the atmosphere. Such a model well applies to the solar photosphere.

The Levenberg–Marquardt algorithm. The Levenberg–Marquardt algorithm can be schematized as follows [28]:

(1) compute a theoretical profile as a function of the N searched for parameters $a_1, a_2, ..., a_N$
(2) compute the χ^2 discrepancy between the observed and theoretical profiles
(3) by using the partial derivatives of the theoretical profile with respect to the parameters $a_1, a_2, ..., a_N$, determine the step to be applied to these parameters to progress in the direction of the χ^2 minimum (see the formulae in [28])
(4) apply the step to $a_1, a_2, ..., a_N$ and go to 1

The Levenberg–Marquardt can be easily applied to the Unno–Rachkowsky solution because the derivatives are analytical. However application of the Levenberg–Marquardt algorithm to numerical derivatives is also possible.

Results of the UNNOFIT checking: the measurements' accuracy. As described by its authors Landolfi et al. [23], UNNOFIT – the straightforward application of the technique outlined above (see also [25]) – provides simultaneous determination of eight free parameters via the fit of the four Stokes profiles.

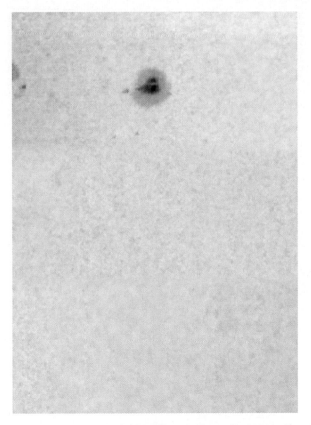

Fig. 22. NOAA 517 observed with THEMIS on 7 December 2003. Continuum near Fe
I 6302.5 Å. Sunspot

The eight free parameters are (1) the line strength η_0; (2) the Zeeman splitting
$\Delta\lambda_H$ that provides the magnetic field strength; (3) the Doppler absorption profile
width $\Delta\lambda_D$; (4) the γ damping parameter of the Voigt function; (5) one single b
parameter describing the Milne–Eddington τ-dependence along the atmosphere
vertical with $b = \mu B_1/B_0$, where B_0 and B_1 are the usual parameters describing
the Milne–Eddington atmosphere (see Eq. (32)) and μ is the cosine of the line
of sight inclination angle; (6) the line central wavelength (providing thus the
Doppler shift); (7) and (8) the magnetic field inclination and azimuth angles. In
[24], we have added a ninth free parameter to be determined by UNNOFIT: the
magnetic filling factor α, which means that the received radiation is the sum of
the magnetic component radiation, weighted α, and of the non-magnetic component one, weighted $1 - \alpha$; that is to say, denoting by 'm' and 'nm' the magnetic
and non-magnetic unresolved contributions, one has

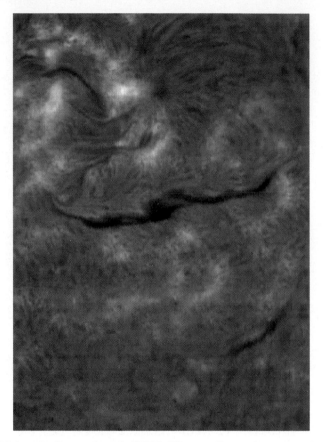

Fig. 23. NOAA 517 observed with THEMIS on 7 December 2003. Hα line centre. Filament and plage regions. From [24]

$$\begin{cases} I = (1 - \alpha)I_{\mathrm{nm}} + \alpha I_{\mathrm{m}} \\ Q = \alpha Q_{\mathrm{m}} \\ U = \alpha U_{\mathrm{m}} \\ V = \alpha V_{\mathrm{m}} \end{cases} \qquad (33)$$

The calculation of the derivatives with respect to the α parameter is straightforward, so that the implementation of this ninth parameter in the algorithm follows. The magnetic and non-magnetic components of the atmosphere have all their physical parameters taken as equal, except the three magnetic field coordinates.

The UNNOFIT code has been checked by computing theoretical profiles for a series of values of the nine parameters, by noising them at the observations noise level and by submitting them to the inverter and comparing the output parameters with the input ones. The detailed results can be found in [24]. The main result is that the magnetic filling factor α and the magnetic field strength B are not individually recovered, but that their product αB, that we call the 'average local magnetic field strength', is correctly determined by the inversion (note that

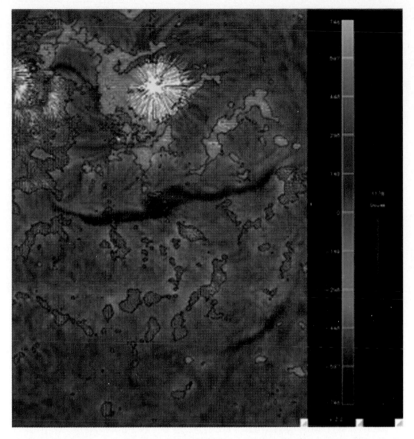

Fig. 24. NOAA 517 observed with THEMIS on 7 December 2003. Network vector magnetic field map. The contours given by $\alpha B = 45$ G separate the network from the internetwork, where the turbulent field is not represented. Vector field: longitudinal component given by colours (cold colours – *green, blue* – for field entering the Sun, warm colours – *yellow, red* – for field coming out of the Sun), transverse component given by *dashes*, without arrow because the fundamental ambiguity is not resolved. From [24]

this average local magnetic field strength is not the local flux which would be the local average longitudinal component of the magnetic field). The determination of αB however saturates at 20 G, weaker values remaining undetermined. The lowering of this threshold requires the reduction of the data polarimetric noise (1.5×10^{-3} per pixel in the present data). Besides, the accuracy of the inversion is obtained, given the polarimetric inaccuracy of 1.5×10^{-3} per pixel: for $\alpha B > 45$ G (network field), αB is obtained ± 5 G, and the inclination and azimuth angles of the field vector are obtained $\pm 5°$. For $\alpha B < 45$ G (internetwork field), αB is obtained ± 5 G, and the inclination and azimuth angles of the field vector are obtained $\pm 20°$.

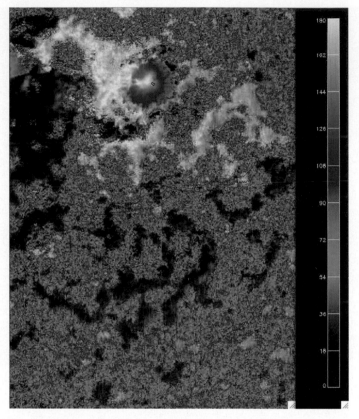

Fig. 25. NOAA 517 observed with THEMIS on 7 December 2003. Map of the field inclination angle: angle between the field and the line of sight which is also nearly the solar vertical, the region being observed near disk centre. In the network the field is nearly vertical, and this draws the network which is the frontiers of the supergranules. In the internetwork the field is turbulent in direction with an horizontal trend. From [24]

4.3 The Photospheric Magnetic Field Structuring in Network and Internetwork

The application of the UNNOFIT inversion to the THEMIS data has led to distinguish two regimes for the photospheric magnetic field, corresponding to two ranges of local average magnetic field strengths: (a) the network, having a field nearly vertical, with a homogeneous azimuth. In this zone the local average field strength is higher than 100 G. The network locates in active regions and plages and at the supergranule frontiers; (b) the internetwork, where the field is turbulent in direction, i.e. changing of direction from pixel to pixel (with, however, a horizontal trend), and where the local average field strength is about 20 G (weaker values are not excluded however due to the insensitivity of UNNOFIT to weaker fields). Thus, the network field appears like a vertical strong field emerging from the carpet of the turbulent weak internetwork field. The solar nature of this turbulence has been proved by an observed correlation between the field directions determined from two different lines, Fe I 6302.5 and Fe I 6301.5, recorded on different camera pixels [24]. An example of maps is given in Figs. 22, 23, 24 and 25, where Fig. 25 shows clearly the two kinds of zones: network and internetwork.

References

1. Bommier, V.: Astron. Astrophys. **87**, 109 (1980)
2. Landi Degl'Innocenti, E.: Sol. Phys. **91**, 1 (1984)
3. Bommier, V., Sahal-Bréchot, S.: Astron. Astrophys. **69**, 57 (1978)
4. Stenflo, J.O.: Sol. Phys. **80**, 209 (1982)
5. Sahal-Bréchot, S.: Space Sci. Rev. **29**, 391 (1981)
6. Tandberg-Hanssen, E.: The Nature of Solar Prominences. Astrophysics and Space Science Library, 199, Kluwer Acad. Pub., Dordrecht (1995)
7. Lyot, B.: Compt. Rend. Acad. Sci. **198**, 249 (1934)
8. Lyot, B.: Compt. Rend. Acad. Sci. **202**, 392 (1936)
9. Hyder, C.L.: Astrophys. J. **141**, 1374 (1965)
10. Leroy, J.L., Ratier, G., Bommier, V.: Astron. Astrophys. **54**, 811 (1977)
11. Athay, R.G., Querfeld, C.W., Smartt, R.N., Landi Degl'Innocenti, E., Bommier, V.: Sol. Phys. **89**, 3 (1983)
12. Querfeld, C.W., Smartt, R.N., Bommier, V., Landi Degl'Innocenti, E., House, L.L.: Sol. Phys. **96**, 277 (1985)
13. Bommier, V., Leroy, J.L., Sahal-Bréchot, S.: Astron. Astrophys. **156**, 79 (1986)
14. Bommier, V.: In: Casini, R., Lites, B.W. (eds.) Solar Polarization 4, ASP Conf. Ser. vol. 358, p. 245 (2006)
15. Bommier, V., Sahal-Bréchot, S.: Ann. Phys. Fr. **16**, 555 (1991)
16. Bommier, V., Leroy, J.L., Sahal-Bréchot, S.: Astron. Astrophys. **156**, 90 (1986)
17. Landi Degl'Innocenti, E., Bommier, V., Sahal-Bréchot, S.: Astron. Astrophys. **186**, 335 (1987)
18. Bommier, V., Landi Degl'Innocenti, E., Leroy, J.L., Sahal-Bréchot, S.: Sol. Phys. **154**, 231 (1994)
19. Bommier, V., Leroy, J.L., Sahal-Bréchot, S.: Astron. Astrophys. **100**, 231 (1981)
20. Leroy, J.L., Bommier, V., Sahal-Bréchot, S.: Astron. Astrophys. **131**, 33 (1984)
21. Sahal-Bréchot, S., Bommier, V., Leroy, J.L.: Astron. Astrophys. **59**, 223 (1977)
22. Arnaud, J., Mein, P., Rayrole, J.: In: Proceedings of "A Crossroads for European Solar & Heliospheric Physics", Tenerife, March 23–27, 1998, ESA SP-417, 213 (1998)
23. Landolfi, M., Landi Degl'Innocenti, E., Arena, P.: Sol. Phys., **93**, 269 (1984)
24. Bommier, V., Landi Degl'Innocenti, E., Landolfi, M., Molodij, G.: Astron. Astrophys. **464**, 323 (2007)
25. Landi Degl'Innocenti, E., Landolfi, M.: Polarization in Spectral Lines. Astrophysics and Space Science Library, 307, Kluwer Acad. Pub., Dordrecht (2004)
26. Landolfi, M., Landi Degl'Innocenti, E.: Sol. Phys. **78**, 355 (1982)
27. Auer, L.H., Heasley, J.N., House, L.L.: Sol. Phys. **55**, 47 (1977)
28. Press, W.H., Flannery, B.P., Teukolsky, S.A., Vetterling, W.T.: Numerical Recipes. The Art of Scientific Computing, Cambridge Univ. Press, New York (1989)

Index

Printing: Krips bv, Meppel, The Netherlands
Binding: Stürtz, Würzburg, Germany